Thermodynamics and Statistical Mechanics

Thermodynamics and Statistical Mechanics

Richard Fitzpatrick
The University of Texas at Austin, USA

World Scientific

NEW JERSEY • LONDON • SINGAPORE • BEIJING • SHANGHAI • HONG KONG • TAIPEI • CHENNAI • TOKYO

Published by

World Scientific Publishing Co. Pte. Ltd.
5 Toh Tuck Link, Singapore 596224
USA office: 27 Warren Street, Suite 401-402, Hackensack, NJ 07601
UK office: 57 Shelton Street, Covent Garden, London WC2H 9HE

Library of Congress Control Number: 2020940890

British Library Cataloguing-in-Publication Data
A catalogue record for this book is available from the British Library.

THERMODYNAMICS AND STATISTICAL MECHANICS

ISBN 978-981-122-335-8 (hardcover)
ISBN 978-981-122-423-2 (paperback)
ISBN 978-981-122-336-5 (ebook for institutions)
ISBN 978-981-122-337-2 (ebook for individuals)

For any available supplementary material, please visit
https://www.worldscientific.com/worldscibooks/10.1142/11912#t=suppl

Preface

The aim of this textbook is to provide a comprehensive exposition of the theory of equilibrium thermodynamics and statistical mechanics at a level suitable for well-prepared undergraduate students. The fundamental message of the book is that all results in equilibrium thermodynamics and statistical mechanics follow from a single unprovable axiom—namely, the principle of equal a priori probabilities—combined with elementary probability theory (see Chapter 1), elementary classical mechanics (see Appendix B), and elementary quantum mechanics (see Appendix C). Furthermore, the derivation of many results in equilibrium thermodynamics and statistical mechanics exploits the fact that macroscopic systems are made up of an extremely large number of atoms and molecules. The major sources for the material presented in this book are [Reif (1965)], [Carter (2001)], and [Schroeder (2000)].

Contents

Chapter 1

Introduction

1.1 Atomic Theory of Matter

According to the well-known *atomic* theory of matter, the familiar objects that make up the visible world, such as tables and chairs, are themselves made up of a great many invisible microscopic particles.

Atomic theory was first proposed by the ancient Greek philosophers Leucippus and Democritus (in works that are, unfortunately, lost), who speculated that the world essentially consists of myriads of tiny indivisible particles, which they called *atoms*, from the Greek ἄτομεν, meaning (an) uncuttable (object). They speculated, further, that the observed properties of everyday materials can be explained either in terms of the different *shapes* of their constituent atoms, or the different *motions* of these atoms. In some respects, modern atomic theory differs substantially from the primitive theory of Leucippus and Democritus. However, the central ideas have remained essentially unchanged. In particular, Leucippus and Democritus were correct to suppose that the properties of materials depend not only on the nature of their constituent atoms or molecules, but also on the relative motions of these particles.

1.2 What is Thermodynamics?

In this book, we shall focus almost exclusively on those physical properties of everyday materials that are associated with the *motions* of their constituent atoms or molecules. In particular, we shall be concerned with the type of motion that we normally call heat. We shall try to establish what controls the flow of heat from one body to another when the bodies are brought into thermal contact. We shall also attempt to understand the relationship between heat and mechanical work. For instance, to what extent does the heat content of a body increase when mechanical work is done on it? More importantly, can we extract heat from a given body in order to do useful work on some other body? This subject area is called thermodynamics, from the Greek roots *θερμόν*, meaning heat, and *δύνᾱμις*, meaning power.

1.3 Need for Statistical Approach

It is necessary to emphasize, from the very outset, that thermodynamics is a difficult subject. In fact, this subject is so difficult that we are forced to adopt a radically different approach to that employed in other areas of physics.

In most areas of physics, we can formulate some exact, or nearly exact, set of equations that governed the system under investigation. For instance, Newton's equations of motion, or Maxwell's equations for electromagnetic fields. We can then analyze the system by solving these equations, either exactly or approximately.

In thermodynamics, we have no problem formulating the governing equations. The motions of atoms and molecules are described exactly by the laws of quantum mechanics. In many cases, they are also described, to a reasonable approximation, by the much simpler laws of classical mechanics. We shall not be dealing with systems sufficiently energetic for atomic nuclei to be disrupted, so we need not consider nuclear forces. Also, in general, the gravitational forces between atoms and molecules are completely negligible. This implies that the forces between atoms and molecules are predominantly electromagnetic in origin, and are, therefore, very well understood. Thus, we can formulate the exact laws of motion for a thermodynamical system, including all of the inter-atomic forces. The problem is the sheer complexity of the resulting system of equations. In one mole of a substance (e.g., in twelve grams of carbon, or eighteen grams of water) there are Avagadro's number of atoms or molecules. That is, about $N_A = 6\times 10^{23}$ particles. To solve the system exactly, we would have to write down approximately 10^{24} coupled equations of motion, with the same number of initial conditions, and then try to integrate the system. This is plainly impossible. It would also be complete overkill. We are generally not interested in knowing the position and velocity of every particle in the system as a function of time. Instead, we want to know things like the volume of the system, the temperature, the pressure, the heat capacity, the coefficient of expansion, et cetera. We would certainly be hard put to specify more than about fifty, say, interesting properties of a thermodynamic system. In other words, the number of pieces of information that we require to characterize the state of the system is absolutely minuscule compared to the number of degrees of freedom of the system. That is, the number of pieces of information needed to completely specify its internal motion. Moreover, the quantities that we are interested in calculating do not depend on the motions of individual particles, or some some small subset of particles, but, instead, depend on the average motions of all the particles in the system. In other words, these quantities depend on the *statistical* properties of the atomic or molecular motion.

The method adopted in thermodynamics is essentially dictated by the enormous complexity of thermodynamic systems. We generally start off with some statistical information about the motions of the constituent atoms or molecules, such as their average kinetic energy, but possess very little information about the motions of individual particles. We then try to deduce some other properties of the system from a statistical treatment of the governing equations of motion. In fact, our approach has to be statistical in nature, because we lack most of the information required to specify the internal state of the system. The best we can do is to provide a few overall constraints, such as the average volume and the average energy.

Thermodynamic systems are ideally suited to a statistical approach because of the enormous numbers of particles that they contain. Statistical arguments actually get more exact as the numbers involved get larger. For example, consider an opinion poll published in a newspaper. Such a poll usually specifies how many people were interviewed. This number is significant because, even if the polling was done without bias, which is extremely unlikely, the laws of statistics say that there is a intrinsic error (in the fraction of people who supposedly support a given opinion) of order one over the square root of the number of people questioned. It follows that if a thousand people were interviewed, which is a typical number, then the error is at least three percent. We can easily appreciate that if we do statistics on a thermodynamic system containing 10^{24} particles then we are going to obtain results that are valid to incredible accuracy. In fact, in most situations, we can forget that the results are statistical at all, and treat them as exact laws of physics. For instance, the familiar equation of state of an ideal gas, $PV = \nu RT$ (see Section 6.2), is actually a statistical result. In other words, it relates the average pressure, and the average volume, to the average temperature. However, for one mole of gas, the relative statistical deviations from average values are only about 10^{-12}, according to the $1/N_A^{1/2}$ law. Of course, it is virtually impossible to measure the pressure, volume, or temperature of a gas to such accuracy, so it is often convenient to forget about the fact that the ideal gas law is a statistical result, and treat it as a law of physics interrelating the actual pressure, volume, and temperature of an ideal gas.

1.4 Microscopic and Macroscopic Systems

It is useful, at this stage, to make a distinction between the different sizes of the systems that we are going to examine. We shall call a system *microscopic* if it is roughly of atomic dimensions, or smaller. On the other hand, we shall call a system *macroscopic* if it is large enough to be visible in the ordinary sense. This is a rather inexact definition. The exact definition depends on the number of particles in the system, which we shall call N. A system is macroscopic if $1/N^{1/2} \ll 1$, which means that statistical arguments can be applied to reasonable accuracy. For instance, if we wish to keep the relative statistical error below one percent then a macroscopic system would have to contain more than about ten thousand particles. Any system containing less than this number of particles would be regarded as essentially microscopic, and, hence, statistical arguments could not be applied to such a system without unacceptable error.

1.5 Classical and Statistical Thermodynamics

In this book, we are going to develop some machinery for interrelating the statistical properties of a system containing a very large number of particles, via a statistical treatment of the laws of atomic or molecular motion. It turns out that, once we have developed this machinery, we can obtain some very general results that do not depend on the exact details of the statistical treatment. These results can be described without reference to the underlying statistical nature of the system, but their validity depends ultimately on statistical arguments. They take the form of general statements regarding heat and work, and are usually referred to as *classical thermodynamics*, or just *thermodynamics*, for short. Historically, classical thermodynamics was the first type of thermodynamics to be discovered. In fact, for many years, the laws of classical thermodynamics seemed rather mysterious, because their statistical justification had yet to be discovered. The strength of classical thermodynamics is its great generality, which comes about because it does not depend on any detailed assumptions about the statistical properties of the system under investigation. This generality is also the principal weakness of classical thermodynamics. Only a relatively few statements can be made on such general grounds, so many interesting properties of the system remain outside the scope of classical thermodynamics.

If we go beyond classical thermodynamics, and start to investigate the statistical machinery that underpins it, then we get all of the results of classical thermodynamics, plus a large number of other results that enable the macroscopic parameters of the system to be calculated from a knowledge of its microscopic constituents. This approach is known as *statistical thermodynamics*, and is extremely powerful. The only drawback is that the further we delve inside the statistical machinery of thermodynamics, the harder it becomes to perform the necessary calculations.

Note that both classical and statistical thermodynamics are only valid for systems in *equilibrium*. If the system is not in equilibrium then the problem becomes considerably more difficult. The thermodynamics of non-equilibrium systems, which is generally known as *irreversible thermodynamics*, is discussed in [Reif (1965)].

1.6 Classical and Quantum Approaches

We mentioned earlier that the motions (by which we really meant the translational motions) of atoms and molecules are described exactly by quantum mechanics, and only approximately by classical mechanics. It turns out that the non-translational motions of molecules, such as their rotation and vibration, are very poorly described by classical mechanics. So, why bother using classical mechanics at all? Unfortunately, quantum mechanics deals with the translational motions of atoms and molecules (via wave mechanics) in a rather awkward manner. The classical approach is far more straightforward, and, under most circumstances, yields the same statistical results. Hence, throughout the first part of this book, we shall use classical mechanics, as much as possible, to describe the translational motion of atoms and molecules, and will reserve quantum mechanics for dealing with non-translational motions. However, towards the end of this book, in Chapter 8, we shall switch to a purely quantum-mechanical approach.

Chapter 2

Probability Theory

2.1 Introduction

This chapter is devoted to a brief, and fairly low-level, introduction to a branch of mathematics known as *probability theory*. In fact, we do not need to know very much about probability theory in order to understand statistical thermodynamics, because the probabilistic calculation that ultimately underpins the latter subject area is extraordinarily simple.

2.2 What is Probability?

What is the scientific definition of probability? Consider an observation made on a general system, S. This can result in any one of a number of different possible outcomes. We want to find the probability of some general outcome, X. In order to ascribe a probability, we have to consider the system as a member of a large set, Σ, of similar systems. Mathematicians call such a group an *ensemble*, which is just the french word for group. So, let us consider an ensemble, Σ, of similar systems, S. The probability of the outcome X is defined as the ratio of the number of systems in the ensemble that exhibit this outcome to the total number of systems, in the limit where the latter number tends to infinity. We can write this symbolically as

$$P(X) = \mathit{lt}_{\,\Omega(\Sigma)\to\infty} \frac{\Omega(X)}{\Omega(\Sigma)}, \tag{2.1}$$

where $\Omega(\Sigma)$ is the total number of systems in the ensemble, and $\Omega(X)$ the number of systems exhibiting the outcome X. We can see that the probability, $P(X)$, must be a number lying between 0 and 1. The probability is zero if no systems exhibit the outcome X, even when the number of systems goes to infinity. This is just another way of saying that the outcome X is impossible. The probability is unity if all systems exhibit the outcome X, in the limit that the number of systems goes to infinity. This is another way of saying that the outcome X is bound to occur.

2.3 Combining Probabilities

Consider two distinct possible outcomes, X and Y, of an observation made on the system S, with probabilities of occurrence $P(X)$ and $P(Y)$, respectively. Let us determine the probability of obtaining either the outcome X or the outcome Y, which we shall denote $P(X \,|\, Y)$. From the basic definition of probability,

$$P(X \,|\, Y) = {}_{lt\, \Omega(\Sigma)\to\infty} \frac{\Omega(X \,|\, Y)}{\Omega(\Sigma)}, \tag{2.2}$$

where $\Omega(X \mid Y)$ is the number of systems in the ensemble that exhibit either the outcome X or the outcome Y. It is clear that

$$\Omega(X \,|\, Y) = \Omega(X) + \Omega(Y) \tag{2.3}$$

if the outcomes X and Y are mutually exclusive (which must be the case if they are two distinct outcomes). Thus,

$$P(X \,|\, Y) = P(X) + P(Y). \tag{2.4}$$

We conclude that the probability of obtaining either the outcome X or the outcome Y is the sum of the individual probabilities of X and Y. For instance, with a six-sided die, the probability of throwing any particular number (one to six) is $1/6$, because all of the possible outcomes are considered to be equally likely. It follows, from the previous discussion, that the probability of throwing either a one or a two is $1/6 + 1/6$, which equals $1/3$.

Let us denote all of the M, say, possible outcomes of an observation made on the system S by X_i, where i runs from 1 to M. Let us determine the probability of obtaining any of these outcomes. This quantity is clearly unity, from the basic definition of probability, because every one of the systems in the ensemble must exhibit one of the possible outcomes. However, this quantity is also equal to the sum of the probabilities of all the individual outcomes, by Equation (2.4), so we conclude that this sum is equal to unity. Thus,

$$\sum_{i=1,M} P(X_i) = 1, \tag{2.5}$$

which is called the *normalization condition*, and must be satisfied by any complete set of probabilities. This condition is equivalent to the self-evident statement that an observation of a system must definitely result in one of its possible outcomes.

There is another way in which we can combine probabilities. Suppose that we make an observation on a system picked at random from the ensemble, and then pick a second system, completely independently, and make another observation. Here, we are assuming that the first observation does not influence the second observation in any way. In other words, the two observations are *statistically independent*. Let us determine the probability of obtaining the outcome X in the first system and the outcome Y in the second system, which we shall denote $P(X \otimes Y)$. In order to determine this probability, we have to form an ensemble of all of the possible pairs of systems that we could choose from the ensemble, Σ. Let us denote this ensemble $\Sigma \otimes \Sigma$. It is obvious that the number of pairs of systems in this new ensemble is just the square of the number of systems in the original ensemble, so

$$\Omega(\Sigma \otimes \Sigma) = \Omega(\Sigma)\,\Omega(\Sigma). \tag{2.6}$$

It is also fairly obvious that the number of pairs of systems in the ensemble $\Sigma \otimes \Sigma$ that exhibit the outcome X in the first system, and Y in the second system, is just the product of the number of systems that exhibit the outcome X, and the number of systems that exhibit the outcome Y, in the original ensemble. Hence,

$$\Omega(X \otimes Y) = \Omega(X)\,\Omega(Y). \tag{2.7}$$

It follows from the basic definition of probability that

$$P(X \otimes Y) = {}_{lt\,\Omega(\Sigma)\to\infty} \frac{\Omega(X \otimes Y)}{\Omega(\Sigma \otimes \Sigma)} = P(X)\,P(Y). \tag{2.8}$$

Thus, the probability of obtaining the outcomes X and Y in two statistically independent observations is the product of the individual probabilities of X and Y. For instance, the probability of throwing a one and then a two on a six-sided die is $1/6 \times 1/6$, which equals $1/36$.

2.4 Two-State System

The simplest non-trivial system that we can investigate using probability theory is one for which there are only two possible outcomes. (There would obviously be little point in investigating a one-outcome system.) Let us suppose that there are two possible outcomes to an observation made on some system, S. Let us denote these outcomes 1 and 2, and let their probabilities of occurrence be

$$P(1) = p, \tag{2.9}$$

$$P(2) = q. \tag{2.10}$$

It follows immediately from the normalization condition, Equation (2.5), that

$$p + q = 1, \tag{2.11}$$

so $q = 1 - p$. The best known example of a two-state system is a tossed coin. The two outcomes are heads and tails, each with equal probabilities 1/2. So, $p = q = 1/2$ for this system.

Suppose that we make N statistically independent observations of S. Let us determine the probability of n_1 occurrences of the outcome 1, and $N - n_1$ occurrences of the outcome 2, with no regard as to the order of these occurrences. Denote this probability $P_N(n_1)$. This type of calculation crops up very often in probability theory. For instance, we might want to know the probability of getting nine heads and only one tails in an experiment where a coin is tossed ten times, or where ten coins are tossed simultaneously.

Consider a simple case in which there are only three observations. Let us try to evaluate the probability of two occurrences of the outcome 1, and one occurrence of the outcome 2. There are three different ways of getting this result. We could get the outcome 1 on the first two observations, and the outcome 2 on the third. Or, we could get the outcome 2 on the first observation, and the outcome 1 on the latter two observations. Or, we could get the outcome 1 on the first and last observations, and the outcome 2 on the middle observation. Writing this symbolically, we have

$$P_3(2) = P(1 \otimes 1 \otimes 2 \,|\, 2 \otimes 1 \otimes 1 \,|\, 1 \otimes 2 \otimes 1). \tag{2.12}$$

Here, the symbolic operator $\otimes$ stands for *and*, whereas the symbolic operator $|$ stands for *or*. This symbolic representation is helpful because of the two basic rules for combining probabilities that we derived earlier in Equations (2.4) and (2.8):

$$P(X \,|\, Y) = P(X) + P(Y), \tag{2.13}$$

$$P(X \otimes Y) = P(X)\, P(Y). \tag{2.14}$$

The straightforward application of these rules gives

$$P_3(2) = p\,p\,q + q\,p\,p + p\,q\,p = 3\,p^2\,q \tag{2.15}$$

for the case under consideration.

The probability of obtaining n_1 occurrences of the outcome 1 in N observations is given by

$$P_N(n_1) = C^{N}_{n_1,\,N-n_1}\, p^{n_1}\, q^{N-n_1}, \tag{2.16}$$

where $C^{N}_{n_1,\,N-n_1}$ is the number of ways of arranging two distinct sets of n_1 and $N - n_1$ indistinguishable objects. Hopefully, this result is, at least, plausible from the previous example. There, the probability of getting two occurrences of the outcome 1, and one occurrence of the outcome 2, was obtained by writing out all of the possible arrangements of two ps (the probability of outcome 1) and one q (the probability of outcome 2), and then adding them all together.

2.5 Combinatorial Analysis

The branch of mathematics that studies the number of different ways of arranging things is called *combinatorial analysis*. We need to know how many different ways there are of arranging N objects that are made up of two groups of n_1 and $N - n_1$ indistinguishable objects. This is a rather difficult problem. Let us start off by tackling a slightly easier problem. How many ways are there of arranging N distinguishable objects? For instance, suppose that we have six pool balls, numbered one through six, and we pot one each into every one of the six pockets of a pool table (that is, top-left, top-right, middle-left, middle-right, bottom-left, and bottom-right). How many different ways are there of doing this? Let us start with the top-left pocket. We could pot any one of the six balls into this pocket, so there are 6 possibilities. For the top-right pocket we only have 5 possibilities, because we have already potted a ball into the top-left pocket, and it cannot be in two pockets simultaneously. So, our 6 original possibilities combined with these 5 new possibilities gives 6×5 ways of potting two balls into the top two pockets. For the middle-left pocket we have 4 possibilities, because we have already potted two balls. These possibilities combined with our 6×5 possibilities gives $6 \times 5 \times 4$ ways of potting three balls into three pockets. At this stage, it should be clear that the final answer is going to be $6 \times 5 \times 4 \times 3 \times 2 \times 1$. The *factorial* of a general positive integer, n, is defined

$$n! = n\,(n-1)\,(n-2)\cdots 3\cdot 2\cdot 1. \tag{2.17}$$

So, $1! = 1$, and $2! = 2 \times 1 = 2$, and $3! = 3 \times 2 \times 1 = 6$, and so on. Clearly, the number of ways of potting six distinguishable pool balls into six pockets is $6!$ (which incidentally equals 720). Because there is nothing special about pool balls, or the number six, we can safely infer that the number of different ways of arranging N distinguishable objects, denoted C^N, is given by

$$C^N = N!\,. \tag{2.18}$$

Suppose that we take the number four ball off the pool table, and replace it by a second number five ball. How many different ways are there of potting the balls now? Consider a previous arrangement in which the number five ball was potted into the top-left pocket, and the number four ball was potted into the top-right pocket, and then consider a second arrangement that only differs from the first because the number four and five balls have been swapped around. These arrangements are now indistinguishable, and are therefore counted as a single arrangement, whereas previously they were counted as two separate arrangements. Clearly, the previous arrangements can be divided into two groups, containing equal numbers of arrangements, that differ only by the permutation of the number four and five balls. Because these balls are now indistinguishable, we conclude that there are only half as many different arrangements as there were before. If we take the number three ball off the table, and replace it by a third number five ball, then we can split the original arrangements into six equal groups of arrangements that differ only by the permutation of the number three, four, and five balls. There are six groups because there are $3! = 6$ separate permutations of these three balls. Because the number three, four, and five balls are now indistinguishable, we conclude that there are only $1/6$ the number of original arrangements. Generalizing this result, we conclude that the number of arrangements of n_1 indistinguishable and $N - n_1$ distinguishable objects is

$$C^N_{n_1} = \frac{N!}{n_1!}. \tag{2.19}$$

We can see that if all the balls on the table are replaced by number five balls then there is only $N!/N! = 1$ possible arrangement. This corresponds, of course, to a number five ball in each pocket. A further straightforward generalization tells us that the number of arrangements of two groups of n_1 and $N - n_1$ indistinguishable objects is

$$C^N_{n_1,\,N-n_1} = \frac{N!}{n_1!\,(N-n_1)!}. \tag{2.20}$$

2.6 Binomial Probability Distribution

It follows from Equations (2.16) and (2.20) that the probability of obtaining n_1 occurrences of the outcome 1 in N statistically independent observations of a two-state system is

$$P_N(n_1) = \frac{N!}{n_1!\,(N-n_1)!}\,p^{n_1}\,q^{N-n_1}. \tag{2.21}$$

This probability function is called the *binomial probability distribution*. The reason for this name becomes obvious if we tabulate the probabilities for the first few possible values of N, as is done in Table 2.1. Of course, we immediately recognize the expressions appearing in the first four rows of this table: they appear in the standard algebraic expansions of $(p + q)$, $(p + q)^2$, $(p + q)^3$, and $(p + q)^4$, respectively. In algebra, the expansion of $(p + q)^N$ is called the *binomial expansion* (hence, the name given to the probability distribution function), and is written

$$(p+q)^N \equiv \sum_{n=0,N} \frac{N!}{n!\,(N-n)!}\,p^n\,q^{N-n}. \tag{2.22}$$

Table 2.1 The binomial probability distribution, $P_N(n_1)$.

		n_1				
		0	1	2	3	4
	1	q	p			
N	2	q^2	$2\,p\,q$	p^2		
	3	q^3	$3\,p\,q^2$	$3\,p^2\,q$	p^3	
	4	q^4	$4\,p\,q^3$	$6\,p^2\,q^2$	$4\,p^3\,q$	p^4

Equations (2.21) and (2.22) can be used to establish the normalization condition for the binomial distribution function:

$$\sum_{n_1=0,N} P_N(n_1) = \sum_{n_1=0,N} \frac{N!}{n_1!\,(N-n_1)!}\,p^{n_1}\,q^{N-n_1} \equiv (p+q)^N = 1, \tag{2.23}$$

because $p + q = 1$. [See Equation (2.11).]

2.7 Mean, Variance, and Standard Deviation

What is meant by the mean, or average, of a quantity? Suppose that we wanted to calculate the average age of undergraduates at the University of Texas at Austin. We could go to the central administration building, and find out how many eighteen year-olds, nineteen year-olds, et cetera, were currently enrolled. We would then write something like

$$\text{Average Age} \simeq \frac{N_{18}\times 18 + N_{19}\times 19 + N_{20}\times 20 + \cdots}{N_{18} + N_{19} + N_{20}\cdots}, \tag{2.24}$$

where N_{18} is the number of enrolled eighteen year-olds, et cetera. Suppose that we were to pick a student at random and then ask "What is the probability of this student being eighteen?" From what we have already discussed, this probability is defined

$$P_{18} = \frac{N_{18}}{N_{\text{students}}}, \tag{2.25}$$

where N_{students} is the total number of enrolled students. We can now see that the average age takes the form

$$\text{Average Age} \simeq P_{18}\times 18 + P_{19}\times 19 + P_{20}\times 20 + \cdots. \tag{2.26}$$

There is nothing special about the age distribution of students at UT Austin. So, for a general variable u, which can take on any one of M possible values, u_1,

$u_2, \cdots, u_M$, with corresponding probabilities $P(u_1)$, $P(u_2), \cdots, P(u_M)$, the mean, or average, value of u, which is denoted $\bar{u}$, is defined as

$$\bar{u} \equiv \sum_{i=1,M} P(u_i)\, u_i. \tag{2.27}$$

Suppose that $f(u)$ is some function of u. For each of the M possible values of u, there is a corresponding value of $f(u)$ that occurs with the same probability. Thus, $f(u_1)$ corresponds to u_1, and occurs with the probability $P(u_1)$, and so on. It follows from our previous definition that the mean value of $f(u)$ is given by

$$\overline{f(u)} \equiv \sum_{i=1,M} P(u_i)\, f(u_i). \tag{2.28}$$

Suppose that $f(u)$ and $g(u)$ are two general functions of u. It follows that

$$\begin{aligned}\overline{f(u)+g(u)} &= \sum_{i=1,M} P(u_i)\,[f(u_i)+g(u_i)] \\ &= \sum_{i=1,M} P(u_i)\, f(u_i) + \sum_{i=1,M} P(u_i)\, g(u_i),\end{aligned} \tag{2.29}$$

so that

$$\overline{f(u)+g(u)} = \overline{f(u)} + \overline{g(u)}. \tag{2.30}$$

Finally, if c is a general constant then it is clear that

$$\overline{c\, f(u)} = c\, \overline{f(u)}. \tag{2.31}$$

We now know how to define the mean value of the general variable u. But, how can we characterize the scatter around the mean value? We could investigate the deviation of u from its mean value, $\bar{u}$, which is denoted

$$\Delta u \equiv u - \bar{u}. \tag{2.32}$$

In fact, this is not a particularly interesting quantity, because its average is obviously zero:

$$\overline{\Delta u} = \overline{(u-\bar{u})} = \bar{u} - \bar{u} = 0. \tag{2.33}$$

This is another way of saying that the average deviation from the mean vanishes. A more interesting quantity is the square of the deviation. The average value of this quantity,

$$\overline{(\Delta u)^2} = \sum_{i=1,M} P(u_i)\,(u_i - \bar{u})^2, \tag{2.34}$$

is usually called the *variance*. The variance is clearly a positive number, unless there is no scatter at all in the distribution, so that all possible values of u correspond to the mean value, $\bar{u}$, in which case it is zero. The following general relation is often useful

$$\overline{(u-\bar{u})^2} = \overline{(u^2 - 2\,u\,\bar{u} + \bar{u}^2)} = \overline{u^2} - 2\,\bar{u}\,\bar{u} + \bar{u}^2, \tag{2.35}$$

giving

$$\overline{(\Delta u)^2} = \overline{u^2} - \bar{u}^2. \tag{2.36}$$

The variance of u is proportional to the square of the scatter of u around its mean value. A more useful measure of the scatter is given by the square root of the variance,

$$\Delta^* u = \left[\overline{(\Delta u)^2}\right]^{1/2}, \tag{2.37}$$

which is usually called the *standard deviation* of u. The standard deviation is essentially the width of the range over which u is distributed around its mean value, $\bar{u}$.

2.8 Application to Binomial Probability Distribution

Let us now apply what we have just learned about the mean, variance, and standard deviation of a general probability distribution function to the specific case of the binomial probability distribution. Recall, from Section 2.6, that if a simple system has just two possible outcomes, denoted 1 and 2, with respective probabilities p and $q = 1 - p$, then the probability of obtaining n_1 occurrences of outcome 1 in N observations is

$$P_N(n_1) = \frac{N!}{n_1!\,(N-n_1)!}\,p^{n_1}\,q^{N-n_1}. \tag{2.38}$$

Thus, making use of Equation (2.27), the mean number of occurrences of outcome 1 in N observations is given by

$$\overline{n_1} = \sum_{n_1=0,N} P_N(n_1)\,n_1 = \sum_{n_1=0,N} \frac{N!}{n_1!\,(N-n_1)!}\,p^{n_1}\,q^{N-n_1}\,n_1. \tag{2.39}$$

We can see that if the final factor n_1 were absent on the right-hand side of the previous expression then it would just reduce to the binomial expansion, which we know how to sum. [See Equation (2.23).] We can take advantage of this fact using a rather elegant mathematical sleight of hand. Observe that because

$$n_1\,p^{n_1} \equiv p\,\frac{\partial}{\partial p}\,p^{n_1}, \tag{2.40}$$

the previous summation can be rewritten as

$$\sum_{n_1=0,N} \frac{N!}{n_1!\,(N-n_1)!}\,p^{n_1}\,q^{N-n_1}\,n_1 \equiv p\,\frac{\partial}{\partial p}\left[\sum_{n_1=0,N} \frac{N!}{n_1!\,(N-n_1)!}\,p^{n_1}\,q^{N-n_1}\right]. \tag{2.41}$$

The term in square brackets is now the familiar binomial expansion, and can be written more succinctly as $(p+q)^N$. Thus,

$$\sum_{n_1=0,N} \frac{N!}{n_1!\,(N-n_1)!}\, p^{n_1}\, q^{N-n_1}\, n_1 = p\,\frac{\partial}{\partial p}\,(p+q)^N = p\,N\,(p+q)^{N-1}. \tag{2.42}$$

However, $p+q=1$ for the case in hand [see Equation (2.11)], so

$$\overline{n_1} = N\,p. \tag{2.43}$$

In fact, we could have guessed the previous result. By definition, the probability, p, is the number of occurrences of the outcome 1 divided by the number of trials, in the limit as the number of trials goes to infinity:

$$p = {}_{lt\,N\to\infty}\,\frac{n_1}{N}. \tag{2.44}$$

If we think carefully, however, we can see that taking the limit as the number of trials goes to infinity is equivalent to taking the mean value, so that

$$p = \overline{\left(\frac{n_1}{N}\right)} = \frac{\overline{n_1}}{N}. \tag{2.45}$$

But, this is just a simple rearrangement of Equation (2.43).

Let us now calculate the variance of n_1. Recall, from Equation (2.36), that

$$\overline{(\Delta n_1)^2} = \overline{(n_1)^2} - (\overline{n_1})^2. \tag{2.46}$$

We already know $\overline{n_1}$, so we just need to calculate $\overline{(n_1)^2}$. This average is written

$$\overline{(n_1)^2} = \sum_{n_1=0,N} \frac{N!}{n_1!\,(N-n_1)!}\, p^{n_1}\, q^{N-n_1}\,(n_1)^2. \tag{2.47}$$

The sum can be evaluated using a simple extension of the mathematical trick that we used previously to evaluate $\overline{n_1}$. Because

$$(n_1)^2\, p^{n_1} \equiv \left(p\,\frac{\partial}{\partial p}\right)^2 p^{n_1}, \tag{2.48}$$

then

$$\begin{aligned}\sum_{n_1=0,N} \frac{N!}{n_1!\,(N-n_1)!}\, p^{n_1}\, q^{N-n_1}\,(n_1)^2 &\equiv \left(p\,\frac{\partial}{\partial p}\right)^2 \sum_{n_1=0,N} \frac{N!}{n_1!\,(N-n_1)!}\, p^{n_1} q^{N-n_1}\\ &= \left(p\,\frac{\partial}{\partial p}\right)^2 (p+q)^N\\ &= \left(p\,\frac{\partial}{\partial p}\right)\left[p\,N\,(p+q)^{N-1}\right]\\ &= p\left[N\,(p+q)^{N-1} + p\,N\,(N-1)\,(p+q)^{N-2}\right].\end{aligned} \tag{2.49}$$

Using $p + q = 1$, we obtain

$$\overline{(n_1)^2} = p\,[N + p\,N\,(N-1)] = N\,p\,[1 + p\,N - p]$$
$$= (N\,p)^2 + N\,p\,q = (\overline{n_1})^2 + N\,p\,q, \tag{2.50}$$

because $\overline{n_1} = N\,p$. [See Equation (2.43).] It follows that the variance of n_1 is given by

$$\overline{(\Delta n_1)^2} = \overline{(n_1)^2} - (\overline{n_1})^2 = N\,p\,q. \tag{2.51}$$

The standard deviation of n_1 is the square root of the variance [see Equation (2.37)], so that

$$\Delta^* n_1 = \sqrt{N\,p\,q}. \tag{2.52}$$

Recall that this quantity is essentially the width of the range over which n_1 is distributed around its mean value. The relative width of the distribution is characterized by

$$\frac{\Delta^* n_1}{\overline{n_1}} = \frac{\sqrt{N\,p\,q}}{N\,p} = \sqrt{\frac{q}{p}}\,\frac{1}{\sqrt{N}}. \tag{2.53}$$

It is clear, from this formula, that the relative width decreases with increasing N like $N^{-1/2}$. So, the greater the number of trials, the more likely it is that an observation of n_1 will yield a result that is relatively close to the mean value, $\overline{n_1}$.

2.9 Gaussian Probability Distribution

Consider a very large number of observations, $N \gg 1$, made on a system with two possible outcomes. Suppose that the probability of outcome 1 is sufficiently large that the average number of occurrences after N observations is much greater than unity: that is,

$$\overline{n_1} = N\,p \gg 1. \tag{2.54}$$

In this limit, the standard deviation of n_1 is also much greater than unity,

$$\Delta^* n_1 = \sqrt{N\,p\,q} \gg 1, \tag{2.55}$$

implying that there are very many probable values of n_1 scattered about the mean value, $\overline{n_1}$. This suggests that the probability of obtaining n_1 occurrences of outcome 1 does not change significantly in going from one possible value of n_1 to an adjacent value. In other words,

$$\frac{|P_N(n_1+1) - P_N(n_1)|}{P_N(n_1)} \ll 1. \tag{2.56}$$

In this situation, it is useful to regard the probability as a smooth function of n_1. Let n be a continuous variable that is interpreted as the number of occurrences of outcome 1 (after N observations) whenever it takes on a positive integer value. The probability that n lies between n and $n + dn$ is defined

$$P(n, n + dn) = \mathcal{P}(n)\, dn, \tag{2.57}$$

where $\mathcal{P}(n)$ is called the *probability density*, and is independent of dn. The probability can be written in this form because $P(n, n + dn)$ can always be expanded as a Taylor series in dn, and must go to zero as $dn \to 0$. We can write

$$\int_{n_1-1/2}^{n_1+1/2} \mathcal{P}(n)\, dn = P_N(n_1), \tag{2.58}$$

which is equivalent to smearing out the discrete probability $P_N(n_1)$ over the range $n_1 \pm 1/2$. Given Equations (2.38) and (2.56), the previous relation can be approximated as

$$\mathcal{P}(n) \simeq P_N(n) = \frac{N!}{n!\,(N-n)!}\, p^n\, q^{N-n}. \tag{2.59}$$

For large N, the relative width of the probability distribution function is small: that is,

$$\frac{\Delta^* n_1}{\overline{n_1}} = \sqrt{\frac{q}{p}}\, \frac{1}{\sqrt{N}} \ll 1. \tag{2.60}$$

This suggests that $\mathcal{P}(n)$ is strongly peaked around the mean value, $\overline{n} = \overline{n_1}$. Suppose that $\ln \mathcal{P}(n)$ attains its maximum value at $n = \tilde{n}$ (where we expect $\tilde{n} \sim \overline{n_1}$). Let us Taylor expand $\ln \mathcal{P}(n)$ around $n = \tilde{n}$. Note that we are expanding the slowly-varying function $\ln \mathcal{P}(n)$, rather than the rapidly-varying function $\mathcal{P}(n)$, because the Taylor expansion of $\mathcal{P}(n)$ does not converge sufficiently rapidly in the vicinity of $n = \tilde{n}$ to be useful. We can write

$$\ln \mathcal{P}(\tilde{n} + \eta) \simeq \ln \mathcal{P}(\tilde{n}) + \eta\, B_1 + \frac{\eta^2}{2}\, B_2 + \cdots, \tag{2.61}$$

where

$$B_k = \left. \frac{d^k \ln \mathcal{P}}{dn^k} \right|_{n=\tilde{n}}. \tag{2.62}$$

By definition,

$$B_1 = 0, \tag{2.63}$$

$$B_2 < 0, \tag{2.64}$$

if $n = \tilde{n}$ corresponds to the maximum value of $\ln \mathcal{P}(n)$.

It follows from Equation (2.59) that

$$\ln \mathcal{P} = \ln N! - \ln n! - \ln (N-n)! + n \ln p + (N-n) \ln q. \tag{2.65}$$

If n is a large integer, such that $n \gg 1$, then $\ln n!$ is almost a continuous function of n, because $\ln n!$ changes by only a relatively small amount when n is incremented by unity. Hence,

$$\frac{d \ln n!}{dn} \simeq \frac{\ln (n+1)! - \ln n!}{1} = \ln \left[\frac{(n+1)!}{n!}\right] = \ln (n+1), \tag{2.66}$$

giving

$$\frac{d \ln n!}{dn} \simeq \ln n, \tag{2.67}$$

for $n \gg 1$. The integral of this relation

$$\ln n! \simeq n \, \ln n - n + O(1), \tag{2.68}$$

valid for $n \gg 1$, is called *Stirling's approximation*, after the Scottish mathematician James Stirling, who first obtained it in 1730.

According to Equations (2.62), (2.65), and (2.67),

$$B_1 = -\ln \tilde{n} + \ln (N - \tilde{n}) + \ln p - \ln q. \tag{2.69}$$

Hence, if $B_1 = 0$ then

$$(N - \tilde{n})\, p = \tilde{n}\, q, \tag{2.70}$$

giving

$$\tilde{n} = N\, p = \overline{n_1}, \tag{2.71}$$

because $p + q = 1$. [See Equations (2.11) and (2.43).] Thus, the maximum of $\ln \mathcal{P}(n)$ occurs exactly at the mean value of n, which equals $\overline{n_1}$.

Further differentiation of Equation (2.69) yields [see Equation (2.62)]

$$B_2 = -\frac{1}{\tilde{n}} - \frac{1}{N - \tilde{n}} = -\frac{1}{N\,p} - \frac{1}{N\,(1-p)} = -\frac{1}{N\,p\,q}, \tag{2.72}$$

because $p + q = 1$. Note that $B_2 < 0$, as required. According to Equation (2.55), the previous relation can also be written

$$B_2 = -\frac{1}{(\Delta^* n_1)^2}. \tag{2.73}$$

It follows, from the previous analysis, that the Taylor expansion of $\ln \mathcal{P}$ can be written

$$\ln \mathcal{P}(\overline{n_1} + \eta) \simeq \ln \mathcal{P}(\overline{n_1}) - \frac{\eta^2}{2\,(\Delta^* n_1)^2} + \cdots. \tag{2.74}$$

Taking the exponential of both sides, we obtain

$$\mathcal{P}(n) \simeq \mathcal{P}(\overline{n_1}) \exp\left[-\frac{(n-\overline{n_1})^2}{2\,(\Delta^* n_1)^2}\right]. \tag{2.75}$$

The constant $\mathcal{P}(\overline{n_1})$ is most conveniently fixed by making use of the normalization condition

$$\sum_{n_1=0,N} P_N(n_1) = 1, \tag{2.76}$$

which becomes

$$\int_0^N \mathcal{P}(n)\,dn \simeq 1 \tag{2.77}$$

for a continuous distribution function. Because we only expect $\mathcal{P}(n)$ to be significant when n lies in the relatively narrow range $\overline{n_1} \pm \Delta^* n_1$, the limits of integration in the previous expression can be replaced by $\pm\infty$ with negligible error. Thus,

$$\mathcal{P}(\overline{n_1}) \int_{-\infty}^{\infty} \exp\left[-\frac{(n-\overline{n_1})^2}{2\,(\Delta^* n_1)^2}\right] dn = \mathcal{P}(\overline{n_1})\,\sqrt{2}\,\Delta^* n_1 \int_{-\infty}^{\infty} \mathrm{e}^{-x^2}\,dx \simeq 1. \tag{2.78}$$

As is well known,

$$\int_{-\infty}^{\infty} \mathrm{e}^{-x^2}\,dx = \sqrt{\pi}. \tag{2.79}$$

(See Exercise 2.1.) It follows from the normalization condition (2.78) that

$$\mathcal{P}(\overline{n_1}) \simeq \frac{1}{\sqrt{2\pi}\,\Delta^* n_1}. \tag{2.80}$$

Finally, we obtain

$$\mathcal{P}(n) \simeq \frac{1}{\sqrt{2\pi}\,\Delta^* n_1} \exp\left[-\frac{(n-\overline{n_1})^2}{2\,(\Delta^* n_1)^2}\right]. \tag{2.81}$$

This is the famous *Gaussian probability distribution*, named after the German mathematician Carl Friedrich Gauss, who discovered it while investigating the distribution of errors in measurements. The Gaussian distribution is only valid in the limits $N \gg 1$ and $\overline{n_1} \gg 1$.

Suppose we were to plot the probability $P_N(n_1)$ against the integer variable n_1, and then fit a continuous curve through the discrete points thus obtained. This curve would be equivalent to the continuous probability density curve $\mathcal{P}(n)$, where n is the continuous version of n_1. According to Equation (2.81), the probability density attains its maximum value when n equals the mean of n_1, and is also symmetric about this point. In fact, when plotted with the appropriate ratio of vertical to horizontal scalings, the Gaussian probability density curve looks rather like the outline of a bell centered on $n = \overline{n_1}$. Hence, this curve is sometimes called a *bell curve*. At one standard deviation away from the mean value—that is $n = \overline{n_1} \pm \Delta^* n_1$—the probability density is about 61% of its peak value. At two standard deviations away from the mean value, the probability density is about 13.5% of its peak value. Finally, at three standard deviations away from the mean value, the probability density is only about 1% of its peak value. We conclude that there is very little chance that n_1 lies more than about three standard deviations away from its mean value. In other words, n_1 is almost certain to lie in the relatively narrow range $\overline{n_1} \pm 3\Delta^* n_1$.

In the previous analysis, we went from a discrete probability function, $P_N(n_1)$, to a continuous probability density, $\mathcal{P}(n)$. The normalization condition becomes

$$1 = \sum_{n_1=0,N} P_N(n_1) \simeq \int_{-\infty}^{\infty} \mathcal{P}(n)\, dn \tag{2.82}$$

under this transformation. Likewise, the evaluations of the mean and variance of the distribution are written

$$\overline{n_1} = \sum_{n_1=0,N} P_N(n_1)\, n_1 \simeq \int_{-\infty}^{\infty} \mathcal{P}(n)\, n\, dn, \tag{2.83}$$

and

$$\overline{(\Delta n_1)^2} \equiv (\Delta^* n_1)^2 = \sum_{n_1=0,N} P_N(n_1)\, (n_1 - \overline{n_1})^2 \simeq \int_{-\infty}^{\infty} \mathcal{P}(n)\, (n - \overline{n_1})^2\, dn, \tag{2.84}$$

respectively. These results follow as simple generalizations of previously established results for the discrete function $P_N(n_1)$. The limits of integration in the previous expressions can be approximated as $\pm\infty$ because $\mathcal{P}(n)$ is only non-negligible in a relatively narrow range of n. Finally, it is easily demonstrated that Equations (2.82)–(2.84) are indeed true by substituting in the Gaussian probability density, Equation (2.81), and then performing a few elementary integrals. (See Exercise 2.3.)

2.10 Central Limit Theorem

It may appear, at first sight, that the Gaussian distribution function is only relevant to two-state systems. In fact, as we shall see, the Gaussian probability distribution is of crucial importance to statistical physics because, under certain circumstances, it applies to all types of system.

Let us briefly review how we obtained the Gaussian distribution function in the first place. We started from a very simple system with only two possible outcomes. Of course, the probability distribution function (for n_1) for this system did not look anything like a Gaussian. However, when we combined very many of these simple systems together, to produce a complicated system with a great number of possible outcomes, we found that the resultant probability distribution function (for n_1) reduced to a Gaussian in the limit that the number of simple systems tended to infinity. We started from a two-outcome system because it was easy to calculate the final probability distribution function when a finite number of such systems were combined together. Clearly, if we had started from a more complicated system then this calculation would have been far more difficult.

Suppose that we start from a general system, with a general probability distribution function (for some measurable quantity x). It turns out that if we combine a sufficiently large number of such systems together then the resultant distribution function (for x) is always Gaussian. This astonishing result is known as the *central limit theorem*. Unfortunately, the central limit theorem is notoriously difficult to prove. A somewhat restricted proof is presented in Sections 1.10 and 1.11 of [Reif (1965)]. A rigorous proof is given in [Khinchin (1949)]. The central limit theorem guarantees that the probability distribution of any measurable quantity is Gaussian, provided that a sufficiently large number of statistically independent observations are made. We can, therefore, confidently predict that Gaussian probability distributions are going to crop up very frequently in statistical thermodynamics.

Exercises

2.1 Let $I_x = \int_{-\infty}^{\infty} \mathrm{e}^{-x^2}\,dx$ and $I_y = \int_{-\infty}^{\infty} \mathrm{e}^{-y^2}\,dy$. Show that

$$I_x\,I_y = \int_0^{\infty} 2\pi\,r\,\mathrm{e}^{-r^2}\,dr,$$

where $r^2 = x^2 + y^2$. Hence, deduce that $I_x = I_y = \pi^{1/2}$.

2.2 Show that

$$\int_{-\infty}^{\infty} \mathrm{e}^{-\beta x^2}\,dx = \sqrt{\frac{\pi}{\beta}}.$$

Hence, deduce that

$$\int_{-\infty}^{\infty} x^2\,\mathrm{e}^{-\beta x^2}\,dx = \frac{\pi^{1/2}}{2\,\beta^{3/2}}.$$

2.3 Confirm via direct integration that

$$\int_{-\infty}^{\infty} \mathcal{P}(n)\,dn = 1,$$

$$\int_{-\infty}^{\infty} \mathcal{P}(n)\,n\,dn = \bar{n},$$

$$\int_{-\infty}^{\infty} \mathcal{P}(n)\,(n-\bar{n})^2\,dn = (\varDelta^* n)^2,$$

where

$$\mathcal{P}(n) = \frac{1}{\sqrt{2\pi}\,\Delta^* n} \exp\left[-\frac{(n-\bar{n})^2}{2\,(\Delta^* n)^2}\right].$$

2.4 Show that the probability of throwing 6 points or less with three (six-sided) dice is $5/54$.

2.5 Consider a game in which six (six-sided) dice are rolled. Find the probability of obtaining:

(a) exactly one ace.
(b) at least one ace.
(c) exactly two aces.

(From [Reif (1965)])

2.6 In the game of Russian roulette, one inserts a single cartridge into the drum of a revolver, leaving the other five chambers of the drum empty. One then spins the drum, aims at one's head, and pulls the trigger.

(a) Show that the probability of still being alive after playing the game N times is $(5/6)^N$.
(b) Show that the probability of surviving $(N-1)$ turns in this game, and then being shot the Nth times one pulls the trigger, is $(5/6)^{N-1}\,(1/6)$.
(c) Show that the mean number of times a player gets to pull the trigger is 6.

(From [Reif (1965)])

2.7 A battery of total emf V is connected to a resistor R. As a result, an amount of power $P = V^2/R$ is dissipated in the resistor. The battery itself consists of N individual cells connected in series, so that V is equal to the sum of the emf's of all these cells. The battery is old, however, so that not all cells are in perfect condition. Thus, there is a probability p that the emf of any individual cell has its normal value v; and a probability $1-p$ that the emf of any individual cell is zero because the cell has become internally shorted. The individual cells are statistically independent of each other. Under these conditions, show that the mean power, $\overline{P}$, dissipated in the resistor, is

$$\overline{P} = \frac{p^2\,V^2}{R}\left[1 + \frac{(1-p)}{N\,p}\right].$$

(From [Reif (1965)])

2.8 A drunk starts out from a lamppost in the middle of a street, taking steps of uniform length l to the right or to the left with equal probability.

(a) Show that the average distance from the lamppost after N steps is zero.

(b) Show that the root-mean-square distance (i.e. the square-root of the mean of the distance squared) from the lamppost after N steps is $\sqrt{N}\, l$.

(c) Show that the probability that the drunk will return to the lamppost after N steps is zero if N is odd, and

$$P_N = \frac{N!}{(N/2)!\,(N/2)!}\left(\frac{1}{2}\right)^N$$

if N is even.

2.9 A molecule in a gas moves equal distances l between collisions with equal probabilities in any direction. Show that, after a total of N such displacements, the mean-square displacement, $\overline{R^2}$, of the molecule from its starting point is $\overline{R^2} = N\, l^2$.

2.10 A penny is tossed 400 times. Find the probability of getting 215 heads. (Hint: Use the Gaussian distribution.) (From [Reif (1965)])

2.11 Suppose that the probability density for the speed s of a car on a highway is given by

$$\rho(s) = A\, s\, \exp\left(\frac{-s}{s_0}\right),$$

where $0 \le s \le \infty$. Here, A and s_0 are positive constants. More explicitly, $\rho(s)\, ds$ gives the probability that a car has a speed between s and $s + ds$.

(a) Determine A in terms of s_0.

(b) What is the mean value of the speed?

(c) What is the most probable speed: that is, the speed for which the probability density has a maximum.

(d) What is the probability that a car has a speed more than three times as large as the mean value?

Chapter 3

Statistical Mechanics

3.1 Introduction

This chapter is devoted to the analysis of the internal motions of a many-particle system using probability theory; a subject area that is known as *statistical mechanics*.

3.2 Specification of State of Many-Particle System

Let us first consider how we might specify the state of a many-particle system. Consider the simplest possible dynamical system, which consists of a single spinless particle moving classically in one dimension. Assuming that we know the particle's equation of motion, the state of the system is fully specified once we simultaneously measure the particle's displacement, q, and momentum, p. In principle, if we know q and p then we can calculate the state of the system at all subsequent times using the equation of motion. In practice, it is impossible to specify q and p exactly, because there is always an intrinsic error in any experimental measurement.

Consider the time evolution of q and p. This can be visualized by plotting the point (q, p) in the q-p plane. This plane is generally known as *phase-space*. In general, as time progresses, the point (q, p) will trace out some very complicated pattern in phase-space. Suppose that we divide phase-space into rectangular cells of uniform dimensions δq and δp. Here, δq is the intrinsic error in the position measurement, and δp the intrinsic error in the momentum measurement. The area of each cell is

$$\delta q\,\delta p = h_0, \tag{3.1}$$

where h_0 is a small constant having the dimensions of angular momentum. The coordinates q and p can now be conveniently specified by indicating the cell in phase-space into which they plot at any given time. This procedure automatically ensures that we do not attempt to specify q and p to an accuracy greater than our experimental error, which would clearly be pointless.

Let us now consider a single spinless particle moving in three dimensions. In order to specify the state of the system, we now need to know three q-p pairs: that is, q_x-p_x, q_y-p_y, and q_z-p_z. Incidentally, the number of q-p pairs needed to specify the state of the system is usually called the *number of degrees of freedom* of the system. (See Section B.1.) Thus, a single particle moving in one dimension constitutes a one degree of freedom system, whereas a single particle moving in three dimensions constitutes a three degree of freedom system.

Consider the time evolution of $\mathbf{q}$ and $\mathbf{p}$, where $\mathbf{q} = (q_x, q_y, q_z)$, et cetera. This can be visualized by plotting the point ($\mathbf{q}$, $\mathbf{p}$) in the six-dimensional $\mathbf{q}$-$\mathbf{p}$ phase-space. Suppose that we divide the q_x-p_x plane into rectangular cells of uniform dimensions δq and δp, and do likewise for the q_y-p_y and q_z-p_z planes. Here, δq and δp are again the intrinsic errors in our measurements of position and momentum, respectively. This is equivalent to dividing phase-space up into regular six-dimensional cells of volume h_0^3. The coordinates $\mathbf{q}$ and $\mathbf{p}$ can now be conveniently specified by indicating the cell in phase-space into which they plot at any given time. Again, this procedure automatically ensures that we do not attempt to specify $\mathbf{q}$ and $\mathbf{p}$ to an accuracy greater than our experimental error.

Finally, let us consider a system consisting of N spinless particles moving classically in three dimensions. In order to specify the state of the system, we need to specify a large number of q-p pairs. The requisite number is simply the number of degrees of freedom, f. For the present case, $f = 3N$, because each particle needs three q-p pairs. Thus, phase-space (i.e., the space of all the q-p pairs) now possesses $2f = 6N$ dimensions. Consider a particular pair of conjugate (see Section B.4) phase-space coordinates, q_i and p_i. As before, we divide the q_i-p_i plane into rectangular cells of uniform dimensions δq and δp. This is equivalent to dividing phase-space into regular $2f$ dimensional cells of volume h_0^f. The state of the system is specified by indicating which cell it occupies in phase-space at any given time.

In principle, we can specify the state of the system to arbitrary accuracy, by taking the limit $h_0 \to 0$. In reality, we know from Heisenberg's uncertainty principle (see Section C.8) that it is impossible to simultaneously measure a coordinate, q_i, and its conjugate momentum, p_i, to greater accuracy than $\delta q_i\, \delta p_i = \hbar/2$. Here, $\hbar$ is Planck's constant divided by 2π. This implies that

$$h_0 \geq \hbar/2. \tag{3.2}$$

In other words, the uncertainty principle sets a lower limit on how finely we can chop up classical phase-space.

In quantum mechanics, we can specify the state of the system by giving its wavefunction at time t,

$$\psi(q_1, \cdots, q_f, s_1, \cdots, s_g, t), \tag{3.3}$$

where f is the number of translational degrees of freedom, and g the number of internal (e.g., spin) degrees of freedom. (See Appendix C.) For instance, if the system consists of N spin-one-half particles then there will be $3N$ translational degrees of freedom, and N spin degrees of freedom (because the spin of each particle can either be directed up or down along the z-axis). Alternatively, if the system is in a *stationary state* (i.e., an eigenstate of the Hamiltonian) then we can just specify $f + g$ quantum numbers. (See Sections C.9 and C.10.) Either way, the future time evolution of the wavefunction is fully determined by Schrödinger's equation. In reality, this approach is not practical because the Hamiltonian of the system is only known approximately. Typically, we are dealing with a system consisting of many weakly-interacting particles. We usually know the Hamiltonian for completely non-interacting particles, but the component of the Hamiltonian associated with particle interactions is either impossibly complicated, or not very well known. We can define approximate stationary eigenstates using the Hamiltonian for non-interacting particles. The state of the system is then specified by the quantum numbers identifying these eigenstates. In the absence of particle interactions, if the system starts off in a stationary state then it stays in that state for ever, so its quantum numbers never change. The interactions allow the system to make transitions between different "stationary" states, causing its quantum numbers to change in time.

3.3 Principle of Equal A Priori Probabilities

We now know how to specify the instantaneous state of a many-particle system. In principle, such a system is completely deterministic. If we know the initial state, and the equations of motion (or the Hamiltonian), then we can evolve the system forward in time, and, thereby, determine all future states. In reality, it is quite impossible to specify the initial state, or the equations of motion, to sufficient accuracy for this method to have any chance of working. Furthermore, even if it were possible, it would still not be a practical proposition to evolve the equations of motion. Remember that we are typically dealing with systems containing Avogadro's number of particles: that is, about 10^{24} particles. We cannot evolve 10^{24} simultaneous differential equations. Even if we could, we would not want to. After all, we are not particularly interested in the motions of individual particles. What we really want is statistical information regarding the motions of all particles in the system.

Clearly, what is required here is a statistical treatment of the problem. Instead of focusing on a single system, let us proceed, in the usual manner, and consider a statistical ensemble consisting of a large number of identical systems. (See Section 2.2.) In general, these systems are distributed over many different states at any given time. In order to evaluate the probability that the system possesses a particular property, we merely need to find the number of systems in the ensemble that exhibit this property, and then divide by the total number of systems, in the limit as the latter number tends to infinity.

We can usually place some general constraints on the system. Typically, we know the total energy, E, the total volume, V, and the total number of particles, N. To be more exact, we can only really say that the total energy lies between E and $E + \delta E$, et cetera, where δE is an experimental error. Thus, we need only concern ourselves with those systems in the ensemble exhibiting states that are consistent with the known constraints. We shall call these the *states accessible to the system*. In general, there are a great many such states.

We now need to calculate the probability of the system being found in each of its accessible states. In fact, the only way that we could calculate these probabilities would be to evolve all of the systems in the ensemble in time, and observe how long, on average, they spend in each accessible state. But, as we have already discussed, such a calculation is completely out of the question. Instead, we shall effectively guess the probabilities.

Let us consider an isolated system in equilibrium. In this situation, we would expect the probability of the system being found in one of its accessible states to be independent of time. This implies that the statistical ensemble does not evolve with time. Individual systems in the ensemble will constantly change state, but the average number of systems in any given state should remain constant. Thus, all macroscopic parameters describing the system, such as the energy and the volume, should also remain constant. There is nothing in the laws of mechanics that would lead us to suppose that the system will be found more often in one of its accessible states than in another. We assume, therefore, that *the system is equally likely to be found in any of its accessible states*. This is called the assumption of *equal a priori probabilities*, and lies at the heart of statistical mechanics. In fact, we use assumptions like this all of the time without really thinking about them. Suppose that we were asked to pick a card at random from a well-shuffled pack of ordinary playing cards. Most people would accept that we have an equal probability of picking any card in the pack. There is nothing that would favor one particular card over all of the others. Hence, because there are fifty-two cards in a normal pack, we would expect the probability of picking the ace of spades, say, to be $1/52$. We could now place some constraints on the system. For instance, we could only count red cards, in which case the probability of picking the ace of hearts, say, would be $1/26$, by the same reasoning. In both cases, we have used the principle of equal a priori probabilities. People really believe that this principle applies to games of chance, such as cards, dice, and roulette. In fact, if the principle were found not to apply to a particular game then most people would conclude that the game was crooked. But, imagine trying to prove that the principle actually does apply to a game of cards. This would be a very difficult task. We would have to show that the way most people shuffle cards is effective at randomizing their order. A convincing study would have to be part mathematics and part psychology.

In statistical mechanics, we treat a many-particle system a little like an extremely large game of cards. Each accessible state corresponds to one of the cards in the pack. The interactions between particles cause the system to continually change state. This is equivalent to constantly shuffling the pack. Finally, an observation of the state of the system is like picking a card at random from the pack. The principle of equal a priori probabilities then boils down to saying that we have an equal chance of choosing any particular card.

It is, unfortunately, impossible to prove with mathematical rigor that the principle of equal a priori probabilities applies to many-particle systems. Over the years, many people have attempted this proof, and all have failed. Not surprisingly, therefore, statistical mechanics was greeted with a great deal of scepticism when it was first proposed in the late 1800's. Nowadays, statistical mechanics is completely accepted into the canon of physics; quite simply because it works.

It is actually possible to formulate a reasonably convincing scientific case for the principle of equal a priori probabilities. To achieve this we have to make use of the so-called H-theorem.

3.4 H-Theorem

Consider a system of weakly-interacting particles. In quantum mechanics, we can write the Hamiltonian for such a system as

$$H = H_0 + H_1, \tag{3.4}$$

where H_0 is the Hamiltonian for completely non-interacting particles, and H_1 a small correction due to the particle interactions. We can define approximate stationary eigenstates of the system using H_0. Thus,

$$H_0\,\Psi_r = E_r\,\Psi_r, \tag{3.5}$$

where the index r labels a state of energy E_r and eigenstate Ψ_r. (See Section C.9.) In general, there are many different eigenstates with the same energy; these are called *degenerate* states. (See Section C.10.)

For example, consider N non-interacting spinless particles of mass m confined in a cubic box of dimension L. According to standard wave-mechanics, the energy levels of the ith particle are given by

$$e_i = \frac{\hbar^2 \pi^2}{2\,m\,L^2}\left(n_{1\,i}^2 + n_{2\,i}^2 + n_{3\,i}^2\right), \tag{3.6}$$

where $n_{1\,i}$, $n_{2\,i}$, and $n_{3\,i}$ are three (positive integer) quantum numbers. (See Section C.10.) Here, $\hbar$ is Planck's constant divided by 2π. The overall energy of the system is the sum of the energies of the individual particles, so that for a general state r,

$$E_r = \sum_{i=1,N} e_i. \tag{3.7}$$

The overall state of the system is thus specified by $3N$ quantum numbers (i.e., three quantum numbers per particle). There are clearly very many different arrangements of these quantum numbers that give the same overall energy.

Consider, now, a statistical ensemble of systems made up of weakly-interacting particles. Suppose that this ensemble is initially very far from equilibrium. For instance, the systems in the ensemble might only be distributed over a very small subset of their accessible states. If each system starts off in a particular stationary state (i.e., with a particular set of quantum numbers) then, in the absence of particle interactions, it will remain in that state for ever. Hence, the ensemble will always stay far from equilibrium, and the principle of equal a priori probabilities will never be applicable. In reality, particle interactions cause each system in the ensemble to make transitions between its accessible "stationary" states. This allows the overall state of the ensemble to change in time.

Let us label the accessible states of our system by the index r. We can ascribe a time-dependent probability, $P_r(t)$, of finding the system in a particular approximate stationary state r at time t. Of course, $P_r(t)$ is proportional to the number of systems in the ensemble in state r at time t. In general, P_r is time dependent because the ensemble is evolving towards an equilibrium state. Let us assume that the probabilities are properly normalized, so that the sum over all accessible states always yields unity: that is,

$$\sum_r P_r(t) = 1. \tag{3.8}$$

Small interactions between particles cause transitions between the approximate stationary states of the system. Thus, there exists some transition probability per unit time, W_{rs}, that a system originally in state r ends up in state s as a result of these interactions. Likewise, there exists a probability per unit time W_{sr} that a system in state s makes a transition to state r. These transition probabilities are meaningful in quantum mechanics provided that the particle interaction strength is sufficiently small, there is a nearly continuous distribution of accessible energy levels, and we consider time intervals that are not too small. These conditions are easily satisfied for the types of systems usually analyzed via statistical mechanics (e.g., nearly-ideal gases). One important conclusion of quantum mechanics is that the forward and backward transition probabilities between two states are the same, so that

$$W_{rs} = W_{sr} \tag{3.9}$$

for any two states r and s. This result follows from the time-reversal symmetry of quantum mechanics. On the microscopic scale of individual particles, all fundamental laws of physics (in particular, classical and quantum mechanics) possess this symmetry. So, if a certain motion of particles satisfies the classical equations of motion (or Schrödinger's equation) then the reversed motion, with all particles starting off from their final positions and then retracing their paths exactly until they reach their initial positions, satisfies these equations just as well.

Suppose that we were to "film" a microscopic process, such as two classical particles approaching one another, colliding, and moving apart. We could then gather an audience together and show them the film. To make things slightly more interesting we could play it either forwards or backwards. Because of the time-reversal symmetry of classical mechanics, the audience would not be able to tell which way the film was running. In both cases, the film would show completely plausible physical events.

We can play the same game for a quantum process. For instance, we could "film" a group of photons impinging on some atoms. Occasionally, one of the atoms will absorb a photon and make a transition to an excited state (i.e., a state with higher than normal energy). We could easily estimate the rate constant for this process by watching the film carefully. If we play the film backwards then it will appear to show excited atoms occasionally emitting a photon, and then decaying back to their unexcited state. If quantum mechanics possesses time-reversal symmetry (which it certainly does) then both films should appear equally plausible. This means that the rate constant for the absorption of a photon to produce an excited state must be the same as the rate constant for the excited state to decay by the emission of a photon. Otherwise, in the backwards film, the excited atoms would appear to emit photons at the wrong rate, and we could then tell that the film was being played backwards. It follows, therefore, that, as a consequence of time-reversal symmetry, the rate constant for any process in quantum mechanics must equal the rate constant for the inverse process.

The probability, P_r, of finding the systems in the ensemble in a particular state r changes with time for two reasons. Firstly, systems in another state s can make transitions to the state r. The rate at which this occurs is P_s, the probability that the systems are in the state s to begin with, multiplied by the rate constant of the transition W_{sr}. Secondly, systems in the state r can make transitions to other states such as s. The rate at which this occurs is clearly P_r times W_{rs}. We can, thus, write a simple differential equation for the time evolution of P_r:

$$\frac{dP_r}{dt} = \sum_{s \neq r} P_s W_{sr} - \sum_{s \neq r} P_r W_{rs}, \tag{3.10}$$

or

$$\frac{dP_r}{dt} = \sum_s W_{rs}(P_s - P_r), \tag{3.11}$$

where use has been made of the symmetry condition (3.9). The summation is over all accessible states.

Consider now the quantity H (from which the H-theorem derives its name), which is defined as the average value of $\ln P_r$ taken over all accessible states:

$$H \equiv \overline{\ln P_r} = \sum_r P_r \ln P_r. \tag{3.12}$$

This quantity changes as the individual probabilities P_r vary in time. Straightforward differentiation of the previous equation yields

$$\frac{dH}{dt} = \sum_r \left(\frac{dP_r}{dt} \ln P_r + \frac{dP_r}{dt} \right) = \sum_r \frac{dP_r}{dt} (\ln P_r + 1). \tag{3.13}$$

According to Equation (3.11), this can be written

$$\frac{dH}{dt} = \sum_r \sum_s W_{rs} (P_s - P_r) (\ln P_r + 1). \tag{3.14}$$

We can now interchange the dummy summations indices r and s to give

$$\frac{dH}{dt} = \sum_r \sum_s W_{sr} (P_r - P_s) (\ln P_s + 1). \tag{3.15}$$

Finally, we can write dH/dt in a more symmetric form by adding the previous two equations, and then making use of Equation (3.9):

$$\frac{dH}{dt} = -\frac{1}{2} \sum_r \sum_s W_{rs} (P_r - P_s) (\ln P_r - \ln P_s). \tag{3.16}$$

Note, however, that $\ln P_r$ is a monotonically increasing function of P_r. It follows that $\ln P_r > \ln P_s$ whenever $P_r > P_s$, and vice versa. Thus, in general, the right-hand side of the previous equation is the sum of many negative contributions. Hence, we conclude that

$$\frac{dH}{dt} \leq 0. \tag{3.17}$$

The equality sign only holds in the special case where all accessible states are equally probable, so that $P_r = P_s$ for all r and s. This result is called the H-theorem.

The H-theorem tells us that if an isolated system is initially not in equilibrium then it will evolve in time, under the influence of particle interactions, in such a manner that the quantity H always decreases. This process will continue until H reaches its minimum possible value, at which point $dH/dt = 0$, and there is no further evolution of the system. According to Equation (3.16), in this final equilibrium state, the system is equally likely to be found in any one of its accessible states. This is, of course, the situation predicted by the principle of equal a priori probabilities.

The previous argument does not constitute a mathematically rigorous proof that the principle of equal a priori probabilities applies to many-particle systems, because we tacitly made an unwarranted assumption. That is, we assumed that the probability of the system making a transition from some state r to another state s is independent of the past history of the system. In general, this is not the case in physical systems, although there are many situations in which it is a fairly good approximation. Thus, the epistemological status of the principle of equal a priori probabilities is that it is plausible, but remains unproven. As we have already mentioned, the ultimate justification for this principle is empirical; it leads to theoretical predictions that are in accordance with experimental observations. The H-theorem, and its implications, are discussed in greater detail in [Tolman (1938)].

3.5 Relaxation Time

The H-theorem guarantees that an isolated many-particle system will eventually reach an equilibrium state, irrespective of its initial state. The typical time required for this process to take place is called the *relaxation time*, and depends, in detail, on the nature of the inter-particle interactions. The principle of equal a priori probabilities is only valid for equilibrium states. It follows that we can only safely apply this principle to systems that have remained undisturbed for many relaxation times since they were setup, or last interacted with the outside world.

The relaxation time for the air in a typical classroom is very much less than one second. This suggests that such air is probably in equilibrium most of the time, and should, therefore, be governed by the principle of equal a priori probabilities. In fact, this is known to be the case.

Consider another example. Our galaxy, the "Milky Way," is an isolated dynamical system made up of about 10^{11} stars. In fact, it can be thought of as a self-gravitating "gas" of stars. At first sight, the Milky Way would seem to be an ideal system on which to test out the ideas of statistical mechanics. Stars in the Milky Way interact via occasional near-miss events in which they exchange energy and momentum. Actual collisions are very rare indeed. Unfortunately, such interactions take place very infrequently, because there is a lot of empty space between the stars. The best estimate for the relaxation time of the Milky Way is about 10^{13} years. This should be compared with the estimated age of the Milky Way, which is only about 10^{10} years. It is clear that, despite its great age, the Milky Way has not been around long enough to reach an equilibrium state. This suggests that the principle of equal a priori probabilities cannot be used to describe stellar dynamics. Not surprisingly, the observed velocity distribution of the stars in the vicinity of the Sun is not governed by this principle.

3.6 Reversibility and Irreversibility

Previously, we mentioned that, on a microscopic level, the laws of physics are invariant under time-reversal. In other words, microscopic phenomena look physically plausible when run in reverse. We usually say that these phenomena are *reversible*. Does this imply that macroscopic phenomena are also reversible? Consider an isolated many-particle system that starts off far from equilibrium. According to the H-theorem, it will evolve towards equilibrium and, as it does so, the macroscopic quantity H will decrease. But, if we run this process backwards in time then the system will appear to evolve away from equilibrium, and the quantity H will increase. This type of behavior is not physical because it violates the H-theorem. In other words, if we saw a film of a macroscopic process then we could very easily tell if it was being run backwards.

For instance, suppose that, by some miracle, we were able to move all of the oxygen molecules in the air in some classroom to one side of the room, and all of the nitrogen molecules to the opposite side. We would not expect this state to persist for very long. Pretty soon the oxygen and nitrogen molecules would start to intermingle, and this process would continue until they were thoroughly mixed together throughout the room. This, of course, is the equilibrium state for air. In reverse, this process appears completely unphysical. We would start off from perfectly normal air, and suddenly, for no good reason, the air's constituent oxygen and nitrogen molecules would appear to separate, and move to opposite sides of the room. This scenario is not impossible, but, from everything we know about the world around us, it is spectacularly unlikely. We conclude, therefore, that macroscopic phenomena are generally *irreversible*, because they appear unphysical when run in reverse.

How does the irreversibility of macroscopic phenomena arise? It certainly does not come from the fundamental laws of physics, because these laws are all reversible. In the previous example, the oxygen and nitrogen molecules intermingled by continually scattering off one another. Each individual scattering event would look perfectly reasonable viewed in reverse. However, the net result of these scattering event appears unphysical when run backwards. How can we obtain an irreversible process from the combined effects of very many reversible processes? This is a vitally important question. Unfortunately, we are not quite at the stage where we can formulate a convincing answer. (We shall answer the question in Section 5.6.) Note, however, that the essential irreversibility of macroscopic phenomena is one of the key results of statistical thermodynamics.

3.7 Probability Calculations

The principle of equal a priori probabilities is fundamental to all of statistical mechanics, and allows a complete description of the properties of macroscopic systems in equilibrium. In principle, statistical mechanics calculations are very simple. Consider a system in equilibrium that is isolated, so that its total energy is known to have a constant value lying somewhere in the range E to $E + \delta E$. In order to make statistical predictions, we focus attention on an ensemble of such systems, all of which have their energy in this range. Let $\Omega(E)$ be the total number of different states in the ensemble with energies in the specified range. Suppose that, among these states, there are a number $\Omega(E; y_k)$ for which some parameter, y, of the system assumes the discrete value y_k. (This discussion can easily be generalized to deal with a parameter that can assume a continuous range of values.) The principle of equal a priori probabilities tells us that all of the $\Omega(E)$ accessible states of the system are equally likely to occur in the ensemble. It follows that the probability, $P(y_k)$, that the parameter y of the system assumes the value y_k is simply

$$P(y_k) = \frac{\Omega(E; y_k)}{\Omega(E)}. \tag{3.18}$$

Clearly, the mean value of y for the system is given by

$$\bar{y} = \frac{\sum_k \Omega(E; y_k)\, y_k}{\Omega(E)}, \tag{3.19}$$

where the sum is over all possible values that y can assume. In the previous formula, it is tacitly assumed that $\Omega(E) \to \infty$, which is generally the case in thermodynamic systems.

It can be seen that, using the principle of equal a priori probabilities, all calculations in statistical mechanics reduce to counting states, subject to various constraints. In principle, this is a fairly straightforward task. In practice, problems arise if the constraints become too complicated. These problems can usually be overcome with a little mathematical ingenuity. Nevertheless, there is no doubt that this type of calculation is far easier than trying to solve the classical equations of motion (or Schrödinger's equation) directly for a many-particle system.

3.8 Behavior of Density of States

Consider an isolated system in equilibrium whose volume is V, and whose energy lies in the range E to $E + \delta E$. Let $\Omega(E, V)$ be the total number of microscopic states that satisfy these constraints. It would be useful if we could estimate how this number typically varies with the macroscopic parameters of the system. The easiest way to do this is to consider a specific example. For instance, an ideal gas made up of spinless monatomic particles. This is a particularly simple example, because, for such a gas, the particles possess translational, but no internal (e.g., vibrational, rotational, or spin), degrees of freedom. By definition, interatomic forces are negligible in an ideal gas. In other words, the individual particles move in an approximately uniform potential. It follows that the energy of the gas is just the total translational kinetic energy of its constituent particles. In other words,

$$E = \frac{1}{2\,m} \sum_{i=1,N} \mathbf{p}_i^{\,2}, \tag{3.20}$$

where m is the particle mass, N the total number of particles, and $\mathbf{p}_i$ the vector momentum of the ith particle.

Consider the system in the limit in which the energy, E, of the gas is much greater than the ground-state energy, so that all of the quantum numbers are large. The classical version of statistical mechanics, in which we divide up phase-space into cells of equal volume, is valid in this limit. The number of states, $\Omega(E, V)$, lying between the energies E and $E + \delta E$ is simply equal to the number of cells in phase-space contained between these energies. In other words, $\Omega(E, V)$ is proportional to the volume of phase-space between these two energies:

$$\Omega(E, V) \propto \int_E^{E+\delta E} d^3\mathbf{r}_1 \cdots d^3\mathbf{r}_N\, d^3\mathbf{p}_1 \cdots d^3\mathbf{p}_N. \tag{3.21}$$

Here, the integrand is the element of volume of phase-space, with

$$d^3\mathbf{r}_i \equiv dx_i\, dy_i\, dz_i, \tag{3.22}$$

$$d^3\mathbf{p}_i \equiv dp_{x\,i}\, dp_{y\,i}\, dp_{z\,i}, \tag{3.23}$$

where (x_i, y_i, z_i) and (p_{xi}, p_{yi}, p_{zi}) are the Cartesian coordinates and momentum components of the ith particle, respectively. The integration is over all coordinates and momenta such that the total energy of the system lies between E and $E + \delta E$.

For an ideal gas, the total energy E does not depend on the positions of the particles. [See Equation (3.20).] This implies that the integration over the position vectors, $\mathbf{r}_i$, can be performed immediately. Because each integral over $\mathbf{r}_i$ extends over the volume of the container (the particles are, of course, not allowed to stray outside the container), $\int d^3\mathbf{r}_i = V$. There are N such integrals, so Equation (3.21) reduces to

$$\Omega(E, V) \propto V^N \chi(E), \tag{3.24}$$

where

$$\chi(E) \propto \int_E^{E+\delta E} d^3\mathbf{p_1} \cdots d^3\mathbf{p_N} \tag{3.25}$$

is a momentum-space integral that is independent of the volume.

The energy of the system can be written

$$E = \frac{1}{2\,m} \sum_{i=1,N} \sum_{\alpha=1,3} p_{\alpha\,i}^2, \tag{3.26}$$

because $\mathbf{p}_i^2 = p_{1\,i}^2 + p_{2\,i}^2 + p_{3\,i}^2$, denoting the (x, y, z) components by (1, 2, 3), respectively. The previous sum contains $3N$ square terms. For E = constant, Equation (3.26) describes the locus of a hyper-sphere of radius $R(E) = (2\,m\,E)^{1/2}$ in the $3N$-dimensional space of the momentum components. Hence, $\chi(E)$ is proportional to the volume of momentum phase-space contained in the region lying between the hyper-sphere of radius $R(E)$, and that of slightly larger radius $R(E + \delta E)$. This volume is proportional to the area of the inner hyper-sphere multiplied by $\delta R \equiv R(E + \delta E) - R(E)$. Because the area varies like R^{3N-1}, and $\delta R \propto \delta E/E^{1/2}$, we have

$$\chi(E) \propto R^{3N-1}/E^{1/2} \propto E^{3N/2-1}. \tag{3.27}$$

Combining this result with Equation (3.24), we obtain

$$\Omega(E, V) \simeq B\,V^N E^{3N/2}, \tag{3.28}$$

where B is a constant independent of V or E, and we have also made use of the fact that $N \gg 1$. Note that, because the number of degrees of freedom of the system is $f = 3N$, the previous relation can be very approximately written

$$\Omega(E, V) \propto V^f E^f. \tag{3.29}$$

In other words, the density of states varies like the extensive macroscopic parameters of the system raised to the power of the number of degrees of freedom. An extensive parameter is one that scales with the size of the system (e.g., the volume). (See Section 8.8.) Because thermodynamic systems generally possess a very large number of degrees of freedom, Equation (3.29) implies that the density of states is an exceptionally rapidly increasing function of the energy and volume. This result, which turns out to be quite general, is of great significance in statistical thermodynamics.

Exercises

3.1 Consider a particle of mass m confined within a cubic box of dimensions $L_x = L_y = L_z$. According to elementary quantum mechanics, the possible energy levels of this particle are given by

$$E = \frac{\hbar^2 \pi^2}{2\,m}\left(\frac{n_x^2}{L_x^2} + \frac{n_y^2}{L_y^2} + \frac{n_z^2}{L_z^2}\right),$$

where n_x, n_y, and n_z are positive integers. (See Section C.10.)

(a) Suppose that the particle is in a given state specified by particular values of the three quantum numbers, n_x, n_y, n_z. By considering how the energy of this state must change when the length, L_x, of the box parallel to the x-axis is very slowly changed by a small amount dL_x, show that the force exerted by a particle in this state on a wall perpendicular to the x-axis is given by $F_x = -\partial E/\partial L_x$.

(b) Explicitly calculate the force per unit area (or pressure) acting on this wall. By averaging over all possible states, find an expression for the mean pressure on this wall. (Hint: exploit the fact that $\overline{n_x^2} = \overline{n_y^2} = \overline{n_z^2}$ must all be equal, by symmetry.) Show that this mean pressure can be written

$$\overline{p} = \frac{2}{3}\frac{\overline{E}}{V},$$

where $\overline{E}$ is the mean energy of the particle, and $V = L_x\,L_y\,L_z$ the volume of the box.

3.2 The state of a system with f degrees of freedom at time t is specified by its generalized coordinates, $q_1, \cdots, q_f$, and conjugate momenta, $p_1, \cdots, p_f$. These evolve according to Hamilton's equations (see Section B.9):

$$\dot{q}_i = \frac{\partial H}{\partial p_i},$$

$$\dot{p}_i = -\frac{\partial H}{\partial q_i}.$$

Here, $H(q_1, \cdots, q_f, p_1, \cdots, p_f, t)$ is the Hamiltonian of the system. Consider a statistical ensemble of such systems. Let

$$\rho(q_1, \cdots, q_f, p_1, \cdots, p_f, t)$$

be the number density of systems in phase-space. In other words, let $\rho(q_1, \cdots, q_f, p_1, \cdots, p_f, t)\, dq_1\, dq_2 \cdots dq_f\, dp_1\, dp_2 \cdots dp_f$ be the number of states with q_1 lying between q_1 and $q_1 + dq_1$, p_1 lying between p_1 and $p_1 + dp_1$, et cetera, at time t.

(a) Show that ρ evolves in time according to *Liouville's theorem*:

$$\frac{\partial \rho}{\partial t} + \sum_{i=1,f} \left(\dot{q}_i \frac{\partial \rho}{\partial q_i} + \dot{p}_i \frac{\partial \rho}{\partial p_i} \right) = 0.$$

[Hint: Consider how the flux of systems into a small volume of phase-space causes the number of systems in the volume to change in time.]

(b) By definition,

$$N = \int_{-\infty}^{\infty} \rho\, dq_1 \cdots dq_f\, dp_1 \cdots dp_f$$

is the total number of systems in the ensemble. The integral is over all of phase-space. Show that Liouville's theorem conserves the total number of systems (i.e., $dN/dt = 0$). You may assume that ρ becomes negligibly small if any of its arguments (i.e., $q_1, \cdots, q_f$ and $p_1, \cdots, p_f$) becomes very large. This is equivalent to assuming that all of the systems are localized to some region of phase-space.

(c) Suppose that H has no explicit time dependence (i.e., $\partial H/\partial t = 0$). Show that the ensemble-averaged energy,

$$\overline{H} = \int_{-\infty}^{\infty} H\, \rho\, dq_1 \cdots dq_f\, dp_1 \cdots dp_f,$$

is a constant of the motion.

(d) Show that if H is also not an explicit function of the coordinate q_j then the ensemble average of the conjugate momentum,

$$\overline{p_j} = \int_{-\infty}^{\infty} p_j\, \rho\, dq_1 \cdots dq_f\, dp_1 \cdots dp_f,$$

is a constant of the motion.

3.3 Consider a system consisting of very many particles. Suppose that an observation of a macroscopic variable, x, can result in any one of a great many closely-spaced values, x_r. Let the (approximately constant) spacing between adjacent values be δx. The probability of occurrence of the value x_r is denoted P_r. The probabilities are assumed to be properly normalized, so that

$$\sum_r P_r \simeq \int_{-\infty}^{\infty} \frac{P_r(x_r)}{\delta x}\,dx_r = 1,$$

where the summation is over all possible values. Suppose that we know the mean and the variance of x, so that

$$\overline{x} = \sum_r x_r\,P_r$$

and

$$\overline{(\Delta x)^2} = \sum_r (x_r - \overline{x})^2\,P_r$$

are both fixed. According to the H-theorem, the system will naturally evolve towards a final equilibrium state in which the quantity

$$H = \sum_r P_r\,\ln P_r$$

is minimized. Used the method of Lagrange multipliers to minimize H with respect to the P_r, subject to the constraints that the probabilities remain properly normalized, and that the mean and variance of x remain constant. (See Section B.6.) Show that the most general form for the P_r which can achieve this goal is

$$P_r(x_r) \simeq \frac{\delta x}{\sqrt{2\pi\,\overline{(\Delta x)^2}}}\left[-\frac{(x_r - \overline{x})^2}{2\,\overline{(\Delta x)^2}}\right].$$

This result demonstrates that the system will naturally evolve towards a final equilibrium state in which all of its macroscopic variables have Gaussian probability distributions, which is in accordance with the central limit theorem. (See Section 2.10.)

Chapter 4

Heat and Work

4.1 Brief History of Heat and Work

In 1789, the French scientist Antoine Lavoisier published a famous treatise on chemistry which, among other things, demolished the then prevalent theory of combustion. This theory, known to history as the *phlogiston theory*, is so extraordinary stupid that it is not even worth describing. In place of phlogiston theory, Lavoisier proposed the first reasonably sensible scientific interpretation of heat. Lavoisier pictured heat as an invisible, tasteless, odorless, weightless fluid, which he called *calorific fluid*. He postulated that hot bodies contain more of this fluid than cold bodies. Furthermore, he suggested that the constituent particles of calorific fluid repel one another, causing heat to flow spontaneously from hot to cold bodies when they are placed in thermal contact.

The modern interpretation of heat is, or course, somewhat different to Lavoisier's *calorific theory*. Nevertheless, there is an important subset of problems, involving heat flow, for which Lavoisier's approach is rather useful. These problems often crop up as examination questions. For example: "A clean dry copper calorimeter contains 100 grams of water at 30° centigrade. A 10 gram block of copper heated to 60° centigrade is added. What is the final temperature of the mixture?". How do we approach this type of problem? According to Lavoisier's theory, there is an analogy between heat flow and incompressible fluid flow under gravity. The same volume of liquid added to containers of different (uniform) cross-sectional area fills them to different heights. If the volume is V, and the cross-sectional area is A, then the height is $h = V/A$. In a similar manner, the same quantity of heat added to different bodies causes them to rise to different temperatures. If Q is the heat and θ is the (absolute) temperature then $\theta = Q/C$, where the constant C is termed the *heat capacity*. [This is a somewhat oversimplified example. In general, the heat capacity is a function of temperature, so that $C = C(\theta)$.] If two containers, filled to different heights, with a free-flowing incompressible fluid are connected together at the bottom, via a small pipe, then fluid will flow under gravity, from one to the other, until the two heights are the same. The final height is easily calculated by equating the total fluid volume in the initial and final states. Thus,

$$h_1 A_1 + h_2 A_2 = h A_1 + h A_2, \tag{4.1}$$

giving

$$h = \frac{h_1 A_1 + h_2 A_2}{A_1 + A_2}. \tag{4.2}$$

Here, h_1 and h_2 are the initial heights in the two containers, A_1 and A_2 are the corresponding cross-sectional areas, and h is the final height. Likewise, if two bodies, initially at different temperatures, are brought into thermal contact then heat will flow, from one to the other, until the two temperatures are the same. The final temperature is calculated by equating the total heat in the initial and final states. Thus,

$$\theta_1 C_1 + \theta_2 C_2 = \theta C_1 + \theta C_2, \tag{4.3}$$

giving

$$\theta = \frac{\theta_1 C_1 + \theta_2 C_2}{C_1 + C_2}, \tag{4.4}$$

where the meaning of the various symbols should be self-evident.

The analogy between heat flow and fluid flow works because, in Lavoisier's theory, heat is a conserved quantity, just like the volume of an incompressible fluid. In fact, Lavoisier postulated that heat was an element. Note that atoms were thought to be indestructible before nuclear reactions were discovered, so the total amount of each element in the cosmos was assumed to be a constant. Thus, if Lavoisier had cared to formulate a law of thermodynamics from his calorific theory then he would have said that the total amount of heat in the universe was a constant.

In 1798, Benjamin Thompson, an Englishman who spent his early years in pre-revolutionary America, was minister for war and police in the German state of Bavaria. One of his jobs was to oversee the boring of cannons in the state arsenal. Thompson was struck by the enormous, and seemingly inexhaustible, amount of heat generated in this process. He simply could not understand where all this heat was coming from. According to Lavoisier's calorific theory, the heat must flow into the cannon from its immediate surroundings, which should, therefore, become colder. The flow should also eventually cease when all of the available heat has been extracted. In fact, Thompson observed that the surroundings of the cannon got hotter, not colder, and that the heating process continued unabated as long as the boring machine was operating. Thompson postulated that some of the mechanical work done on the cannon by the boring machine was being converted into heat. At the time, this was quite a revolutionary concept, and most people were not ready to accept it. This is somewhat surprising, because, by the end of the eighteenth century, the conversion of heat into work, by steam engines, was quite commonplace. Nevertheless, the conversion of work into heat did not gain broad acceptance until 1849, when an English physicist called James Prescott Joule published the results of a long and painstaking series of experiments. Joule confirmed that work could indeed be converted into heat. Moreover, he found that the same amount of work always generates the same quantity of heat. This is true regardless of the nature of the work (e.g., mechanical, electrical, et cetera). Joule was able to formulate what became known as the *work equivalent of heat*. Namely, that 1 newton meter of work is equivalent to 0.241 calories of heat. A calorie is the amount of heat required to raise the temperature of 1 gram of water by 1 degree centigrade. Nowadays, we measure both heat and work using the same units, so that one newton meter, or joule, of work is equivalent to one joule of heat.

In 1850, the German physicist Rudolf Clausius correctly postulated that the essential conserved quantity is neither heat nor work, but some combination of the two which quickly became known as *energy*, from the ancient greek word ἐνέϱγεια, which means *activity* or *action*. According to Clausius, the change in the internal energy of a macroscopic body can be written

$$\Delta E = Q - W, \tag{4.5}$$

where Q is the heat absorbed from the surroundings, and W is the work done on the surroundings. This relation is known as the *first law of thermodynamics*.

4.2 Macrostates and Microstates

In describing a system made up of a great many particles, it is usually possible to specify some macroscopically measurable independent parameters, $x_1, x_2, \cdots, x_n$, that affect the particles' equations of motion. These parameters are termed the *external parameters* of the system. Examples of such parameters are the volume (this gets into the equations of motion because the potential energy becomes infinite when a particle strays outside the available volume), and any applied electric and magnetic fields. A *microstate* of the system is defined as a state for which the motions of the individual particles are completely specified (subject, of course, to the unavoidable limitations imposed by the uncertainty principle of quantum mechanics). In general, the overall energy of a given microstate, r, is a function of the external parameters:

$$E_r \equiv E_r(x_1, x_2, \cdots, x_n). \tag{4.6}$$

A *macrostate* of the system is defined by specifying the external parameters, and any other constraints to which the system is subject. For example, if we are dealing with an isolated system (i.e., one that can neither exchange heat with, nor do work on, its surroundings) then the macrostate might be specified by giving the values of the volume and the constant total energy. For a many-particle system, there are generally a very great number of microstates that are consistent with a given macrostate.

4.3 Microscopic Interpretation of Heat and Work

Consider a macroscopic system, A, that is known to be in a given macrostate. To be more exact, consider an ensemble of similar macroscopic systems, A, where each system in the ensemble is in one of the many microstates consistent with the given macrostate. There are two fundamentally different ways in which the average energy of A can change due to interaction with its surroundings. If the external parameters of the system remain constant then the interaction is termed a purely *thermal* interaction. Any change in the average energy of the system is attributed to an exchange of heat with its environment. Thus,

$$\Delta\overline{E} = Q, \tag{4.7}$$

where Q is the heat absorbed by the system. On a microscopic level, the energies of the individual microstates are unaffected by the absorption of heat. In fact, it is the distribution of the systems in the ensemble over the various microstates that is modified.

Suppose that the system A is thermally insulated from its environment. This can be achieved by surrounding it by an *adiabatic* envelope (i.e., an envelope fabricated out of a material that is a poor conductor of heat, such a fiber glass). Incidentally, the term adiabatic is derived from the ancient greek word *ἀδίαβᾶτος*, which means *impassable*. In scientific terminology, an adiabatic process is one in which there is no exchange of heat. The system A is still capable of interacting with its environment via its external parameters. This type of interaction is termed *mechanical* interaction, and any change in the average energy of the system is attributed to work done on it by its surroundings. Thus,

$$\Delta\overline{E} = -W, \tag{4.8}$$

where W is the work done by the system on its environment. On a microscopic level, the energy of the system changes because the energies of the individual microstates are functions of the external parameters. [See Equation (4.6).] Thus, if the external parameters are changed then, in general, the energies of all of the systems in the ensemble are modified (because each is in a specific microstate). Such a modification usually gives rise to a redistribution of the systems in the ensemble over the accessible microstates (without any heat exchange with the environment). Clearly, from a microscopic viewpoint, performing work on a macroscopic system is quite a complicated process. Nevertheless, macroscopic work is a quantity that is easy to measure experimentally. For instance, if the system A exerts a force $\mathbf{F}$ on its immediate surroundings, and the change in external parameters corresponds to a displacement $\mathbf{x}$ of the center of mass of the system, then the work done by A on its surroundings is simply

$$W = \mathbf{F}\cdot\mathbf{x}: \tag{4.9}$$

that is, the product of the force and the displacement along the line of action of the force. In a general interaction of the system A with its environment there is both heat exchange and work performed. We can write

$$Q \equiv \Delta\overline{E} + W, \tag{4.10}$$

which serves as the general definition of the absorbed heat Q. (Hence, the equivalence sign.) The quantity Q is simply the change in the mean energy of the system that is not due to the modification of the external parameters. Note that the notion of a quantity of heat has no independent meaning apart from Equation (4.10). The mean energy, $\overline{E}$, and work performed, W, are both physical quantities that can be determined experimentally, whereas Q is merely a derived quantity.

4.4 Quasi-Static Processes

Consider the special case of an interaction of the system A with its surroundings that is carried out so slowly that A remains arbitrarily close to equilibrium at all times. Such a process is said to be *quasi-static* for the system A. In practice, a quasi-static process must be carried out on a timescale that is much longer than the relaxation time of the system. Recall that the relaxation time is the typical timescale for the system to return to equilibrium after being suddenly disturbed. (See Section 3.5.)

A finite quasi-static change can be built up out of many infinitesimal changes. The infinitesimal heat, $đQ$, absorbed by the system when infinitesimal work, $đW$, is done on its environment, and its average energy changes by $d\overline{E}$, is given by

$$đQ \equiv d\overline{E} + đW. \tag{4.11}$$

The special symbols $đW$ and $đQ$ are introduced to emphasize that the work done, and the heat absorbed, are infinitesimal quantities that do not correspond to the difference between two works or two heats. Instead, the work done, and the heat absorbed, depend on the interaction process itself. Thus, it makes no sense to talk about the work in the system before and after the process, or the difference between these.

If the external parameters of the system have the values $x_1, \cdots, x_n$ then the energy of the system in a definite microstate, r, can be written

$$E_r = E_r(x_1, \cdots, x_n). \tag{4.12}$$

Hence, if the external parameters are changed by infinitesimal amounts, so that $x_\alpha \to x_\alpha + dx_\alpha$ for α in the range 1 to n, then the corresponding change in the energy of the microstate is

$$dE_r = \sum_{\alpha=1,n} \frac{\partial E_r}{\partial x_\alpha}\, dx_\alpha. \tag{4.13}$$

The work, $đW$, done by the system when it remains in this particular state r is [see Equation (4.8)]

$$đW = \sum_{\alpha=1,n} X_{\alpha\, r}\, dx_\alpha, \tag{4.14}$$

where

$$X_{\alpha\, r} \equiv -\frac{\partial E_r}{\partial x_\alpha} \tag{4.15}$$

is termed the *generalized force* (conjugate to the external parameter x_α) in the state r. (See Section B.2.) Note that if x_α is a displacement then $X_{\alpha\, r}$ is an ordinary force.

Consider, now, an ensemble of systems. Provided that the external parameters of the system are changed quasi-statically, the generalized forces $X_{\alpha\, r}$ have well-defined mean values that are calculable from the distribution of systems in the ensemble characteristic of the instantaneous macrostate. The macroscopic work, $đW$, resulting from an infinitesimal quasi-static change of the external parameters is obtained by calculating the decrease in the mean energy resulting from the parameter change. Thus,

$$đW = \sum_{\alpha=1,n} \overline{X}_\alpha\, dx_\alpha, \tag{4.16}$$

where

$$\overline{X}_\alpha \equiv -\overline{\frac{\partial E_r}{\partial x_\alpha}} \tag{4.17}$$

is the mean generalized force conjugate to x_α. The mean value is calculated from the equilibrium distribution of systems in the ensemble corresponding to the external parameter values x_α. The macroscopic work, W, resulting from a finite quasi-static change of external parameters can be obtained by integrating Equation (4.16).

The most well-known example of quasi-static work in thermodynamics is that done by pressure when the volume changes. For simplicity, suppose that the volume V is the only external parameter of any consequence. The work done in changing the volume from V to $V + dV$ is simply the product of the force and the displacement (along the line of action of the force). By definition, the mean equilibrium pressure, $\bar{p}$, of a given macrostate is equal to the normal force per unit area acting on any surface element. Thus, the normal force acting on a surface element $d\mathbf{S}_i$ is $\bar{p}\ d\mathbf{S}_i$. Suppose that the surface element is subject to a displacement $d\mathbf{x}_i$. The work done by the element is $\bar{p}\ d\mathbf{S}_i\cdot d\mathbf{x}_i$. The total work done by the system is obtained by summing over all of the surface elements. Thus,

$$đW = \bar{p}\ dV, \tag{4.18}$$

where

$$dV = \sum_i d\mathbf{S}_i\cdot d\mathbf{x}_i \tag{4.19}$$

is the infinitesimal volume change due to the displacement of the surface. It follows from Equations (4.16) and (4.18) that

$$\bar{p} = -\frac{\partial \overline{E}}{\partial V}. \tag{4.20}$$

Thus, the mean pressure is the generalized force conjugate to the volume, V.

Suppose that a quasi-static process is carried out in which the volume is changed from V_i to V_f. In general, the mean pressure, $\bar{p}$, is a function of the volume, so $\bar{p} = \bar{p}(V)$. It follows that the macroscopic work done by the system is given by

$$W_{if} = \int_{V_i}^{V_f} đW = \int_{V_i}^{V_f} \bar{p}(V)\, dV. \tag{4.21}$$

This quantity is just the area under the curve in a plot of $\bar{p}(V)$ versus V.

4.5 Exact and Inexact Differentials

In our investigation of heat and work, we have come across various infinitesimal objects, such as $d\overline{E}$ and $đW$. It is instructive to examine these infinitesimals more closely.

Consider the purely mathematical problem where $F(x, y)$ is some general function of two independent variables, x and y. Consider the change in F in going from the point (x, y) in the x-y plane to the neighboring point $(x + dx, y + dy)$. This is given by

$$dF = F(x + dx, y + dy) - F(x, y), \tag{4.22}$$

which can also be written

$$dF = X(x, y)\, dx + Y(x, y)\, dy, \tag{4.23}$$

where $X = \partial F/\partial x$ and $Y = \partial F/\partial y$. Clearly, dF is simply the infinitesimal difference between two adjacent values of the function F. This type of infinitesimal quantity is termed an *exact differential* to distinguish it from another type to be discussed presently. If we move in the x-y plane from an initial point $i \equiv (x_i, y_i)$ to a final point $f \equiv (x_f, y_f)$ then the corresponding change in F is given by

$$\Delta F = F_f - F_i = \int_i^f dF = \int_i^f (X\, dx + Y\, dy). \tag{4.24}$$

Note that because the difference on the left-hand side depends only on the initial and final points, the integral on the right-hand side can only depend on these points as well. In other words, the value of the integral is independent of the path taken in going from the initial to the final point. This is the distinguishing feature of an exact differential. Consider an integral taken around a closed circuit in the x-y plane. In this case, the initial and final points correspond to the same point, so the difference $F_f - F_i$ is clearly zero. It follows that the integral of an exact differential over a closed circuit is always zero:

$$\oint dF \equiv 0. \tag{4.25}$$

Of course, not every infinitesimal quantity is an exact differential. Consider the infinitesimal object

$$đG \equiv X'(x, y)\, dx + Y'(x, y)\, dz, \tag{4.26}$$

where X' and Y' are two general functions of x and y. It is easy to test whether or not an infinitesimal quantity is an exact differential. Consider the expression (4.23). It is clear that because $X = \partial F/\partial x$ and $Y = \partial F/\partial y$ then

$$\frac{\partial X}{\partial y} = \frac{\partial Y}{\partial x} = \frac{\partial^2 F}{\partial x\, \partial y}. \tag{4.27}$$

Thus, if

$$\frac{\partial X'}{\partial y} \neq \frac{\partial Y'}{\partial x} \tag{4.28}$$

(as is assumed to be the case), then $đG$ cannot be an exact differential, and is instead termed an *inexact differential*. The special symbol $đ$ is used to denote an inexact differential. Consider the integral of $đG$ over some path in the x-y plane. In general, it is not true that

$$\int_i^f đG = \int_i^f (X'\,dx + Y'\,dy) \tag{4.29}$$

is independent of the path taken between the initial and final points. This is the distinguishing feature of an inexact differential. In particular, the integral of an inexact differential around a closed circuit is not necessarily zero, so

$$\oint đG \neq 0. \tag{4.30}$$

Consider, for the moment, the solution of

$$đG = 0, \tag{4.31}$$

which reduces to the ordinary differential equation

$$\frac{dy}{dx} = -\frac{X'}{Y'}. \tag{4.32}$$

Because the right-hand side is a known function of x and y, the previous equation defines a definite direction (i.e., gradient) at each point in the x-y plane. The solution simply consists of drawing a system of curves in the x-y plane such that, at any point, the tangent to the curve is as specified by Equation (4.32). This defines a set of curves that can be written $\sigma(x, y) = c$, where c is a labeling parameter. It follows that, on a particular curve,

$$\frac{d\sigma}{dx} \equiv \frac{\partial\sigma}{\partial x} + \frac{\partial\sigma}{\partial y}\frac{dy}{dx} = 0. \tag{4.33}$$

The elimination of dy/dx between Equations (4.32) and (4.33) yields

$$Y'\,\frac{\partial\sigma}{\partial x} = X'\,\frac{\partial\sigma}{\partial y} = \frac{X'\,Y'}{\tau}, \tag{4.34}$$

where $\tau(x, y)$ is function of x and y. The previous equation could equally well be written

$$X' = \tau\,\frac{\partial\sigma}{\partial x}, \qquad Y' = \tau\,\frac{\partial\sigma}{\partial y}. \tag{4.35}$$

Inserting Equation (4.35) into Equation (4.26) gives

$$đG = \tau\left(\frac{\partial\sigma}{\partial x}\,dx + \frac{\partial\sigma}{\partial y}\,dy\right) = \tau\,d\sigma, \tag{4.36}$$

or

$$\frac{đG}{\tau} = d\sigma. \tag{4.37}$$

Thus, dividing the inexact differential $đG$ by τ yields the exact differential $d\sigma$. A factor τ that possesses this property is termed an *integrating factor*. Because the previous analysis is quite general, it is clear that an inexact differential involving two independent variables always admits of an integrating factor. Note, however, this is not generally the case for inexact differentials involving more than two variables.

4.6 Heat and Work

After the mathematical excursion in the previous section, let us return to a physical situation of interest. The macrostate of a macroscopic system can be determined by specifying the values of the external parameters (e.g., the volume), and the mean energy, $\overline{E}$. This, in turn, fixes other parameters, such as the mean pressure, $\bar{p}$. Alternatively, we can specify the external parameters and the mean pressure, which fixes the mean energy. Quantities such as $d\bar{p}$ and $d\overline{E}$ are infinitesimal differences between well-defined quantities: that is, they are exact differentials. For example, $d\overline{E} = \overline{E}_f - \overline{E}_i$ is just the difference between the mean energy of the system in the final macrostate f and the initial macrostate i, in the limit where these two states are nearly the same. It follows that if the system is taken from an initial macrostate i to any final macrostate f then the mean energy change is given by

$$\Delta\overline{E} = \overline{E}_f - \overline{E}_i = \int_i^f d\overline{E}. \tag{4.38}$$

However, because the mean energy is just a function of the macrostate under consideration, $\overline{E}_f$ and $\overline{E}_i$ depend only on the initial and final states, respectively. Thus, the integral $\int d\overline{E}$ depends only on the initial and final states, and not on the particular process used to get between them.

Consider, now, the infinitesimal work done by the system in going from some initial macrostate i to some neighbouring final macrostate f. In general, $đW = \sum \overline{X}_\alpha\, dx_\alpha$ is not the difference between two numbers referring to the properties of two neighboring macrostates. Instead, it is merely an infinitesimal quantity characteristic of the process of going from state i to state f. In other words, the work $đW$ is in general an inexact differential. The total work done by the system in going from any macrostate i to some other macrostate f can be written as

$$W_{if} = \int_i^f đW, \tag{4.39}$$

where the integral represents the sum of the infinitesimal amounts of work $đW$ performed at each stage of the process. In general, the value of the integral does depend on the particular process used in going from macrostate i to macrostate f.

Recall that, in going from macrostate i to macrostate f, the change in energy, $\Delta\overline{E}$, does not depend on the process used, whereas the work done, W, in general, does. Thus, it follows from the first law of thermodynamics, Equation (4.10), that the heat absorbed, Q, in general, also depends on the process used. It follows that

$$đQ \equiv d\overline{E} + đW \tag{4.40}$$

is an inexact differential. However, by analogy with the mathematical example in the previous section, there may exist some integrating factor, T (say), that converts the inexact differential $đQ$ into an exact differential. So,

$$\frac{đQ}{T} \equiv dS. \tag{4.41}$$

It will be interesting to discover the physical quantities that correspond to the functions T and S. (See Section 5.5.)

Suppose that the system is thermally insulated, so that $đQ = 0$. In this case, the first law of thermodynamics implies that

$$W_{if} = -\Delta\overline{E}. \tag{4.42}$$

Thus, in this special case, the work done depends only on the energy difference between the initial and final states, and is independent of the process. In fact, when Clausius first formulated the first law of thermodynamics, in 1850, he wrote:

> If a thermally isolated system is brought from some initial to some final state then the work done by the system is independent of the process used.

If the external parameters of the system are kept fixed, so that no work is done, then $đW = 0$, and Equation (4.11) reduces to

$$đQ = d\overline{E}, \tag{4.43}$$

and $đQ$ becomes an exact differential. The amount of heat, Q, absorbed in going from one macrostate to another depends only on the mean energy difference between them, and is independent of the process used to effect the change. In this situation, heat is a conserved quantity, and acts very much like the invisible indestructible fluid of Lavoisier's calorific theory.

Exercises

4.1 The mean pressure, $\bar{p}$, of a thermally insulated gas varies with volume according to the relation

$$\bar{p}\,V^{\gamma} = K,$$

where $\gamma > 1$ and K are positive constants. Show that the work done by this gas in a quasi-static process in which the state of the gas evolves from an initial macrostate with pressure $\bar{p}_i$ and volume V_i to a final macrostate with pressure $\bar{p}_f$ and volume V_f is

$$W_{if} = \frac{1}{\gamma - 1}\left(\bar{p}_i\,V_i - \bar{p}_f\,V_f\right).$$

4.2 Consider the infinitesimal quantity

$$đF \equiv (x^2 - y)\,dx + x\,dy.$$

Is this an exact differential? If not, find the integrating factor that converts it into an exact differential.

4.3 A system undergoes a quasi-static process that appears as a closed curve in a diagram of mean pressure, $\bar{p}$, versus volume, V. Such a process is termed *cyclic*, because the system ends up in a final macrostate that is identical to its initial macrostate. Show that the work done by the system is given by the area contained within the closed curve in the $\bar{p}$-V plane.

Chapter 5

Statistical Thermodynamics

5.1 Introduction

Let us briefly review the material that we have covered in this book so far. We started off by studying the mathematics of probability. (See Chapter 2.) We then used probabilistic reasoning to analyze the dynamics of many-particle systems, a subject area known as *statistical mechanics*. (See Chapter 3.) Next, we explored the physics of heat and work, the study of which is termed *thermodynamics*. (See Chapter 4.) The final step in our investigation is to combine statistical mechanics with thermodynamics. In other words, to investigate heat and work via statistical arguments. This discipline is called *statistical thermodynamics*, the study of which will form the central subject matter of this book. This chapter constitutes an introduction to the fundamental concepts of statistical thermodynamics. The remaining chapters will then explore the various applications of these concepts.

5.2 Thermal Interaction Between Macrosystems

Let us begin our investigation of statistical thermodynamics by examining a purely thermal interaction between two macroscopic systems, A and A', from a microscopic point of view. Suppose that the energies of these two systems are E and E', respectively. The external parameters are held fixed, so that A and A' cannot do work on one another. However, we shall assume that the systems are free to exchange heat energy (i.e., they are in thermal contact). It is convenient to divide the energy scale into small subdivisions of width δE. The number of microstates of A consistent with a macrostate in which the energy lies in the range E to $E+\delta E$ is denoted $\Omega(E)$. Likewise, the number of microstates of A' consistent with a macrostate in which the energy lies between E' and $E' + \delta E$ is denoted $\Omega'(E')$.

The combined system $A^{(0)} = A + A'$ is assumed to be isolated (i.e., it neither does work on, nor exchanges heat with, its surroundings). It follows from the first law of thermodynamics that the total energy, $E^{(0)}$, is constant. When speaking of thermal contact between two distinct systems, we usually assume that the mutual interaction is sufficiently weak for the energies to be additive. Thus,

$$E + E' \simeq E^{(0)} = \text{constant}. \tag{5.1}$$

Of course, in the limit of zero interaction, the energies are strictly additive. However, a small residual interaction is always required to enable the two systems to exchange heat energy, and, thereby, eventually reach thermal equilibrium. (See Section 3.4.) In fact, if the interaction between A and A' is too strong for the energies to be additive then it makes little sense to consider each system in isolation, because the presence of one system clearly strongly perturbs the other, and vice versa. In this case, the smallest system that can realistically be examined in isolation is $A^{(0)}$.

According to Equation (5.1), if the energy of A lies in the range E to $E + \delta E$ then the energy of A' must lie between $E^{(0)} - E - \delta E$ and $E^{(0)} - E$. Thus, the number of microstates accessible to each system is given by $\Omega(E)$ and $\Omega'(E^{(0)} - E)$, respectively. Because every possible state of A can be combined with every possible state of A' to form a distinct microstate, the total number of distinct states accessible to $A^{(0)}$ when the energy of A lies in the range E to $E + \delta E$ is

$$\Omega^{(0)} = \Omega(E)\,\Omega'(E^{(0)} - E). \tag{5.2}$$

Consider an ensemble of pairs of thermally interacting systems, A and A', that are left undisturbed for many relaxation times, so that they can attain thermal equilibrium. The principle of equal a priori probabilities is applicable to this situation. (See Section 3.3.) According to this principle, the probability of occurrence of a given macrostate is proportional to the number of accessible microstates, because all microstates are equally likely. Thus, the probability that the system A has an energy lying in the range E to $E + \delta E$ can be written

$$P(E) = C\,\Omega(E)\,\Omega'(E^{(0)} - E), \tag{5.3}$$

where C is a constant that is independent of E.

We know, from Section 3.8, that the typical variation of the number of accessible states with energy is of the form

$$\Omega \propto E^{f}, \tag{5.4}$$

where f is the number of degrees of freedom. For a macroscopic system, f is an exceedingly large number. It follows that the probability, $P(E)$, in Equation (5.3) is the product of an extremely rapidly increasing function of E, and an extremely rapidly decreasing function of E. Hence, we would expect the probability to exhibit a very pronounced maximum at some particular value of the energy.

Let us Taylor expand the logarithm of $P(E)$ in the vicinity of its maximum value, which is assumed to occur at $E = \tilde{E}$. We expand the relatively slowly-varying logarithm, rather than the function itself, because the latter varies so rapidly with the energy that the radius of convergence of its Taylor series is too small for this expansion to be of any practical use. The expansion of $\ln \Omega(E)$ yields

$$\ln \Omega(E) = \ln \Omega(\tilde{E}) + \beta(\tilde{E})\,\eta - \frac{1}{2}\,\lambda(\tilde{E})\,\eta^2 + \cdots\,, \tag{5.5}$$

where

$$\eta = E - \tilde{E}, \tag{5.6}$$

$$\beta = \frac{\partial \ln \Omega}{\partial E}, \tag{5.7}$$

$$\lambda = -\frac{\partial^2 \ln \Omega}{\partial E^2} = -\frac{\partial \beta}{\partial E}. \tag{5.8}$$

Now, because $E' = E^{(0)} - E$, we have

$$E' - \tilde{E}' = -(E - \tilde{E}) = -\eta. \tag{5.9}$$

Here, $\tilde{E}' = E^{(0)} - \tilde{E}$. It follows that

$$\ln \Omega'(E') = \ln \Omega'(\tilde{E}') + \beta'(\tilde{E}')\,(-\eta) - \frac{1}{2}\,\lambda'(\tilde{E}')\,(-\eta)^2 + \cdots\,, \tag{5.10}$$

where β' and λ' are defined in an analogous manner to the parameters β and λ. Equations (5.5) and (5.10) can be combined to give

$$\ln\,[\Omega(E)\,\Omega'(E')] = \ln\,[\Omega(\tilde{E})\,\Omega'(\tilde{E}')] + [\beta(\tilde{E}) - \beta'(\tilde{E}')]\,\eta - \frac{1}{2}\,[\lambda(\tilde{E}) + \lambda'(\tilde{E}')]\,\eta^2 + \cdots\,. \tag{5.11}$$

At the maximum of $\ln\,[\Omega(E)\,\Omega'(E')]$, the linear term in the Taylor expansion must vanish, so

$$\beta(\tilde{E}) = \beta'(\tilde{E}'), \tag{5.12}$$

which enables us to determine $\tilde{E}$. It follows that

$$\ln P(E) = \ln P(\tilde{E}) - \frac{1}{2}\,\lambda_0\,\eta^2, \tag{5.13}$$

or

$$P(E) = P(\tilde{E})\exp\left[-\frac{1}{2}\,\lambda_0\,(E - \tilde{E})^2\right], \tag{5.14}$$

where

$$\lambda_0 = \lambda(\tilde{E}) + \lambda'(\tilde{E}'). \tag{5.15}$$

Now, the parameter λ_0 must be positive, otherwise the probability $P(E)$ does not exhibit a pronounced maximum value. That is, the combined system, $A^{(0)}$, does not possess a well-defined equilibrium state as, physically, we know it must. It is clear that $\lambda(\tilde{E})$ must also be positive, because we could always choose for A' a system with a negligible contribution to λ_0, in which case the constraint $\lambda_0 > 0$ would effectively correspond to $\lambda(\tilde{E}) > 0$. [A similar argument can be used to show that $\lambda'(\tilde{E}')$ must be positive.] The same conclusion also follows from the estimate $\Omega \propto E^f$, which implies that

$$\lambda(\tilde{E}) \sim \frac{f}{\tilde{E}^2} > 0. \tag{5.16}$$

According to Equation (5.14), the probability distribution function, $P(E)$, is Gaussian. (See Section 2.9.) This is hardly surprising, because the central limit theorem ensures that the probability distribution for any macroscopic variable, such as E, is Gaussian in nature. (See Section 2.10.) It follows that the mean value of E corresponds to the situation of maximum probability (i.e., the peak of the Gaussian curve), so that

$$\overline{E} = \tilde{E}. \tag{5.17}$$

The standard deviation of the distribution is

$$\Delta^* E = \lambda_0^{-1/2} \sim \frac{\overline{E}}{\sqrt{f}}, \tag{5.18}$$

where use has been made of Equation (5.16) (assuming that system A makes the dominant contribution to λ_0). It follows that the fractional width of the probability distribution function is given by

$$\frac{\Delta^* E}{\overline{E}} \sim \frac{1}{\sqrt{f}}. \tag{5.19}$$

Hence, if A contains 1 mole of particles then $f \sim N_A \simeq 10^{24}$, and $\Delta^* E/\overline{E} \sim 10^{-12}$. Clearly, the probability distribution for E has an exceedingly sharp maximum. Experimental measurements of this energy will almost always yield the mean value, and the underlying statistical nature of the distribution may not be apparent.

5.3 Temperature

Suppose that the systems A and A' are initially thermally isolated from one another, with respective energies E_i and E_i'. (Because the energy of an isolated system cannot fluctuate, we do not have to bother with mean energies here.) If the two systems are subsequently placed in thermal contact, so that they are free to exchange heat energy, then, in general, the resulting state is an extremely improbable one [i.e., $P(E_i)$ is much less than the peak probability]. The configuration will, therefore, tend to evolve in time until the two systems attain final mean energies, $\overline{E}_f$ and $\overline{E}_f'$, which are such that

$$\beta_f = \beta_f', \tag{5.20}$$

where $\beta_f \equiv \beta(\overline{E}_f)$ and $\beta_f' \equiv \beta'(\overline{E}_f')$. This corresponds to the state of maximum probability. (See Section 5.2.) In the special case in which the initial energies, E_i and E_i', lie very close to the final mean energies, $\overline{E}_f$ and $\overline{E}_f'$, respectively, there is no change in the two systems when they are brought into thermal contact, because the initial state already corresponds to a state of maximum probability.

It follows from energy conservation that

$$\overline{E}_f + \overline{E}_f' = E_i + E_i'. \tag{5.21}$$

The mean energy change in each system is simply the net heat absorbed, so that

$$Q \equiv \overline{E}_f - E_i, \tag{5.22}$$

$$Q' \equiv \overline{E}_f' - E_i'. \tag{5.23}$$

The conservation of energy then reduces to

$$Q + Q' = 0. \tag{5.24}$$

In other words, the heat given off by one system is equal to the heat absorbed by the other. (In our notation, absorbed heat is positive, and emitted heat is negative.)

It is clear that if the systems A and A' are suddenly brought into thermal contact then they will only exchange heat, and evolve towards a new equilibrium state, if the final state is more probable than the initial one. In other words, the system will evolve if

$$P(\overline{E}_f) > P(E_i), \tag{5.25}$$

or

$$\ln P(\overline{E}_f) > \ln P(E_i), \tag{5.26}$$

because the logarithm is a monotonic function. The previous inequality can be written

$$\ln \Omega(\overline{E}_f) + \ln \Omega'(\overline{E}_f') > \ln \Omega(E_i) + \ln \Omega'(E_i'), \tag{5.27}$$

with the aid of Equation (5.3). Taylor expansion to first order yields

$$\frac{\partial \ln \Omega(E_i)}{\partial E}\,(\overline{E}_f - E_i) + \frac{\partial \ln \Omega'(E_i')}{\partial E'}\,(\overline{E}_f' - E_i') > 0, \tag{5.28}$$

which finally gives

$$(\beta_i - \beta_i')\,Q > 0, \tag{5.29}$$

where $\beta_i \equiv \beta(E_i)$, $\beta_i' \equiv \beta'(E_i')$, and use has been made of Equations (5.22)–(5.24).

It is clear, from the previous analysis, that the parameter β, defined

$$\beta = \frac{\partial \ln \Omega}{\partial E}, \tag{5.30}$$

has the following properties:

(1) If two systems separately in equilibrium have the same value of β then the systems will remain in equilibrium when brought into thermal contact with one another.
(2) If two systems separately in equilibrium have different values of β then the systems will not remain in equilibrium when brought into thermal contact with one another. Instead, the system with the higher value of β will absorb heat from the other system until the two β values are the same. [See Equation (5.29).]

Incidentally, a partial derivative is used in Equation (5.30) because, in a purely thermal interaction, the external parameters of the system are held constant as the energy changes.

Let us define the dimensionless parameter T, such that

$$\frac{1}{k\,T} \equiv \beta = \frac{\partial \ln \Omega}{\partial E}, \tag{5.31}$$

where k is a positive constant having the dimensions of energy. The parameter T is termed the *thermodynamic temperature*, and controls heat flow in much the same manner as a conventional temperature. Thus, if two isolated systems in equilibrium possess the same thermodynamic temperature then they will remain in equilibrium when brought into thermal contact. However, if the two systems have different thermodynamic temperatures then heat will flow from the system with the higher temperature (i.e., the "hotter" system) to the system with the lower temperature, until the temperatures of the two systems are the same. In addition, suppose that we have three systems A, B, and C. We know that if A and B remain in equilibrium when brought into thermal contact then their temperatures are the same, so that $T_A = T_B$. Similarly, if B and C remain in equilibrium when brought into thermal contact, then $T_B = T_C$. But, we can then conclude that $T_A = T_C$, so systems A and C will also remain in equilibrium when brought into thermal contact. Thus, we arrive at the following statement, which is sometimes called the *zeroth law of thermodynamics*:

> If two systems are separately in thermal equilibrium with a third system then they must also be in thermal equilibrium with one another.

The thermodynamic temperature of a macroscopic body, as defined in Equation (5.31), depends only on the rate of change of the number of accessible microstates with the total energy. Thus, it is possible to define a thermodynamic temperature for systems with radically different microscopic structures (e.g., matter and radiation). The thermodynamic, or *absolute*, scale of temperature is measured in degrees kelvin. The parameter k is chosen to make this temperature scale accord as much as possible with more conventional temperature scales. The choice

$$k = 1.381 \times 10^{-23} \text{ joules/kelvin}, \tag{5.32}$$

ensures that there are 100 degrees kelvin between the freezing and boiling points of water at atmospheric pressure (the two temperatures are 273.15 K and 373.15 K, respectively). Here, k is known as the *Boltzmann constant*. In fact, the Boltzmann constant is fixed by international convention so as to make the *triple point* of water (i.e., the unique temperature at which the three phases of water co-exist in thermal equilibrium) exactly 273.16 K. Note that the zero of the thermodynamic scale, the so called *absolute zero* of temperature, does not correspond to the freezing point of water, but to some far more physically significant temperature that we shall discuss presently. (See Section 5.9.)

The familiar $\Omega \propto E^f$ scaling for translational degrees of freedom (see Section 3.8) yields

$$kT \sim \frac{\overline{E}}{f}, \tag{5.33}$$

using Equation (5.31), so kT is a rough measure of the mean energy associated with each degree of freedom in the system. In fact, for a classical system (i.e., one in which quantum effects are unimportant) it is possible to show that the mean energy associated with each degree of freedom is exactly $(1/2)\,kT$. This result, which is known as the *equipartition theorem*, will be discussed in more detail later on in this book. (See Section 8.10.)

The absolute temperature, T, is usually positive, because $\Omega(E)$ is ordinarily a very rapidly increasing function of energy. In fact, this is the case for all conventional systems in which the kinetic energy of the particles is taken into account, because there is no upper bound on the possible energy of the system, and $\Omega(E)$

consequently increases roughly like E^f. It is, however, possible to envisage a situation in which we ignore the translational degrees of freedom of a system, and concentrate only on its spin degrees of freedom. In this case, there is an upper bound to the possible energy of the system (i.e., all spins lined up anti-parallel to an applied magnetic field). Consequently, the total number of states available to the system is finite. In this situation, the density of spin states, $\Omega_{\rm spin}(E)$, first increases with increasing energy, as in conventional systems, but then reaches a maximum and decreases again. Thus, it is possible to get absolute spin temperatures that are negative, as well as positive. (See Exercise 5.2.)

In Lavoisier's calorific theory, the basic mechanism that forces heat to spontaneously flow from hot to cold bodies is the supposed mutual repulsion of the constituent particles of calorific fluid. In statistical mechanics, the explanation is far less contrived. Heat flow occurs because statistical systems tend to evolve towards their most probable states, subject to the imposed physical constraints. When two bodies at different temperatures are suddenly placed in thermal contact, the initial state corresponds to a spectacularly improbable state of the overall system. For systems containing of order 1 mole of particles, the only reasonably probable final equilibrium states are such that the two bodies differ in temperature by less than 1 part in 10^{12}. Thus, the evolution of the system towards these final states (i.e., towards thermal equilibrium) is effectively driven by probability.

5.4 Mechanical Interaction Between Macrosystems

Let us now examine a purely mechanical interaction between macrostates, where one or more of the external parameters is modified, but there is no exchange of heat energy. Consider, for the sake of simplicity, a situation where only one external parameter, x, of the system is free to vary. In general, the number of microstates accessible to the system when the overall energy lies between E and $E + \delta E$ depends on the particular value of x, so we can write $\Omega \equiv \Omega(E, x)$.

When x is changed by the amount dx, the energy $E_r(x)$ of a given microstate r changes by $(\partial E_r/\partial x)\,dx$. The number of states, $\sigma(E, x)$, whose energy is changed from a value less than E to a value greater than E, when the parameter changes from x to $x + dx$, is given by the number of microstates per unit energy range multiplied by the average shift in energy of the microstates. Hence,

$$\sigma(E, x) = \frac{\Omega(E, x)}{\delta E}\,\overline{\frac{\partial E_r}{\partial x}}\,dx, \tag{5.34}$$

where the mean value of $\partial E_r/\partial x$ is taken over all accessible microstates (i.e., all states for which the energy lies between E and $E + \delta E$, and the external parameter takes the value x). The previous equation can also be written

$$\sigma(E, x) = -\frac{\Omega(E, x)}{\delta E}\,\overline{X}\,dx, \tag{5.35}$$

where

$$\overline{X}(E, x) = -\overline{\frac{\partial E_r}{\partial x}} \tag{5.36}$$

is the mean generalized force conjugate to the external parameter x. (See Section 4.4.)

Consider the total number of microstates whose energies lies between E and $E+\delta E$. When the external parameter changes from x to $x+dx$, the number of states in this energy range changes by $(\partial\Omega/\partial x)\,dx$. This change is due to the difference between the number of states that enter the range because their energy is changed from a value less than E to one greater than E, and the number that leave because their energy is changed from a value less than $E+\delta E$ to one greater than $E+\delta E$. In symbols,

$$\frac{\partial\Omega(E, x)}{\partial x}\,dx = \sigma(E) - \sigma(E+\delta E) \simeq -\frac{\partial\sigma}{\partial E}\,\delta E, \tag{5.37}$$

which yields

$$\frac{\partial\Omega}{\partial x} = \frac{\partial(\Omega\,\overline{X})}{\partial E}, \tag{5.38}$$

where use has been made of Equation (5.35). Dividing both sides by Ω gives

$$\frac{\partial\ln\Omega}{\partial x} = \frac{\partial\ln\Omega}{\partial E}\,\overline{X} + \frac{\partial\overline{X}}{\partial E}. \tag{5.39}$$

However, according to the usual estimate $\Omega \propto E^f$ (see Section 3.8), the first term on the right-hand side is of order $(f/\overline{E})\,\overline{X}$, whereas the second term is only of order $\overline{X}/\overline{E}$. Clearly, for a macroscopic system with many degrees of freedom, the second term is utterly negligible, so we have

$$\frac{\partial\ln\Omega}{\partial x} = \frac{\partial\ln\Omega}{\partial E}\,\overline{X} = \beta\,\overline{X}, \tag{5.40}$$

where use has been made of Equation (5.30).

When there are several external parameters, $x_1, \cdots, x_n$, so that $\Omega \equiv \Omega(E, x_1, \cdots, x_n)$, the previous derivation is valid for each parameter taken in isolation. Thus,

$$\frac{\partial\ln\Omega}{\partial x_\alpha} = \beta\,\overline{X}_\alpha, \tag{5.41}$$

where $\overline{X}_\alpha$ is the mean generalized force conjugate to the parameter x_α. (See Section B.2.)

5.5 General Interaction Between Macrosystems

Consider two systems, A and A', that can interact by exchanging heat energy and doing work on one another. Let the system A have energy E, and adjustable external parameters $x_1, \cdots, x_n$. Likewise, let the system A' have energy E', and adjustable external parameters $x'_1, \cdots, x'_n$. The combined system $A^{(0)} = A + A'$ is assumed to be isolated. It follows from the first law of thermodynamics that

$$E + E' = E^{(0)} = \text{constant.} \tag{5.42}$$

Thus, the energy E' of system A' is determined once the energy E of system A is given, and vice versa. In fact, E' could be regarded as a function of E. Furthermore, if the two systems can interact mechanically then, in general, the parameters x' are some function of the parameters x. As a simple example, if the two systems are separated by a movable partition, in an enclosure of fixed volume $V^{(0)}$, then

$$V + V' = V^{(0)} = \text{constant,} \tag{5.43}$$

where V and V' are the volumes of systems A and A', respectively.

The total number of microstates accessible to $A^{(0)}$ is clearly a function of E, and the parameters x_α (where α runs from 1 to n), so $\Omega^{(0)} \equiv \Omega^{(0)}(E, x_1, \cdots, x_n)$. We have already demonstrated (in Section 5.2) that $\Omega^{(0)}$ exhibits a very pronounced maximum at one particular value of the energy, $E = \tilde{E}$, when E is varied but the external parameters are held constant. This behavior comes about because of the very strong,

$$\Omega \propto E^f, \tag{5.44}$$

increase in the number of accessible microstates of A (or A') with energy. However, according to Section 3.8, the number of accessible microstates exhibits a similar strong increase with the volume, which is a typical external parameter, so that

$$\Omega \propto V^f. \tag{5.45}$$

It follows that the variation of $\Omega^{(0)}$ with a typical parameter, x_α, when all the other parameters and the energy are held constant, also exhibits a very sharp maximum at some particular value, $x_\alpha = \tilde{x}_\alpha$. The equilibrium situation corresponds to the configuration of maximum probability, in which virtually all systems $A^{(0)}$ in the ensemble have values of E and x_α very close to $\tilde{E}$ and $\tilde{x}_\alpha$, respectively. The mean values of these quantities are thus given by $\overline{E} = \tilde{E}$ and $\bar{x}_\alpha = \tilde{x}_\alpha$.

Consider a quasi-static process in which the system A is brought from an equilibrium state described by $\overline{E}$ and $\bar{x}_\alpha$ to an infinitesimally different equilibrium state described by $\overline{E} + d\overline{E}$ and $\bar{x}_\alpha + d\bar{x}_\alpha$. Let us calculate the resultant change in the number of microstates accessible to A. Because $\Omega \equiv \Omega(E, x_1, \cdots, x_n)$, the change in $\ln \Omega$ follows from standard mathematics:

$$d \ln \Omega = \frac{\partial \ln \Omega}{\partial E}\, d\overline{E} + \sum_{\alpha=1,n} \frac{\partial \ln \Omega}{\partial x_\alpha}\, d\bar{x}_\alpha. \tag{5.46}$$

However, we have previously demonstrated that

$$\beta = \frac{\partial \ln \Omega}{\partial E}, \quad \beta\, \overline{X}_\alpha = \frac{\partial \ln \Omega}{\partial x_\alpha} \tag{5.47}$$

[in Equations (5.30) and (5.41), respectively], so Equation (5.46) can be written

$$d \ln \Omega = \beta \left(d\overline{E} + \sum_\alpha \overline{X}_\alpha\, d\bar{x}_\alpha \right). \tag{5.48}$$

Note that the temperature parameter, β, and the mean conjugate forces, $\overline{X}_\alpha$, are only well defined for equilibrium states. This is why we are only considering quasi-static changes in which the two systems are always arbitrarily close to equilibrium.

Let us rewrite Equation (5.48) in terms of the thermodynamic temperature, T, using the relation $\beta \equiv 1/k\,T$. We obtain

$$dS = \left(d\overline{E} + \sum_\alpha \overline{X}_\alpha\, d\bar{x}_\alpha \right) \Big/ T, \tag{5.49}$$

where

$$S = k \ln \Omega. \tag{5.50}$$

Equation (5.49) is a differential relation that enables us to calculate the quantity S as a function of the mean energy, $\overline{E}$, and the mean external parameters, $\bar{x}_\alpha$, assuming that we can calculate the temperature, T, and mean conjugate forces, $\overline{X}_\alpha$, for each equilibrium state. The function $S(\overline{E}, \bar{x}_\alpha)$ is termed the *entropy* of system A. The word entropy is derived from the ancient greek word ἐντροπή, which means *a turning towards* or *tendency*. The reason for this etymology will become apparent presently. It can be seen from Equation (5.50) that the entropy is merely a parameterization of the number of accessible microstates. Hence, according to statistical mechanics, $S(\overline{E}, \bar{x}_\alpha)$ is essentially a measure of the relative probability of a state characterized by values of the mean energy and mean external parameters $\overline{E}$ and $\bar{x}_\alpha$, respectively.

According to Equation (4.16), the net amount of work performed during a quasi-static change is given by

$$đW = \sum_\alpha \overline{X}_\alpha\, d\bar{x}_\alpha. \tag{5.51}$$

It follows from Equation (5.49) that

$$dS = \frac{d\overline{E} + đW}{T} = \frac{đQ}{T}. \tag{5.52}$$

Thus, the thermodynamic temperature, T, is the integrating factor for the first law of thermodynamics,

$$đQ = d\overline{E} + đW, \tag{5.53}$$

which converts the inexact differential $đQ$ into the exact differential dS. (See Section 4.5.) It follows that the entropy difference between any two macrostates i and f can be written

$$S_f - S_i = \int_i^f dS = \int_i^f \frac{đQ}{T}, \tag{5.54}$$

where the integral is evaluated for any process through which the system is brought *quasi-statically* via a sequence of near-equilibrium configurations from its initial to its final macrostate. The process has to be quasi-static because the temperature, T, which appears in the integrand, is only well defined for an equilibrium state. Because the left-hand side of the previous equation only depends on the initial and final states, it follows that the integral on the right-hand side is independent of the particular sequence of quasi-static changes used to get from i to f. Thus, $\int_i^f đQ/T$ is independent of the process (provided that it is quasi-static).

All of the concepts that we have encountered up to now in this book, such as temperature, heat, energy, volume, pressure, et cetera, have been fairly familiar to us from other branches of physics. However, entropy, which turns out to be of crucial importance in thermodynamics, is something quite new. Let us consider the following questions. What does the entropy of a thermodynamic system actually signify? What use is the concept of entropy?

5.6 Entropy

Consider an isolated system whose energy is known to lie in a narrow range. Let Ω be the number of accessible microstates. According to the principle of equal a priori probabilities, the system is equally likely to be found in any one of these states when it is in thermal equilibrium. The accessible states are just that set of microstates that are consistent with the macroscopic constraints imposed on the system. These constraints can usually be quantified by specifying the values of some parameters, $y_1, \cdots, y_n$, that characterize the macrostate. Note that these parameters are not necessarily external in nature. For example, we could specify either the volume (an external parameter) or the mean pressure (the mean force conjugate to the volume). The number of accessible states is clearly a function of the chosen parameters, so we can write $\Omega \equiv \Omega(y_1, \cdots, y_n)$ for the number of microstates consistent with a macrostate in which the general parameter y_α lies in the range y_α to $y_\alpha + dy_\alpha$.

Suppose that we start from a system in thermal equilibrium. According to statistical mechanics, each of the Ω_i, say, accessible states are equally likely. Let us now remove, or relax, some of the constraints imposed on the system. Clearly, all of the microstates formally accessible to the system are still accessible, but many additional states will, in general, become accessible. Thus, removing or relaxing constraints can only have the effect of increasing, or possibly leaving unchanged, the number of microstates accessible to the system. If the final number of accessible states is Ω_f then we can write

$$\Omega_f \geq \Omega_i. \tag{5.55}$$

Immediately after the constraints are relaxed, the systems in the ensemble are not in any of the microstates from which they were previously excluded. So, the systems only occupy a fraction

$$P_i = \frac{\Omega_i}{\Omega_f} \tag{5.56}$$

of the Ω_f states now accessible to them. This is clearly not a equilibrium situation. Indeed, if $\Omega_f \gg \Omega_i$ then the configuration in which the systems are only distributed over the original Ω_i states is an extremely unlikely one. In fact, its probability of occurrence is given by Equation (5.56). According to the H-theorem (see Section 3.4), the ensemble will evolve in time until a more probable final state is reached in which the systems are evenly distributed over the Ω_f available states.

As a simple example, consider a system consisting of a box divided into two regions of equal volume. Suppose that, initially, one region is filled with gas, and the other is empty. The constraint imposed on the system is, thus, that the coordinates of all of the gas molecules must lie within the filled region. In other words, the volume accessible to the system is $V = V_i$, where V_i is half the volume of the box. The constraints imposed on the system can be relaxed by removing the partition, and allowing gas to flow into both regions. The volume accessible to the gas is now $V = V_f = 2\,V_i$. Immediately after the partition is removed, the system is in an extremely improbable state. We know, from Section 3.8, that, at constant energy, the variation of the number of accessible states of an ideal gas with the volume is

$$\Omega \propto V^N, \tag{5.57}$$

where N is the number of particles. Thus, the probability of observing the state immediately after the partition is removed in an ensemble of equilibrium systems with volume $V = V_f$ is

$$P_i = \frac{\Omega_i}{\Omega_f} = \left(\frac{V_i}{V_f}\right)^N = \left(\frac{1}{2}\right)^N. \tag{5.58}$$

If the box contains of order 1 mole of molecules then $N \sim 10^{24}$, and this probability is fantastically small:

$$P_i \sim \exp\left(-10^{24}\right). \tag{5.59}$$

Clearly, the system will evolve towards a more probable state.

This discussion can also be phrased in terms of the parameters, $y_1, \cdots, y_n$, of the system. Suppose that a constraint is removed. For instance, one of the parameters, y, say, which originally had the value $y = y_i$, is now allowed to vary. According to statistical mechanics, all states accessible to the system are equally likely. So, the probability $P(y)$ of finding the system in equilibrium with the parameter in the range y to $y + \delta y$ is proportional to the number of microstates in this interval: that is,

$$P(y) \propto \Omega(y). \tag{5.60}$$

Usually, $\Omega(y)$ has a very pronounced maximum at some particular value $\tilde{y}$. (See Section 5.2.) This means that practically all systems in the final equilibrium ensemble have values of y close to $\tilde{y}$. Thus, if $y_i \neq \tilde{y}$, initially, then the parameter y will change until it attains a final value close to $\tilde{y}$, where Ω is maximum. This discussion can be summed up in a single phrase:

> If some of the constraints of an isolated system are removed then the parameters of the system tend to readjust themselves in such a way that
>
> $$\Omega(y_1, \cdots, y_n) \to \text{maximum}.$$

Suppose that the final equilibrium state has been reached, so that the systems in the ensemble are uniformly distributed over the Ω_f accessible final states. If the original constraints are reimposed then the systems in the ensemble still occupy these Ω_f states with equal probability. Thus, if $\Omega_f > \Omega_i$ then simply restoring the constraints does not restore the initial situation. Once the systems are randomly distributed over the Ω_f states, they cannot be expected to spontaneously move out of some of these states, and occupy a more restricted class of states, merely in response to the reimposition of a constraint. The initial condition can also not be restored by removing further constraints. This could only lead to even more states becoming accessible to the system.

Suppose that a process occurs in which an isolated system goes from some initial configuration to some final configuration. If the final configuration is such that the imposition or removal of constraints cannot by itself restore the initial condition then the process is deemed *irreversible*. On the other hand, if it is such that the imposition or removal of constraints can restore the initial condition then the process is deemed *reversible*. From what we have already said, an irreversible process is clearly one in which the removal of constraints leads to a situation where $\Omega_f > \Omega_i$. A reversible process corresponds to the special case where the removal of constraints does not change the number of accessible states, so that $\Omega_f = \Omega_i$. In this situation, the systems remain distributed with equal probability over these states irrespective of whether the constraints are imposed or not.

Our microscopic definition of irreversibility is in accordance with the macroscopic definition discussed in Section 3.6. Recall that, on a macroscopic level, an irreversible process is one that appears unphysical when viewed in reverse. On a microscopic level, it is clearly plausible that a system should spontaneously evolve from an improbable to a probable configuration in response to the relaxation of some constraint. However, it is quite manifestly implausible that a system should ever spontaneously evolve from a probable to an improbable configuration. Let us consider our example again. If a gas is initially restricted to one half of a box, via a partition, then the flow of gas from one side of the box to the other when the partition is removed is an irreversible process. This process is irreversible on a microscopic level because the initial configuration cannot be recovered by simply replacing the partition. It is irreversible on a macroscopic level because it is obviously unphysical for the molecules of a gas to spontaneously distribute themselves in such a manner that they only occupy half of the available volume.

It is actually possible to quantify irreversibility. In other words, in addition to stating that a given process is irreversible, we can also give some indication of how irreversible it is. The parameter that measures irreversibility is the number of accessible states, Ω. Thus, if Ω for an isolated system spontaneously increases then the process is irreversible, the degree of irreversibility being proportional to the amount of the increase. If Ω stays the same then the process is reversible. Of course, it is unphysical for Ω to ever spontaneously decrease. In symbols, we can write

$$\Omega_f - \Omega_i \equiv \Delta\Omega \geq 0, \tag{5.61}$$

for any physical process operating on an isolated system. In practice, Ω itself is a rather unwieldy parameter with which to measure irreversibility. For instance, in the previous example, where an ideal gas doubles in volume (at constant energy) due to the removal of a partition, the fractional increase in Ω is

$$\frac{\Omega_f}{\Omega_i} \simeq 10^{2\,\nu\times 10^{23}}, \tag{5.62}$$

where ν is the number of moles. This is an extremely large number. It is far more convenient to measure irreversibility in terms of $\ln \Omega$. If Equation (5.61) is true then it is certainly also true that

$$\ln \Omega_f - \ln \Omega_i \equiv \varDelta \ln \Omega \geq 0 \tag{5.63}$$

for any physical process operating on an isolated system. The increase in $\ln \Omega$ when an ideal gas doubles in volume (at constant energy) is

$$\ln \Omega_f - \ln \Omega_i = \nu\, N_A\, \ln 2, \tag{5.64}$$

where $N_A = 6 \times 10^{23}$. This is a far more manageable number. Because we usually deal with particles by the mole in laboratory physics, it makes sense to pre-multiply our measure of irreversibility by a number of order $1/N_A$. For historical reasons, the number that is generally used for this purpose is the Boltzmann constant, k, which can be written

$$k = \frac{R}{N_A} \quad \text{joules/kelvin}, \tag{5.65}$$

where

$$R = 8.3143 \quad \text{joules/kelvin/mole} \tag{5.66}$$

is the ideal gas constant that appears in the well-known equation of state for an ideal gas, $P\,V = \nu\,R\,T$. Thus, the final form for our measure of irreversibility is

$$S = k\, \ln \Omega. \tag{5.67}$$

This quantity is termed entropy, and is measured in joules per degree kelvin. The increase in entropy when an ideal gas doubles in volume (at constant energy) is

$$S_f - S_i = \nu R\, \ln 2, \tag{5.68}$$

which is order unity for laboratory-scale systems (i.e., those containing about one mole of particles). The essential irreversibility of macroscopic phenomena can be summed up as follows:

$$S_f - S_i \equiv \varDelta S \geq 0, \tag{5.69}$$

for a process acting on an isolated system. [This formula is equivalent to Equations (5.61) and (5.63).] Thus:

> The entropy of an isolated system tends to increase with time, and can never decrease.

This proposition is known as the *second law of thermodynamics*.

One way of thinking of the number of accessible states, Ω, is that it is a measure of the disorder associated with a macrostate. For a system exhibiting a high degree of order, we would expect a strong correlation between the motions of the individual particles. For instance, in a fluid there might be a strong tendency for the particles to move in one particular direction, giving rise to an ordered flow of the system in that direction. On the other hand, for a system exhibiting a low degree of order, we expect far less correlation between the motions of individual particles. It follows that, all other things being equal, an ordered system is more constrained than a disordered system, because the former is excluded from microstates in which there is not a strong correlation between individual particle motions, whereas the latter is not. Another way of saying this is that an ordered system has less accessible microstates than a corresponding disordered system. Thus, entropy is effectively a measure of the disorder in a system (the disorder increases with increasing S). With this interpretation, the second law of thermodynamics reduces to the statement that isolated systems tend to become more disordered with time, and can never become more ordered.

Note that the second law of thermodynamics only applies to isolated systems. The entropy of a non-isolated system can decrease. For instance, if a gas expands (at constant energy) to twice its initial volume after the removal of a partition, we can subsequently recompress the gas to its original volume. The energy of the gas will increase because of the work done on it during compression, but if we absorb some heat from the gas then we can restore it to its initial state. Clearly, in restoring the gas to its original state, we have restored its original entropy. This appears to violate the second law of thermodynamics, because the entropy should have increased in what is obviously an irreversible process. However, if we consider a new system consisting of the gas plus the compression and heat absorption machinery then it is still true that the entropy of this system (which is assumed to be isolated) must increase in time. Thus, the entropy of the gas is only kept the same at the expense of increasing the entropy of the rest of the system, and the total entropy is increased. If we consider the system of everything in the universe, which is certainly an isolated system because there is nothing outside it with which it could interact, then the second law of thermodynamics becomes:

> The disorder of the universe tends to increase with time, and can never decrease.

5.7 Properties of Entropy

Entropy, as we have defined it, has some dependence on the resolution, δE, to which the energy of macrostates is measured. Recall that $\Omega(E)$ is the number of accessible microstates with energy in the range E to $E + \delta E$. Suppose that we choose a new resolution $\delta^* E$, and define a new density of states, $\Omega^*(E)$, which is the number of states with energy in the range E to $E + \delta^* E$. It can easily be seen that

$$\Omega^*(E) = \frac{\delta^* E}{\delta E}\,\Omega(E). \tag{5.70}$$

It follows that the new entropy, $S^* = k\ln\Omega^*$, is related to the previous entropy $S = k\ln\Omega$ via

$$S^* = S + k\ln\left(\frac{\delta^* E}{\delta E}\right). \tag{5.71}$$

Now, our usual estimate that $\Omega \sim E^f$ (see Section 3.8) gives $S \sim k f$, where f is the number of degrees of freedom. It follows that even if $\delta^* E$ were to differ from δE by of order f (i.e., twenty four orders of magnitude), which is virtually inconceivable, the second term on the right-hand side of the previous equation is still only of order $k\ln f$, which is utterly negligible compared to $k f$. It follows that

$$S^* = S \tag{5.72}$$

to an excellent approximation, so our definition of entropy is completely insensitive to the resolution to which we measure energy (or any other macroscopic parameter).

Note that, like the temperature, the entropy of a macrostate is only well defined if the macrostate is in equilibrium. The crucial point is that it only makes sense to talk about the number of accessible states if the systems in the ensemble are given sufficient time to thoroughly explore all of the possible microstates consistent with the known macroscopic constraints. In other words, we can only be sure that a given microstate is inaccessible when the systems in the ensemble have had ample opportunity to move into it, and yet have not done so. For an equilibrium state, the entropy is just as well defined as more familiar quantities such as the temperature and the mean pressure.

Consider, again, two systems, A and A', that are in thermal contact, but can do no work on one another. (See Section 5.2.) Let E and E' be the energies of the two systems, and $\Omega(E)$ and $\Omega'(E')$ the respective densities of states. Furthermore, let $E^{(0)}$ be the conserved energy of the system as a whole, and $\Omega^{(0)}$ the corresponding density of states. We have from Equation (5.2) that

$$\Omega^{(0)}(E) = \Omega(E)\,\Omega'(E'), \tag{5.73}$$

where $E' = E^{(0)} - E$. In other words, the number of states accessible to the whole system is the product of the numbers of states accessible to each subsystem, because every microstate of A can be combined with every microstate of A' to form a distinct microstate of the whole system. We know, from Section 5.2, that in equilibrium the mean energy of A takes the value $\overline{E} = \tilde{E}$ for which $\Omega^{(0)}(E)$ is maximum, and the temperatures of A and A' are equal. The distribution of E around the mean value is of order $\Delta^* E = \tilde{E}/\sqrt{f}$, where f is the number of degrees of freedom. It follows that the total number of accessible microstates is approximately the number of states which lie within $\Delta^* E$ of $\tilde{E}$. Thus,

$$\Omega^{(0)}_{\rm tot} \simeq \frac{\Omega^{(0)}(\tilde{E})}{\delta E}\,\Delta^* E. \tag{5.74}$$

The entropy of the whole system is given by

$$S^{(0)} = k\,\ln\Omega^{(0)}_{\rm tot} = k\,\ln\Omega^{(0)}(\tilde{E}) + k\,\ln\left(\frac{\Delta^* E}{\delta E}\right). \tag{5.75}$$

According to our usual estimate, $\Omega \sim E^f$ (see Section 3.8), the first term on the right-hand side of the previous equation is of order $k\,f$, whereas the second term is of order $k\,\ln(\tilde{E}/\sqrt{f}\,\delta E)$. Any reasonable choice for the energy subdivision, δE, should be greater than $\tilde{E}/f$, otherwise there would be less than one microstate per subdivision. It follows that the second term is less than, or of order, $k\,\ln f$, which is utterly negligible compared to $k\,f$. Thus,

$$S^{(0)} = k\,\ln\Omega^{(0)}(\tilde{E}) = k\,\ln[\Omega(\tilde{E})\,\Omega(\tilde{E}')] = k\,\ln\Omega(\tilde{E}) + k\,\ln\Omega'(\tilde{E}') \tag{5.76}$$

to an excellent approximation, giving

$$S^{(0)} = S(\tilde{E}) + S'(\tilde{E}'). \tag{5.77}$$

It can be seen that the probability distribution for $\Omega^{(0)}(E)$ is so strongly peaked around its maximum value that, for the purpose of calculating the entropy, the total number of states is equal to the maximum number of states [i.e., $\Omega^{(0)}_{\rm tot} \sim \Omega^{(0)}(\tilde{E})$]. One consequence of this is that entropy possesses the simple additive property illustrated in Equation (5.77). Thus, the total entropy of two thermally interacting systems in equilibrium is the sum of the entropies of each system taken in isolation.

5.8 Uses of Entropy

We have defined a new function called entropy, denoted S, that parameterizes the amount of disorder in a macroscopic system. The entropy of an equilibrium macrostate is related to the number of accessible microstates, Ω, via

$$S = k\,\ln\Omega. \tag{5.78}$$

On a macroscopic level, the increase in entropy due to a quasi-static change in which an infinitesimal amount of heat, $đQ$, is absorbed by the system is given by

$$dS = \frac{đQ}{T}, \tag{5.79}$$

where T is the absolute temperature of the system. The second law of thermodynamics states that the entropy of an isolated system can never spontaneously decrease. Let us now briefly examine some consequences of these results.

Consider two bodies, A and A', that are in thermal contact, but can do no work on one another. We know what is supposed to happen in this situation. Heat flows from the hotter to the colder of the two bodies until their temperatures become the same. Consider a quasi-static exchange of heat between the two bodies. According to the first law of thermodynamics, if an infinitesimal amount of heat, $đQ$, is absorbed by A then infinitesimal heat, $đQ' = -đQ$, is absorbed by A'. The increase in the entropy of system A is $dS = đQ/T$, and the corresponding increase in the entropy of A' is $dS' = đQ'/T'$. Here, T and T' are the temperatures of the two systems, respectively. Note that $đQ$ is assumed to the sufficiently small that the heat transfer does not substantially modify the temperatures of either system. The change in entropy of the whole system is

$$dS^{(0)} = dS + dS' = \left(\frac{1}{T} - \frac{1}{T'}\right) đQ. \tag{5.80}$$

This change must be positive or zero, according to the second law of thermodynamics, so $dS^{(0)} \geq 0$. It follows that $đQ$ is positive (i.e., heat flows from A' to A) when $T' > T$, and vice versa. The spontaneous flow of heat only ceases when $T = T'$. Thus, the direction of spontaneous heat flow is a consequence of the second law of thermodynamics. Note that the spontaneous flow of heat between bodies at different temperatures is always an irreversible process that increases the entropy, or disorder, of the universe.

Consider, now, the slightly more complicated situation in which the two systems can exchange heat, and also do work on one another via a movable partition. Suppose that the total volume is invariant, so that

$$V^{(0)} = V + V' = \text{constant}, \tag{5.81}$$

where V and V' are the volumes of A and A', respectively. Consider a quasi-static change in which system A absorbs an infinitesimal amount of heat, $đQ$, and its volume simultaneously increases by an infinitesimal amount, dV. The infinitesimal amount of work done by system A is $đW = \bar{p}\, dV$ (see Section 4.4), where $\bar{p}$ is the mean pressure of A. According to the first law of thermodynamics,

$$đQ = dE + đW = dE + \bar{p}\, dV, \tag{5.82}$$

where dE is the change in the internal energy of A. Because $dS = đQ/T$, the increase in entropy of system A is written

$$dS = \frac{dE + \bar{p}\,dV}{T}. \tag{5.83}$$

Likewise, the increase in entropy of system A' is given by

$$dS' = \frac{dE' + \bar{p}'\,dV'}{T'}. \tag{5.84}$$

According to Equation (5.83),

$$\frac{1}{T} = \left(\frac{\partial S}{\partial E}\right)_V, \tag{5.85}$$

$$\frac{\bar{p}}{T} = \left(\frac{\partial S}{\partial V}\right)_E, \tag{5.86}$$

where the subscripts are to remind us what is held constant in the partial derivatives. We can write an analogous pair of equations for the system A'.

The overall system is assumed to be isolated, so conservation of energy gives $dE + dE' = 0$. Furthermore, Equation (5.81) implies that $dV + dV' = 0$. It follows that the total change in entropy is given by

$$dS^{(0)} = dS + dS' = \left(\frac{1}{T} - \frac{1}{T'}\right) dE + \left(\frac{\bar{p}}{T} - \frac{\bar{p}'}{T'}\right) dV. \tag{5.87}$$

The equilibrium state is the most probable state. (See Section 5.2.) According to statistical mechanics, this is equivalent to the state with the largest number of accessible microstates. Finally, Equation (5.78) implies that this is the maximum entropy state. The system can never spontaneously leave a maximum entropy state, because this would imply a spontaneous reduction in entropy, which is forbidden by the second law of thermodynamics. A maximum or minimum entropy state must satisfy $dS^{(0)} = 0$ for arbitrary small variations of the energy and external parameters. It follows from Equation (5.87) that

$$T = T', \tag{5.88}$$

$$\bar{p} = \bar{p}', \tag{5.89}$$

for such a state. This corresponds to a maximum entropy state (i.e., an equilibrium state) provided that

$$\left(\frac{\partial^2 S}{\partial E^2}\right)_V < 0, \tag{5.90}$$

$$\left(\frac{\partial^2 S}{\partial V^2}\right)_E < 0, \tag{5.91}$$

with a similar pair of inequalities for system A'. The usual estimate $\Omega \propto E^f V^f$ (see Section 3.8), giving $S = k f \ln E + k f \ln V + \cdots$, ensures that the previous inequalities are satisfied in conventional macroscopic systems. In the maximum entropy state, the systems A and A' have equal temperatures (i.e., they are in thermal equilibrium), and equal pressures (i.e., they are in mechanical equilibrium). The second law of thermodynamics implies that the two interacting systems will evolve towards this state, and will then remain in it indefinitely (if left undisturbed).

5.9 Entropy and Quantum Mechanics

The entropy, S, of a system is defined in terms of the number, Ω, of accessible microstates consistent with an overall energy in the range E to $E + \delta E$ via

$$S = k \ln \Omega. \tag{5.92}$$

We have already demonstrated that this definition is utterly insensitive to the resolution, δE, to which the macroscopic energy is measured (See Section 5.7.) In classical mechanics, if a system possesses f degrees of freedom then phase-space is conventionally subdivided into cells of arbitrarily chosen volume h_0^f. (See Section 3.2.) The number of accessible microstates is equivalent to the number of these cells in the volume of phase-space consistent with an overall energy of the system lying in the range E to $E + \delta E$. Thus,

$$\Omega = \frac{1}{h_0^f} \int \cdots \int dq_1 \cdots dq_f \, dp_1 \cdots dp_f, \tag{5.93}$$

giving

$$S = k \ln\left(\int \cdots \int dq_1 \cdots dq_f \, dp_1 \cdots dp_f\right) - k f \ln h_0. \tag{5.94}$$

Thus, in classical mechanics, the entropy is undetermined to an arbitrary additive constant that depends on the size of the cells in phase-space. In fact, S increases as the cell size decreases. The second law of thermodynamics is only concerned with changes in entropy, and is, therefore, unaffected by an additive constant. Likewise, macroscopic thermodynamical quantities, such as the temperature and pressure, that can be expressed as partial derivatives of the entropy with respect to various macroscopic parameters [see Equations (5.85) and (5.86)] are unaffected by such a constant. So, in classical mechanics, the entropy is rather like a gravitational potential; it is undetermined to an additive constant, but this does not affect any physical laws.

The non-unique value of the entropy comes about because there is no limit to the precision to which the state of a classical system can be specified. In other words, the cell size, h_0, can be made arbitrarily small, which corresponds to specifying the particle coordinates and momenta to arbitrary accuracy. However, in quantum mechanics, the uncertainty principle sets a definite limit to how accurately the particle coordinates and momenta can be specified. (See Section C.8.) In general,

$$\delta q_i\,\delta p_i \geq \hbar/2, \tag{5.95}$$

where $\hbar$ is Planck's constant over 2π, p_i is the momentum conjugate to the generalized coordinate q_i, and δq_i, δp_i are the uncertainties in these quantities, respectively. In fact, in quantum mechanics, the number of accessible quantum states with the overall energy in the range E to $E + \delta E$ is completely determined. This implies that, in reality, the entropy of a system has a unique and unambiguous value. Quantum mechanics can often be simulated in classical mechanics by setting the cell size in phase-space equal to Planck's constant, so that $h_0 = h$. This automatically enforces the most restrictive form of the uncertainty principle, $\delta q_i\,\delta p_i \simeq \hbar/2$. In many systems, the substitution $h_0 \rightarrow h$ in Equation (5.94) gives the same, unique value for S as that obtained from a full quantum-mechanical calculation. (See Section 10.10.)

Consider a simple quantum-mechanical system consisting of N non-interacting spinless particles of mass m confined in a cubic box of dimension L. (See Section C.10.) The energy levels of the ith particle are given by

$$e_i = \frac{\hbar^2\pi^2}{2\,m\,L^2}\left(n_{1\,i}^2 + n_{2\,i}^2 + n_{3\,i}^2\right), \tag{5.96}$$

where $n_{1\,i}$, $n_{2\,i}$, and $n_{3\,i}$ are three (positive) quantum numbers. The overall energy of the system is the sum of the energies of the individual particles, so that for a general state r,

$$E_r = \sum_{i=1,N} e_i. \tag{5.97}$$

The overall state of the system is completely specified by $3N$ quantum numbers, so the number of degrees of freedom is $f = 3N$. The classical limit corresponds to the situation where all of the quantum numbers are much greater than unity. In this limit, the number of accessible states varies with energy according to our usual estimate $\Omega \propto E^f$. (See Section 3.8.) The lowest possible energy state of the system, the so-called ground-state, corresponds to the situation where all quantum numbers take their lowest possible value, unity. Thus, the ground-state energy, E_0, is given by

$$E_0 = \frac{f\,\hbar^2\,\pi^2}{2\,m\,L^2}. \tag{5.98}$$

There is only one accessible microstate at the ground-state energy (i.e., that where all quantum numbers are unity), so by our usual definition of entropy,

$$S(E_0) = k \ln 1 = 0. \tag{5.99}$$

In other words, there is no disorder in the system when all the particles are in their ground-states.

Clearly, as the energy approaches the ground-state energy, the number of accessible states becomes far less than the usual classical estimate E^f. This is true for all quantum mechanical systems. In general, the number of microstates varies roughly like

$$\Omega(E) \sim 1 + C\,(E - E_0)^f, \tag{5.100}$$

where C is a positive constant. According to Equation (5.31), the temperature varies approximately like

$$T \sim \frac{E - E_0}{k\,f}, \tag{5.101}$$

provided $\Omega \gg 1$. Thus, as the absolute temperature of a system approaches zero, the internal energy approaches a limiting value E_0 (the quantum-mechanical ground-state energy), and the entropy approaches the limiting value zero. This proposition is known as the *third law of thermodynamics*.

At low temperatures, great care must be taken to ensure that equilibrium thermodynamical arguments are applicable, because the rate of attaining equilibrium may be very slow. Another difficulty arises when dealing with a system in which the atoms possess nuclear spins. Typically, when such a system is brought to a very low temperature, the entropy associated with the degrees of freedom not involving nuclear spins becomes negligible. Nevertheless, the number of microstates, Ω_s, corresponding to the possible nuclear spin orientations may be very large. Indeed, it may be just as large as the number of states at room temperature. The reason for this is that nuclear magnetic moments are extremely small, and, therefore, have extremely weak mutual interactions. Thus, it only takes a tiny amount of heat energy in the system to completely randomize the spin orientations. Typically, a temperature as small as 10^{-3} degrees kelvin is sufficient to randomize the spins.

Suppose that the system consists of N atoms with nuclei of spin $1/2$. Each spin can have two possible orientations. If there is enough residual heat energy in the system to randomize the spins then each orientation is equally likely. It follows that there are $\Omega_s = 2^N$ accessible spin states. The entropy associated with these states is $S_0 = k \ln \Omega_s = \nu R \ln 2$. Below some critical temperature, T_0, the interaction between the nuclear spins becomes significant, and the system settles down in some unique quantum-mechanical ground-state (e.g., with all spins aligned). In this situation, $S \to 0$, in accordance with the third law of thermodynamics. However, for temperatures which are small, but not small enough to freeze out the nuclear spin degrees of freedom, the entropy approaches a limiting value, S_0, that depends only on the kinds of atomic nuclei in the system. This limiting value is independent of the spatial arrangement of the atoms, or the interactions between them. Thus, for most practical purposes, the third law of thermodynamics can be written

$$\text{as } T \to 0_+, \quad S \to S_0, \tag{5.102}$$

where 0_+ denotes a temperature that is very close to absolute zero, but still much larger than T_0. This modification of the third law is useful because it can be applied at temperatures that are not prohibitively low.

5.10 Laws of Thermodynamics

We have now come to the end of our investigation of the fundamental postulates of classical and statistical thermodynamics. The remainder of this book is devoted to the application of the ideas that we have just discussed to various situations of interest in physics. Before we proceed, however, it is useful to summarize the results of our investigations. Our summary takes the form of a number of general statements regarding macroscopic variables that are usually referred to as the laws of thermodynamics:

Zeroth Law: If two systems are separately in thermal equilibrium with a third system then they must be in thermal equilibrium with one another. (See Section. 5.3.)

First Law: The change in internal energy of a system in going from one macrostate to another is the difference between the net heat absorbed by the system from its surroundings, and the net work done by the system on its surroundings. (See Section 4.1.)

Second Law: The entropy of an isolated system can never spontaneously decrease. (See Section 5.6.)

Third Law: In the limit as the absolute temperature tends to zero, the entropy also tends to zero. (See Section 5.9.)

Exercises

5.1 A box is separated by a partition that divides its volume in the ratio 3:1. The larger portion of the box contains 1000 molecules of Ne gas, and the smaller, 100 molecules of He gas. A small hole is punctured in the partition, and the system is allowed to settle down and attain an equilibrium state.

(a) Find the mean number of molecules of each type on either side of the partition.
(b) What is the probability of finding 1000 molecules of Ne gas in the larger partition, and 100 molecules of He gas in the smaller (in other words, the same distribution as in the initial system)?

(From [Reif (1965)])

5.2 Consider an isolated system consisting of a large number, N, of very weakly-interacting localized particles of spin $1/2$. Each particle has a magnetic moment, μ, that can point either parallel or antiparallel to an applied magnetic field, B. The energy, E, of the system is then $E = -(n_1 - n_2)\,\mu B$, where n_1 is the number of spins aligned parallel to B, and n_2 the number of antiparallel spins.

(a) Consider the energy range between E and $E + \delta E$, where δE is very small compared to E, but is microscopically large, so that $\delta E \gg \mu B$. Show that the total number of states lying in this energy range is

$$\Omega(E) = \frac{N!}{(N/2 - E/2\,\mu B)!\,(N/2 + E/2\,\mu B)!}\,\frac{\delta E}{2\,\mu B}.$$

(b) Write down an expression for $\ln \Omega(E)$. Simplify this expression using Stirling's approximation (i.e., $n! \simeq n \ln n - n$) to obtain

$$\ln \Omega(E) \simeq N\,\ln N - \left(\frac{N}{2} - \frac{E}{2\,\mu B}\right)\ln\left(\frac{N}{2} - \frac{E}{2\,\mu B}\right) - \left(\frac{N}{2} + \frac{E}{2\,\mu B}\right)\ln\left(\frac{N}{2} + \frac{E}{2\,\mu B}\right).$$

(c) Use the definition $\beta = \partial \ln \Omega/\partial E$ (where $\beta = 1/k\,T$) to show that the energy of the system can be written

$$E \simeq -N\,\mu\,B\,\tanh\left(\frac{\mu B}{k\,T}\right).$$

(d) Under what circumstances is the temperature negative?
(e) Suppose that two otherwise identical spin systems with equal and opposite temperatures are brought into thermal contact. What is the final temperature of the overall system?

5.3 A glass bulb contains air at room temperature and at a pressure of 1 atmosphere. It is placed in a far larger chamber filled with helium gas at

1 atmosphere and at room temperature. A few months later, the experimenter happens to read in a journal that the particular glass of which the bulb is made is quite permeable to helium, but not to any other gas. Assuming that equilibrium has been reached by this time, what gas pressure will the experimenter measure inside the bulb when he/she goes back to check? (From [Reif (1965)])

5.4 The heat absorbed by a mole of ideal gas in a quasi-static process in which the temperature, T, changes by dT, and the volume, V, by dV, is given by

$$đQ = c\,dT + p\,dV,$$

where c is its constant molar specific heat at constant volume, and $p = RT/V$ is its pressure. Find an expression for the change in entropy of the gas in a quasi-static process which takes it from the initial values of temperature and volume T_i and V_i, respectively, to the final values T_f and V_f, respectively. Does the answer depend on the process involved in going from the initial to the final state? What is the relationship between the temperature and the volume in an adiabatic process (i.e. a quasi-static process in which no heat is absorbed)? What is the change in entropy in an adiabatic process in which the volume changes from an initial value V_i to a final value V_f? (From [Reif (1965)])

5.5 A solid contains N magnetic atoms having spin $1/2$. At sufficiently high temperatures, each spin is completely randomly oriented. In other words, it is equally likely to be in either one of two possible states. But at sufficiently low temperature, the interactions between the magnetic atoms causes them to exhibit ferromagnetism, with the result that their spins become oriented in the same direction. A very crude approximation suggests that the spin-dependent contribution, $C(T)$, to the heat capacity of the solid has an approximate temperature dependence given by

$$C(T) = C_1\left(2\,\frac{T}{T_1} - 1\right)$$

for $T_1/2 < T < T_1$, and

$$C(T) = 0,$$

otherwise. The abrupt increase in specific heat as T is reduced below T_1 is due to the onset of ferromagnetic behavior. Find two expressions for the increase in entropy as the temperature of the system is raised from a value below $T_1/2$ to one above T_1. By equating these two expressions, show that

$$C_1 = 2.26\,N\,k.$$

(From [Reif (1965)])

Chapter 6

Classical Thermodynamics

6.1 Introduction

We have already learned that macroscopic quantities, such as energy, temperature, and pressure, are, in fact, statistical in nature. In other words, in an equilibrium state, they exhibit random fluctuations about some mean value. If we were to plot the probability distribution for the energy (say) of a system in thermal equilibrium with its surroundings then we would obtain a Gaussian distribution with a very small fractional width. In fact, we expect

$$\frac{\Delta^* E}{\overline{E}} \sim \frac{1}{\sqrt{f}}, \tag{6.1}$$

where the number of degrees of freedom, f, is about 10^{24} for laboratory-scale systems. This implies that the statistical fluctuations of macroscopic quantities about their mean values are typically only about 1 part in 10^{12}.

Because the statistical fluctuations of equilibrium quantities are so small, it is an excellent approximation to neglect them altogether, and, thereby, to replace macroscopic quantities, such as energy, temperature, and pressure, by their mean values. In other words, $E \to \overline{E}$, $T \to \overline{T}$, $p \to \bar{p}$, et cetera. In the following discussion, we shall drop the overbars, so that E should be understood to represent the mean energy, $\overline{E}$, et cetera. This prescription, which is the essence of classical thermodynamics, is equivalent to replacing all statistically-varying quantities by their most probable values.

Although, formally, there are four laws of thermodynamics (i.e., the zeroth to the third), the zeroth law is really a consequence of the second law, and the third law is only important at temperatures close to absolute zero. So, for most practical purposes, the two laws that actually matter are the first law and the second law.

For an infinitesimal process, the first law of thermodynamics is written

$$đQ = dE + đW, \tag{6.2}$$

where dE is the change in internal energy of the system, $đQ$ the heat absorbed by the system from its surroundings, and $đW$ the work done by the system on its surroundings. Note that this particular formulation of the first law is merely a convention. We could equally well write the first law in terms of the heat emitted by the system, or the work done on the system. It does not really matter, as long as we are consistent in our definitions. The second law of thermodynamics implies that

$$đQ = T\,dS, \tag{6.3}$$

for an infinitesimal quasi-static process, where T is the thermodynamic temperature, and dS the change in entropy of the system. Furthermore, for systems in which the only external parameter is the volume (e.g., gases), the work done on the environment is

$$đW = p\,dV, \tag{6.4}$$

where p is the pressure, and dV the change in volume. Thus, the first and second laws of thermodynamics can be combined to give the fundamental thermodynamic relation

$$T\,dS = dE + p\,dV. \tag{6.5}$$

6.2 Ideal Gas Equation of State

Let us commence our discussion of classical thermodynamics by considering the simplest possible macroscopic system, which is an ideal gas. All of the thermodynamic properties of an ideal gas are summed up in its equation of state, which specifies the relationship between its pressure, volume, and temperature. Unfortunately, classical thermodynamics is unable determine this equation of state from first principles. In fact, classical thermodynamics cannot determine anything from first principles. We always need to provide some initial information before classical thermodynamics can generate new results. This initial information may come from statistical physics (i.e., from our knowledge of the microscopic structure of the system under consideration), but, more usually, it is empirical in nature (i.e., it is obtained from experiments). In fact, the ideal gas law was first discovered empirically by Robert Boyle. Nowadays, however, it can be justified via statistical arguments, as is described in the following.

Recall (from Section 3.8) that the number of accessible states of a monotonic ideal gas varies like

$$\Omega \propto V^N \chi(E), \tag{6.6}$$

where N is the number of atoms, and $\chi(E)$ depends only on the energy of the gas (and is, therefore, independent of the volume). We obtained this result by integrating over the volume of accessible phase-space. Because the energy of an ideal gas is independent of the atomic coordinates (given that there are no inter-atomic forces in an ideal gas), the integrals over these coordinates just reduced to N simultaneous volume integrals, giving the V^N factor in the previous expression. The integrals over the particle momenta were more complicated, but were clearly independent of V, giving the $\chi(E)$ factor in the previous expression. Consider the following statistical result [see Equation (5.41)]:

$$X_\alpha = \frac{1}{\beta}\frac{\partial \ln \Omega}{\partial x_\alpha}, \tag{6.7}$$

where X_α is the mean force conjugate to the external parameter x_α (i.e., $đW = \sum_\alpha X_\alpha\, dx_\alpha$), and $\beta = 1/(k\,T)$. For an ideal gas, the only external parameter is the volume, and its conjugate force is the pressure (because $đW = p\,dV$). So, we can write

$$p = \frac{1}{\beta}\frac{\partial \ln \Omega}{\partial V}. \tag{6.8}$$

It immediately follows from Equation (6.6) that

$$p = \frac{N\,k\,T}{V}. \tag{6.9}$$

However, $N = \nu\,N_A$, where ν is the number of moles, and N_A is Avagadro's number. Also, $k\,N_A = R$, where R is the molar ideal gas constant. This allows us to write the equation of state in its standard form

$$p\,V = \nu\,R\,T. \tag{6.10}$$

Incidentally, the fact that $\Omega = \Omega(V, E)$ in Equation (6.6) suggests that the macroscopic state of an ideal gas can be uniquely specified by giving the values of two independent parameters (e.g., the energy and the volume, the pressure and the volume, the temperature and the volume, et cetera).

The previous derivation of the ideal gas equation of state is rather elegant. It is certainly far easier to obtain the equation of state in this manner than to treat the atoms that make up the gas as tiny billiard balls continually bouncing off the walls of a container. The latter derivation is difficult to perform correctly, because it is necessary to average over all possible directions of atomic motion. (See Section 8.16 and Exercise 8.17.) It is clear, from the previous derivation, that the crucial element needed to obtain the ideal gas equation of state is the absence of interatomic forces. This automatically gives rise to a variation of the number of accessible states with E and V of the form (6.6), which, in turn, implies the ideal gas law. So, the ideal gas law should also apply to polyatomic gases with no inter-molecular forces. Polyatomic gases are more complicated that monatomic gases because the molecules can rotate and vibrate, giving rise to extra degrees of freedom, in addition to the translational degrees of freedom of a monatomic gas. In other words, $\chi(E)$, in Equation (6.6), becomes much more complicated in polyatomic gases. However, as long as there are no inter-molecular forces, the volume dependence of Ω is still V^N, and the ideal gas law should still hold true. In fact, we shall discover that the extra degrees of freedom of polyatomic gases manifest themselves by increasing the specific heat capacity. (See Section 8.13.)

There is one other conclusion that we can draw from Equation (6.6). The statistical definition of temperature is

$$\frac{1}{kT} = \frac{\partial \ln \Omega}{\partial E}. \tag{6.11}$$

[See Equation (5.31).] It follows that

$$\frac{1}{kT} = \frac{\partial \ln \chi}{\partial E}. \tag{6.12}$$

We can see that, because χ is a function of the energy, but not of the volume, the temperature must also be a function of the energy, but not the volume. This implies that

$$E = E(T). \tag{6.13}$$

In other words, the internal energy of an ideal gas depends only on the temperature of the gas, and is independent of the volume. This is a fairly obvious result, because if there are no inter-molecular forces then increasing the volume, which effectively increases the mean separation between molecules, is not going to affect the molecular energies. Hence, the energy of the whole gas is unaffected.

The volume independence of the internal energy can also be derived directly from the ideal gas equation of state. The internal energy of a gas can be considered to be a general function of the temperature and volume, so that

$$E = E(T, V). \tag{6.14}$$

It follows from mathematics that

$$dE = \left(\frac{\partial E}{\partial T}\right)_V dT + \left(\frac{\partial E}{\partial V}\right)_T dV, \tag{6.15}$$

where the subscript V reminds us that the first partial derivative is taken at constant volume, and the subscript T reminds us that the second partial derivative is taken at constant temperature. The first and second laws of thermodynamics imply that for a quasi-static change of parameters,

$$T\,dS = dE + p\,dV. \tag{6.16}$$

[See Equation (6.5).] The ideal gas equation of state, (6.10), can be used to express the pressure in term of the volume and the temperature in the previous expression:

$$dS = \frac{1}{T}\,dE + \frac{\nu R}{V}\,dV. \tag{6.17}$$

Using Equation (6.15), this becomes

$$dS = \frac{1}{T}\left(\frac{\partial E}{\partial T}\right)_V dT + \left[\frac{1}{T}\left(\frac{\partial E}{\partial V}\right)_T + \frac{\nu R}{V}\right] dV. \tag{6.18}$$

However, dS is the exact differential of a well-defined state function, S. This means that we can consider the entropy to be a function of the temperature and volume. Thus, $S = S(T, V)$, and mathematics immediately yields

$$dS = \left(\frac{\partial S}{\partial T}\right)_V dT + \left(\frac{\partial S}{\partial V}\right)_T dV. \tag{6.19}$$

The previous expression is valid for all small values of dT and dV, so a comparison with Equation (6.18) gives

$$\left(\frac{\partial S}{\partial T}\right)_V = \frac{1}{T}\left(\frac{\partial E}{\partial T}\right)_V, \tag{6.20}$$

$$\left(\frac{\partial S}{\partial V}\right)_T = \frac{1}{T}\left(\frac{\partial E}{\partial V}\right)_T + \frac{\nu R}{V}. \tag{6.21}$$

A well-known property of partial differentials is the equality of second derivatives, irrespective of the order of differentiation, so

$$\frac{\partial^2 S}{\partial V\,\partial T} = \frac{\partial^2 S}{\partial T\,\partial V}. \tag{6.22}$$

This implies that

$$\left(\frac{\partial}{\partial V}\right)_T\left(\frac{\partial S}{\partial T}\right)_V = \left(\frac{\partial}{\partial T}\right)_V\left(\frac{\partial S}{\partial V}\right)_T. \tag{6.23}$$

The previous expression can be combined with Equations (6.20) and (6.21) to give

$$\frac{1}{T}\left(\frac{\partial^2 E}{\partial V\,\partial T}\right) = \left[-\frac{1}{T^2}\left(\frac{\partial E}{\partial V}\right)_T + \frac{1}{T}\left(\frac{\partial^2 E}{\partial T\,\partial V}\right)\right]. \tag{6.24}$$

Because second derivatives are equivalent, irrespective of the order of differentiation, the previous relation reduces to

$$\left(\frac{\partial E}{\partial V}\right)_T = 0, \tag{6.25}$$

which implies that the internal energy is independent of the volume for any gas obeying the ideal equation of state. This result was confirmed experimentally by James Joule in the middle of the nineteenth century.[??]

6.3 Specific Heat

Suppose that a body absorbs an amount of heat ΔQ, and its temperature consequently rises by ΔT. The usual definition of the heat capacity, or *specific heat*, of the body is

$$C = \frac{\Delta Q}{\Delta T}. \tag{6.26}$$

If the body consists of ν moles of some substance then the *molar specific heat* (i.e., the specific heat of one mole of this substance) is defined

$$c = \frac{1}{\nu}\frac{\Delta Q}{\Delta T}. \tag{6.27}$$

In writing the previous expressions, we have tacitly assumed that the specific heat of a body is independent of its temperature. In general, this is not true. We can overcome this problem by only allowing the body in question to absorb a very small amount of heat, so that its temperature only rises slightly, and its specific heat remains approximately constant. In the limit that the amount of absorbed heat becomes infinitesimal, we obtain

$$c = \frac{1}{\nu}\frac{đQ}{dT}. \tag{6.28}$$

In classical thermodynamics, it is usual to define two molar specific heats. Firstly, the molar specific heat at constant volume, denoted

$$c_V = \frac{1}{\nu}\left(\frac{đQ}{dT}\right)_V, \tag{6.29}$$

and, secondly, the molar specific heat at constant pressure, denoted

$$c_p = \frac{1}{\nu}\left(\frac{đQ}{dT}\right)_p. \tag{6.30}$$

Consider the molar specific heat at constant volume of an ideal gas. Because $dV = 0$, no work is done by the gas on its surroundings, and the first law of thermodynamics reduces to

$$đQ = dE. \tag{6.31}$$

It follows from Equation (6.29) that

$$c_V = \frac{1}{\nu}\left(\frac{\partial E}{\partial T}\right)_V. \tag{6.32}$$

Now, for an ideal gas, the internal energy is volume independent. [See Equation (6.25).] Thus, the previous expression implies that the specific heat at constant volume is also volume independent. Because E is a function of T only, we can write

$$dE = \left(\frac{\partial E}{\partial T}\right)_V dT. \tag{6.33}$$

The previous two expressions can be combined to give

$$dE = \nu\, c_V\, dT \tag{6.34}$$

for an ideal gas.

Let us now consider the molar specific heat at constant pressure of an ideal gas. In general, if the pressure is kept constant then the volume changes, and so the gas does work on its environment. According to the first law of thermodynamics,

$$đQ = dE + p\, dV = \nu\, c_V\, dT + p\, dV, \tag{6.35}$$

where use has been made of Equation (6.34). The equation of state of an ideal gas, (6.10), implies that if the volume changes by dV, the temperature changes by dT, and the pressure remains constant, then

$$p\, dV = \nu R\, dT. \tag{6.36}$$

The previous two equations can be combined to give

$$đQ = \nu\, c_V\, dT + \nu R\, dT. \tag{6.37}$$

Now, by definition,

$$c_p = \frac{1}{\nu}\left(\frac{đQ}{dT}\right)_p, \tag{6.38}$$

so we obtain

$$c_p = c_V + R \tag{6.39}$$

for an ideal gas. Note that, at constant volume, all of the heat absorbed by the gas goes into increasing its internal energy, and, hence, its temperature, whereas, at constant pressure, some of the absorbed heat is used to do work on the environment as the volume increases. This means that, in the latter case, less heat is available to increase the temperature of the gas. Thus, we expect the specific heat at constant pressure to exceed that at constant volume, as indicated by the previous formula.

The ratio of the two specific heats, c_p/c_V, is conventionally denoted γ. We have

$$\gamma \equiv \frac{c_p}{c_V} = 1 + \frac{R}{c_V} \tag{6.40}$$

for an ideal gas. In fact, γ is easy to measure experimentally because the speed of sound in an ideal gas takes the form

$$u = \sqrt{\frac{\gamma\, p}{\rho}}, \tag{6.41}$$

where ρ is the mass density. (See Exercise 6.8.) Table 6.1 lists some experimental measurements of c_V and γ for common gases. The extent of the agreement between γ calculated from Equation (6.40) and the experimental γ is remarkable.

Table 6.1 Molar specific heats of common gases in joules/mole/degree (at 15°C and 1 atmosphere). [Reif (1965); Partington and Shilling (1924)]

Gas	Symbol	c_V (experiment)	γ (experiment)	γ (theory)
Helium	He	12.5	1.666	1.666
Argon	Ar	12.5	1.666	1.666
Nitrogen	N_2	20.6	1.405	1.407
Oxygen	O_2	21.1	1.396	1.397
Carbon Dioxide	CO_2	28.2	1.302	1.298
Ethane	C_2H_6	39.3	1.220	1.214

6.4 Calculation of Specific Heats

Now that we know the relationship between the molar specific heats at constant volume and constant pressure for an ideal gas, it would be interesting if we could calculate either one of these quantities from first principles. Classical thermodynamics cannot help us here. However, it is straightforward to calculate the specific heat at constant volume using our knowledge of statistical physics. Recall that the variation of the number of accessible states of an ideal gas with energy and volume is written

$$\Omega(E, V) \propto V^N \chi(E). \tag{6.42}$$

In fact, for the specific case of a monatomic ideal gas, we worked out a more exact expression for Ω in Section 3.8: namely,

$$\Omega(E, V) = B\, V^N E^{3N/2}, \tag{6.43}$$

where B is a constant independent of the energy and volume. It follows that

$$\ln \Omega = \ln B + N \ln V + \frac{3N}{2} \ln E. \tag{6.44}$$

The absolute temperature is given by [see Equation (5.30)]

$$\beta \equiv \frac{1}{k T} = \frac{\partial \ln \Omega}{\partial E} = \frac{3N}{2}\frac{1}{E}, \tag{6.45}$$

so

$$E = \frac{3}{2} N k T. \tag{6.46}$$

Because $N = \nu N_A$, and $N_A k = R$, we can rewrite the previous expression as

$$E = \frac{3}{2} \nu R T, \tag{6.47}$$

where $R = 8.3143$ joules/mole/degree is the molar ideal gas constant. The previous formula tells us exactly how the internal energy of a monatomic ideal gas depends on its temperature.

The molar specific heat at constant volume of a monatomic ideal gas is clearly

$$c_V = \frac{1}{\nu}\left(\frac{\partial E}{\partial T}\right)_V = \frac{3}{2}\,R. \tag{6.48}$$

This has the numerical value

$$c_V = 12.47 \text{ joules/mole/degree.} \tag{6.49}$$

Furthermore, we have

$$c_p = c_V + R = \frac{5}{2}\,R, \tag{6.50}$$

and

$$\gamma \equiv \frac{c_p}{c_V} = \frac{5}{3} = 1.667. \tag{6.51}$$

We can see from Table 6.1 that these predictions are borne out fairly well for the monatomic gases helium and argon. Note that the specific heats of polyatomic gases are larger than those of monatomic gases, because polyatomic molecules can rotate around their centers of mass, as well as translate, so polyatomic gases can store energy in the rotational, as well as the translational, energy states of their constituent particles. We shall analyze this effect in greater detail later on in this book. (See Section 8.13.)

6.5 Isothermal and Adiabatic Expansion

Suppose that the temperature of an ideal gas is held constant by keeping the gas in thermal contact with a heat reservoir. If the gas is allowed to expand quasi-statically under these so-called *isothermal* conditions then the ideal gas equation of state tells us that

$$p\,V = \text{constant}. \tag{6.52}$$

This result is known as the *isothermal gas law*.

Suppose, now, that the gas is thermally isolated from its surroundings. If the gas is allowed to expand quasi-statically under these so-called *adiabatic* conditions then it does work on its environment, and, hence, its internal energy is reduced, and its temperature changes. Let us calculate the relationship between the pressure and volume of the gas during adiabatic expansion. According to the first law of thermodynamics,

$$đQ = \nu\,c_V\,dT + p\,dV = 0, \tag{6.53}$$

in an adiabatic process (in which no heat is absorbed). [See Equation (6.35).] The ideal gas equation of state, (6.10), can be differentiated, yielding

$$p\,dV + V\,dp = \nu\,R\,dT. \tag{6.54}$$

The temperature increment, dT, can be eliminated between the previous two expressions to give

$$0 = \frac{c_V}{R}\,(p\,dV + V\,dp) + p\,dV = \left(\frac{c_V}{R} + 1\right) p\,dV + \frac{c_V}{R}\,V\,dp, \tag{6.55}$$

which reduces to

$$(c_V + R)\,p\,dV + c_V\,V\,dp = 0. \tag{6.56}$$

Dividing through by $c_V\,p\,V$ yields

$$\gamma\,\frac{dV}{V} + \frac{dp}{p} = 0, \tag{6.57}$$

where

$$\gamma \equiv \frac{c_p}{c_V} = \frac{c_V + R}{c_V}. \tag{6.58}$$

It turns out that c_V is a slowly-varying function of temperature in most gases. Consequently, it is usually a good approximation to treat the ratio of specific heats, γ, as a constant, at least over a limited temperature range. If γ is constant then we can integrate Equation (6.57) to give

$$\gamma \ln V + \ln p = \text{constant}, \tag{6.59}$$

or

$$p\,V^{\gamma} = \text{constant}. \tag{6.60}$$

This result is known as the *adiabatic gas law*. It is straightforward to obtain analogous relationships between V and T, and between p and T, during adiabatic expansion or contraction. In fact, because $p = \nu\,R\,T/V$, the previous formula also implies that

$$T\,V^{\gamma-1} = \text{constant}, \tag{6.61}$$

and

$$p^{\,1-\gamma}\,T^{\,\gamma} = \text{constant}. \tag{6.62}$$

Equations (6.60)–(6.62) are all completely equivalent.

6.6 Hydrostatic Equilibrium of Atmosphere

The gas that we are most familiar with in everyday life is, of course, the Earth's atmosphere. It turns out that we can use the isothermal and adiabatic gas laws to explain most of the observed features of the atmosphere.

Let us, first of all, consider the hydrostatic equilibrium of the atmosphere. Consider a thin vertical slice of the atmosphere, of cross-sectional area A, that starts at height z above ground level, and extends to height $z + dz$. The upward force exerted on this slice by the gas below it is $p(z)\,A$, where $p(z)$ is the pressure at height z. Likewise, the downward force exerted by the gas above the slice is $p(z+dz)\,A$. The net upward force is $[p(z)-p(z+dz)]\,A$. In equilibrium, this upward force must be balanced by the downward force due to the weight of the slice, which is $\rho\,A\,dz\,g$, where ρ is the mass density of the gas, and g the acceleration due to gravity. It follows that the force balance condition can be written

$$[p(z) - p(z+dz)]\,A = \rho\,A\,dz\,g, \tag{6.63}$$

which reduces to

$$\frac{dp}{dz} = -\rho\,g. \tag{6.64}$$

This result is known as the *equation of hydrostatic equilibrium* for the atmosphere.

We can express the mass density of a gas in the following form,

$$\rho = \frac{\nu\,\mu}{V}, \tag{6.65}$$

where μ is the *molecular weight* of the gas, and is equal to the mass of one mole of gas particles. For instance, the molecular weight of nitrogen gas is 28×10^{-3} kg. The previous formula for the mass density of a gas, combined with the ideal gas law, $p\,V = \nu\,R\,T$, yields

$$\rho = \frac{p\,\mu}{R\,T}. \tag{6.66}$$

It follows that the equation of hydrostatic equilibrium can be rewritten

$$\frac{dp}{p} = -\frac{\mu\,g}{R\,T}\,dz. \tag{6.67}$$

6.7 Isothermal Atmosphere

As a first approximation, let us assume that the temperature of the atmosphere is uniform. In such an *isothermal atmosphere*, we can directly integrate the previous equation to give

$$p = p_0 \exp\left(-\frac{z}{z_0}\right). \tag{6.68}$$

Here, p_0 is the pressure at ground level ($z = 0$), which is generally about 1 bar ($10^5\ \mathrm{N m^{-2}}$ in SI units). The quantity

$$z_0 = \frac{R\,T}{\mu\,g} \tag{6.69}$$

is known as the *isothermal scale-height* of the atmosphere. At ground level, the atmospheric temperature is, on average, about 15°C, which is 288 K on the absolute scale. The mean molecular weight of air at sea level is 29×10^{-3} kg (i.e., the molecular weight of a gas made up of 78% nitrogen, 21% oxygen, and 1% argon). The mean acceleration due to gravity is $9.81\ \mathrm{m\,s^{-2}}$ at ground level. Also, the molar ideal gas constant is 8.314 joules/mole/degree. Combining all of this information, the isothermal scale-height of the atmosphere comes out to be about 8.4 kilometers.

We have discovered that, in an isothermal atmosphere, the pressure decreases exponentially with increasing height. Because the temperature is assumed to be constant, and $\rho \propto p/T$ [see Equation (6.66)], it follows that the density also decreases exponentially with the same scale-height as the pressure. According to Equation (6.68), the pressure, or the density, of the atmosphere decreases by a factor 10 every $\ln 10\, z_0$, or 19.3 kilometers, increase in altitude above sea level. Clearly, the effective height of the atmosphere is very small compared to the Earth's radius, which is about 6,400 kilometers. In other words, the atmosphere constitutes a relatively thin layer covering the surface of the Earth. Incidentally, this justifies our neglect of the decrease of g with increasing altitude.

One of the highest points in the United States of America is the peak of Mount Elbert in Colorado. This peak lies 14,432 feet, or about 4.4 kilometers, above sea level. At this altitude, Equation (6.68) predicts that the air pressure should be about 0.6 atmospheres. Surprisingly enough, after a few days acclimatization, people can survive quite comfortably at this sort of pressure. In the highest inhabited regions of the Andes and Tibet, the air pressure falls to about 0.5 atmospheres. Humans can just about survive at such pressures. However, people cannot survive for any extended period in air pressures below half an atmosphere. This sets an upper limit on the altitude of permanent human habitation, which is about 19,000 feet, or 5.8 kilometers, above sea level. Incidentally, this is also the maximum altitude at which a pilot can fly an unpressurized aircraft without requiring additional oxygen.

The highest point in the world is, of course, the peak of Mount Everest in Nepal. This peak lies at an altitude of 29,028 feet, or 8.85 kilometers, above sea level, where we expect the air pressure to be a mere 0.35 atmospheres. This explains why Mount Everest was only conquered after lightweight portable oxygen cylinders were invented. Admittedly, some climbers have subsequently ascended

Mount Everest without the aid of additional oxygen, but this is a very foolhardy venture, because, above 19,000 feet, the climbers are slowly dying.

Commercial airliners fly at a cruising altitude of 32,000 feet. At this altitude, we expect th e air pressure to be only 0.3 atmospheres, which explains why airline cabins are pressurized. In fact, the cabins are only pressurized to 0.85 atmospheres (which accounts for the popping of passengers ears during air travel). The reason for this partial pressurization is quite simple. At 32,000 feet, the pressure difference between the air in the cabin and that outside is about half an atmosphere. Clearly, the walls of the cabin must be strong enough to support this pressure difference, which means that they must be of a certain thickness, and, hence, that the aircraft must be of a certain weight. If the cabin were fully pressurized then the pressure difference at cruising altitude would increase by about 30%, which means that the cabin walls would have to be much thicker, and, thus, the aircraft would have to be substantially heavier. So, a fully pressurized aircraft would be more comfortable to fly in (because your ears would not pop), but it would also be far less economical to operate.

6.8 Adiabatic Atmosphere

Of course, we know that the atmosphere is not isothermal. In fact, air temperature falls quite noticeably with increasing altitude. In ski resorts, the general rule of thumb is that the temperature drops by about 1 degree per 100 meters increase in altitude. Many people cannot understand why the atmosphere gets colder with increasing height. They reason that because higher altitudes are closer to the Sun they ought to be hotter. In fact, the explanation is quite simple. It depends on three important properties of air. The first property is that air is transparent to most, but by no means all, of the electromagnetic spectrum. In particular, most infrared radiation, which carries heat energy, passes straight through the lower atmosphere, and heats the ground. In other words, the lower atmosphere is heated from below, not from above. The second important property of air is that it is constantly in motion. In fact, the lower 20 kilometers of the atmosphere (the so-called *troposphere*) are fairly thoroughly mixed. You might think that this would imply that the atmosphere is isothermal. However, this is not the case because of the final important property of air; namely, it is a very poor conductor of heat. This, of course, is why woolly sweaters work; they trap a layer of air close to the body, and, because air is such a poor conductor of heat, you stay warm.

Imagine a packet of air that is swirling around in the atmosphere. We would expect it to always remain at the same pressure as its surroundings, otherwise it would be mechanically unstable. It is also plausible that the packet moves around too quickly to effectively exchange heat with its surroundings, because air is very a poor heat conductor, and heat flow is consequently quite a slow process. So, to a first approximation, the air in the packet is adiabatic. In a steady-state atmosphere, we expect that, as the packet moves upwards, expands due to the reduced pressure, and cools adiabatically, its temperature always remains the same as that of its immediate surroundings. This means that we can use the adiabatic gas law to characterize the cooling of the atmosphere with increasing altitude. In this particular case, the most useful manifestation of the adiabatic law is

$$p^{1-\gamma}\,T^{\gamma} = \text{constant}, \tag{6.70}$$

giving

$$\frac{dp}{p} = \frac{\gamma}{\gamma-1}\,\frac{dT}{T}. \tag{6.71}$$

Combining the previous expression with the equation of hydrostatic equilibrium, (6.67), we obtain

$$\frac{\gamma}{\gamma-1}\,\frac{dT}{T} = -\frac{\mu\,g}{R\,T}\,dz, \tag{6.72}$$

or

$$\frac{dT}{dz} = -\frac{\gamma-1}{\gamma}\,\frac{\mu\,g}{R}. \tag{6.73}$$

Now, the ratio of specific heats for air (which is effectively a diatomic gas) is about 1.4. (See Table 6.1.) Hence, we can deduce, from the previous expression, that the temperature of the atmosphere decreases with increasing height at a constant rate of 9.8°C per kilometer. This value is called the (dry) *adiabatic lapse rate* of the atmosphere. Our calculation accords well with the 1 degree colder per 100 meters higher rule of thumb used in ski resorts. The basic reason that air is colder at higher altitudes is that it expands as its pressure decreases with height. It, therefore, does work on its environment, without absorbing any heat (because of its low thermal conductivity), so its internal energy, and, hence, its temperature decreases.

According to the adiabatic lapse rate calculated previously, the air temperature at the cruising altitude of airliners (32,000 feet) should be about −80°C (assuming a sea level temperature of 15°C). In fact, this is somewhat of an underestimate. A more realistic value is about −60°C. The explanation for this discrepancy is the presence of water vapor in the atmosphere. As air rises, expands, and cools, water vapor condenses out, releasing latent heat, which prevents the temperature from falling as rapidly with height as the adiabatic lapse rate would predict. (See Exercise 7.6.) In fact, in the tropics, where the air humidity is very high, the lapse rate of the atmosphere (i.e., the rate of decrease of temperature with altitude) is significantly less than the adiabatic value. The adiabatic lapse rate is only observed when the humidity is low. This is the case in deserts, in the arctic (where water vapor is frozen out of the atmosphere), and, of course, in ski resorts.

Suppose that the lapse rate of the atmosphere differs from the adiabatic value. Let us ignore the complication of water vapor, and assume that the atmosphere is dry. Consider a packet of air that moves slightly upwards from its equilibrium height. The temperature of the packet will decrease with altitude according to the adiabatic lapse rate, because its expansion is adiabatic. We shall assume that the packet always maintains pressure balance with its surroundings. It follows that because $\rho T \propto p$, according to the ideal gas law,

$$(\rho T)_{\rm packet} = (\rho T)_{\rm atmosphere}. \tag{6.74}$$

If the atmospheric lapse rate is less than the adiabatic value then $T_{\rm atmosphere} > T_{\rm packet}$ implying that $\rho_{\rm packet} > \rho_{\rm atmosphere}$. So, the packet will be denser than its immediate surroundings, and will, therefore, tend to fall back to its original height. Clearly, an atmosphere whose lapse rate is less than the adiabatic value is vertically stable. On the other hand, if the atmospheric lapse rate exceeds the adiabatic value then, after rising a little way, the packet will be less dense than its immediate surroundings, and will, therefore, continue to rise due to buoyancy effects. Clearly, an atmosphere whose lapse rate is greater than the adiabatic value is vertically unstable. This effect is of great importance in meteorology. The normal stable state of the atmosphere is for the lapse rate to be slightly less than the adiabatic value. Occasionally, however, the lapse rate exceeds the adiabatic value, and this is always associated with extremely disturbed weather patterns.

Let us consider the temperature, pressure, and density profiles in an adiabatic atmosphere. We can directly integrate Equation (6.73) to give

$$T = T_0\left(1 - \frac{\gamma - 1}{\gamma}\frac{z}{z_0}\right), \tag{6.75}$$

where T_0 is the ground-level temperature, and

$$z_0 = \frac{R\,T_0}{\mu\,g} \tag{6.76}$$

the isothermal scale-height calculated using this temperature. The pressure profile is easily calculated from the adiabatic gas law $p^{1-\gamma}\,T^{\gamma} = \text{constant}$, or $p \propto T^{\gamma/(\gamma-1)}$. It follows that

$$p = p_0\left(1 - \frac{\gamma - 1}{\gamma}\frac{z}{z_0}\right)^{\gamma/(\gamma-1)}. \tag{6.77}$$

Consider the limit $\gamma \rightarrow 1$. In this limit, Equation (6.75) yields T independent of height (i.e., the atmosphere becomes isothermal). We can evaluate Equation (6.77) in the limit as $\gamma \rightarrow 1$ using the mathematical identity

$$\mathrm{lt}_{m\rightarrow 0}\,(1+m\,x)^{1/m} \equiv \exp(x). \tag{6.78}$$

We obtain

$$p = p_0 \exp\left(-\frac{z}{z_0}\right), \tag{6.79}$$

which, not surprisingly, is the predicted pressure variation in an isothermal atmosphere. In reality, the ratio of specific heats of the atmosphere is not unity, but is about 1.4 (i.e., the ratio for diatomic gases), which implies that in the real atmosphere

$$p = p_0 \left(1 - \frac{z}{3.5\,z_0}\right)^{3.5}. \tag{6.80}$$

In fact, this formula gives very similar results to the isothermal formula, Equation (6.79), for heights below one scale-height (i.e., $z < z_0$). For heights above one scale-height, the isothermal formula tends to predict too high a pressure. (See Figure 6.1.) So, in an adiabatic atmosphere, the pressure falls off more quickly with altitude than in an isothermal atmosphere, but this effect is only noticeable at pressures significantly below one atmosphere. In fact, the isothermal formula is a fairly good approximation below altitudes of about 10 kilometers. Because $\rho \propto p/T$, the variation of density with height is

$$\rho = \rho_0 \left(1 - \frac{\gamma - 1}{\gamma}\frac{z}{z_0}\right)^{1/(\gamma-1)} = \rho_0 \left(1 - \frac{z}{3.5\,z_0}\right)^{2.5}, \tag{6.81}$$

where ρ_0 is the density at ground level. Thus, the density falls off more rapidly with altitude than the temperature, but less rapidly than the pressure.

Note that an adiabatic atmosphere has a sharp upper boundary. Above height $z_1 = [\gamma/(\gamma - 1)]\,z_0$, the temperature, pressure, and density are all zero. In other words, there is no atmosphere. For real air, with $\gamma = 1.4$, the upper boundary of an adiabatic atmosphere lies at height $z_1 \simeq 3.5\,z_0 \simeq 29.4$ kilometers above sea level. This behavior is quite different to that of an isothermal atmosphere, which has a diffuse upper boundary. In reality, there is no sharp upper boundary to the atmosphere. The adiabatic gas law does not apply above about 20 kilometers (i.e., in the *stratosphere*) because, at these altitudes, the air is no longer strongly mixed. Thus, in the stratosphere, the pressure falls off exponentially with increasing height.

In conclusion, we have demonstrated that the temperature of the lower atmosphere should decrease approximately linearly with increasing height above

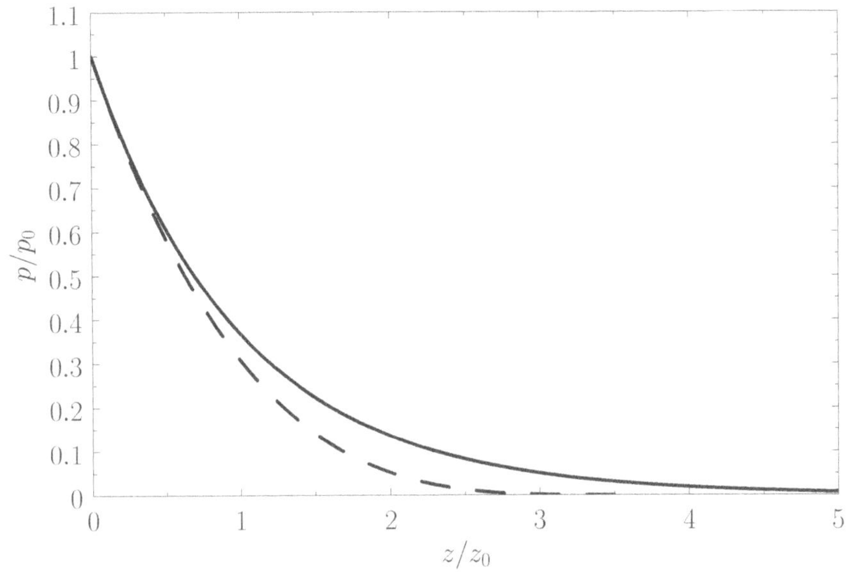

Figure 6.1 The solid curve shows the variation of pressure (normalized to the pressure at ground level) with altitude (normalized to the isothermal scale-height at ground level) in an isothermal atmosphere. The dashed curve shows the variation of pressure with altitude in an adiabatic atmosphere.

ground level, while the pressure should decrease far more rapidly than the temperature, and the density should decrease at some intermediate rate. We have also shown that the lapse rate of the temperature should be about 10°C per kilometer in dry air, but somewhat less than this in wet air. In fact, all of these predictions are, more or less, correct. It is amazing that such accurate predictions can be obtained from the two simple laws; $p\,V =$ constant for an isothermal gas, and $p\,V^{\gamma} =$ constant for an adiabatic gas.

6.9 Internal Energy

Equation (6.5) can be rearranged to give

$$dE = T\,dS - p\,dV. \tag{6.82}$$

This equation shows how the internal energy, E, depends on independent variations of the entropy, S, and the volume, V. If S and V are considered to be the two independent parameters that specify the system then

$$E = E(S, V). \tag{6.83}$$

It immediately follows that

$$dE = \left(\frac{\partial E}{\partial S}\right)_V dS + \left(\frac{\partial E}{\partial V}\right)_S dV. \tag{6.84}$$

Now, Equations (6.82) and (6.84) must be equivalent for all possible values of dS and dV. Hence, we deduce that

$$\left(\frac{\partial E}{\partial S}\right)_V = T, \tag{6.85}$$

$$\left(\frac{\partial E}{\partial V}\right)_S = -p. \tag{6.86}$$

We also know that

$$\frac{\partial^2 E}{\partial V\,\partial S} = \frac{\partial^2 E}{\partial S\,\partial V}, \tag{6.87}$$

or

$$\left(\frac{\partial}{\partial V}\right)_S \left(\frac{\partial E}{\partial S}\right)_V = \left(\frac{\partial}{\partial S}\right)_V \left(\frac{\partial E}{\partial V}\right)_S. \tag{6.88}$$

In fact, this is a necessary condition for dE in Equation (6.84) to be an exact differential. (See Section 4.5.) It follows from Equations (6.85) and (6.86) that

$$\left(\frac{\partial T}{\partial V}\right)_S = -\left(\frac{\partial p}{\partial S}\right)_V. \tag{6.89}$$

6.10 Enthalpy

The analysis in the previous section is based on the premise that S and V are the two independent parameters that specify the system. Suppose, however, that we choose S and p to be the two independent parameters. Because

$$p\,dV = d(p\,V) - V\,dp, \tag{6.90}$$

we can rewrite Equation (6.82) in the form

$$dH = T\,dS + V\,dp, \tag{6.91}$$

where

$$H = E + p\,V \tag{6.92}$$

is termed the *enthalpy*. The name is derived from the ancient greek verb ἐνθάλπω, which means *to warm in* (something).

The analysis now proceeds in an analogous manner to that in the preceding section. First, we write

$$H = H(S, p), \tag{6.93}$$

which implies that

$$dH = \left(\frac{\partial H}{\partial S}\right)_p dS + \left(\frac{\partial H}{\partial p}\right)_S dp. \tag{6.94}$$

Comparison of Equations (6.91) and (6.94) yields

$$\left(\frac{\partial H}{\partial S}\right)_p = T, \tag{6.95}$$

$$\left(\frac{\partial H}{\partial p}\right)_S = V. \tag{6.96}$$

We also know that

$$\frac{\partial^2 H}{\partial p\,\partial S} = \frac{\partial^2 H}{\partial S\,\partial p}, \tag{6.97}$$

or

$$\left(\frac{\partial}{\partial p}\right)_S \left(\frac{\partial H}{\partial S}\right)_p = \left(\frac{\partial}{\partial S}\right)_p \left(\frac{\partial H}{\partial p}\right)_S. \tag{6.98}$$

Thus, it follows from Equations (6.95) and (6.96) that

$$\left(\frac{\partial T}{\partial p}\right)_S = \left(\frac{\partial V}{\partial S}\right)_p. \tag{6.99}$$

6.11 Helmholtz Free Energy

Suppose that T and V are the are the two independent parameters that specify the system. Because

$$T\,dS = d(T\,S) - S\,dT, \tag{6.100}$$

we can rewrite Equation (6.82) in the form

$$dF = -S\,dT - p\,dV, \tag{6.101}$$

where

$$F = E - T\,S \tag{6.102}$$

is termed the *Helmholtz free energy*.

Proceeding as before, we write

$$F = F(T, V), \tag{6.103}$$

which implies that

$$dF = \left(\frac{\partial F}{\partial T}\right)_V dT + \left(\frac{\partial F}{\partial V}\right)_T dV. \tag{6.104}$$

Comparison of Equations (6.101) and (6.104) yields

$$\left(\frac{\partial F}{\partial T}\right)_V = -S, \tag{6.105}$$

$$\left(\frac{\partial F}{\partial V}\right)_T = -p. \tag{6.106}$$

We also know that

$$\frac{\partial^2 F}{\partial V\,\partial T} = \frac{\partial^2 F}{\partial T\,\partial V}, \tag{6.107}$$

or

$$\left(\frac{\partial}{\partial V}\right)_T\left(\frac{\partial F}{\partial T}\right)_V = \left(\frac{\partial}{\partial T}\right)_V\left(\frac{\partial F}{\partial V}\right)_T. \tag{6.108}$$

Thus, it follows from Equations (6.105) and (6.106) that

$$\left(\frac{\partial S}{\partial V}\right)_T = \left(\frac{\partial p}{\partial T}\right)_V. \tag{6.109}$$

6.12 Gibbs Free Energy

Suppose, finally, that T and p are the are the two independent parameters that specify the system. Because

$$T\,dS = d(T\,S) - S\,dT, \tag{6.110}$$

we can rewrite Equation (6.91) in the form

$$dG = -S\,dT + V\,dp, \tag{6.111}$$

where

$$G = H - T\,S = E - T\,S + p\,V \tag{6.112}$$

is termed the *Gibbs free energy*.

Proceeding as before, we write

$$G = G(T, p), \tag{6.113}$$

which implies that

$$dG = \left(\frac{\partial G}{\partial T}\right)_p dT + \left(\frac{\partial G}{\partial p}\right)_T dp. \tag{6.114}$$

Comparison of Equations (6.111) and (6.114) yields

$$\left(\frac{\partial G}{\partial T}\right)_p = -S, \tag{6.115}$$

$$\left(\frac{\partial G}{\partial p}\right)_T = V. \tag{6.116}$$

We also know that

$$\frac{\partial^2 G}{\partial p\,\partial T} = \frac{\partial^2 G}{\partial T\,\partial p}, \tag{6.117}$$

or

$$\left(\frac{\partial}{\partial p}\right)_T \left(\frac{\partial G}{\partial T}\right)_p = \left(\frac{\partial}{\partial T}\right)_p \left(\frac{\partial G}{\partial p}\right)_T. \tag{6.118}$$

Thus, it follows from Equations (6.115) and (6.116) that

$$-\left(\frac{\partial S}{\partial p}\right)_T = \left(\frac{\partial V}{\partial T}\right)_p. \tag{6.119}$$

Equations (6.89), (6.99), (6.109), and (6.119) are known collectively as *Maxwell relations.*

6.13 General Relation Between Specific Heats

Consider a general homogeneous substance (not necessarily a gas) whose volume, V, is the only relevant external parameter. Let us find the general relationship between this substance's molar specific heat at constant volume, c_V, and its molar specific heat at constant pressure, c_p.

The heat capacity at constant volume is given by

$$C_V = \left(\frac{đQ}{dT}\right)_V = T\left(\frac{\partial S}{\partial T}\right)_V. \tag{6.120}$$

Likewise, the heat capacity at constant pressure is written

$$C_p = \left(\frac{đQ}{dT}\right)_p = T\left(\frac{\partial S}{\partial T}\right)_p. \tag{6.121}$$

Experimentally, the parameters that are most easily controlled are the temperature, T, and the pressure, p. Let us consider these as the independent variables. Thus, $S = S(T, p)$, which implies that

$$đQ = T\,dS = T\left[\left(\frac{\partial S}{\partial T}\right)_p dT + \left(\frac{\partial S}{\partial p}\right)_T dp\right] \tag{6.122}$$

in an infinitesimal quasi-static process in which an amount of heat $đQ$ is absorbed. It follows from Equation (6.121) that

$$đQ = T\,dS = C_p\,dT + T\left(\frac{\partial S}{\partial p}\right)_T dp. \tag{6.123}$$

Suppose that $p = p(T, V)$. The previous equation can be written

$$đQ = T\,dS = C_p\,dT + T\left(\frac{\partial S}{\partial p}\right)_T \left[\left(\frac{\partial p}{\partial T}\right)_V dT + \left(\frac{\partial p}{\partial V}\right)_T dV\right]. \tag{6.124}$$

At constant volume, $dV = 0$. Hence, Equation (6.120) gives

$$C_V = T\left(\frac{\partial S}{\partial T}\right)_V = C_p + T\left(\frac{\partial S}{\partial p}\right)_T\left(\frac{\partial p}{\partial T}\right)_V. \tag{6.125}$$

This is the general relationship between C_V and C_p. Unfortunately, it contains quantities on the right-hand side that are not readily measurable.

Consider $(\partial S/\partial p)_T$. According to the Maxwell relation (6.119),

$$\left(\frac{\partial S}{\partial p}\right)_T = -\left(\frac{\partial V}{\partial T}\right)_p. \tag{6.126}$$

Now, the quantity

$$\alpha_V \equiv \frac{1}{V}\left(\frac{\partial V}{\partial T}\right)_p, \tag{6.127}$$

which is known as the *volume coefficient of expansion*, is easily measured experimentally. Hence, it is convenient to make the substitution

$$\left(\frac{\partial S}{\partial p}\right)_T = -V\,\alpha_V \tag{6.128}$$

in Equation (6.125).

Consider the quantity $(\partial p/\partial T)_V$. Writing $V = V(T, p)$, we obtain

$$dV = \left(\frac{\partial V}{\partial T}\right)_p dT + \left(\frac{\partial V}{\partial p}\right)_T dp. \tag{6.129}$$

At constant volume, $dV = 0$, so we obtain

$$\left(\frac{\partial p}{\partial T}\right)_V = -\left(\frac{\partial V}{\partial T}\right)_p \bigg/ \left(\frac{\partial V}{\partial p}\right)_T. \tag{6.130}$$

The (usually positive) quantity

$$\kappa_T \equiv -\frac{1}{V}\left(\frac{\partial V}{\partial p}\right)_T, \tag{6.131}$$

which is known as the *isothermal compressibility*, is easily measured experimentally. Hence, it is convenient to make the substitution

$$\left(\frac{\partial p}{\partial T}\right)_V = \frac{\alpha_V}{\kappa_T} \tag{6.132}$$

in Equation (6.125). It follows that

$$C_p - C_V = V\,T\,\frac{\alpha_V^2}{\kappa_T}, \tag{6.133}$$

and

$$c_p - c_V = v\,T\,\frac{\alpha_V^2}{\kappa_T}, \tag{6.134}$$

where $v = V/\nu$ is the *molar volume*.

As an example, consider an ideal gas, for which

$$p\,V = \nu\,R\,T. \tag{6.135}$$

At constant p, we have

$$p\,dV = \nu\,R\,dT. \tag{6.136}$$

Hence,

$$\left(\frac{dV}{dT}\right)_p = \frac{\nu\,R}{p} = \frac{V}{T}, \tag{6.137}$$

and the expansion coefficient defined in Equation (6.127) becomes

$$\alpha_V = \frac{1}{T}. \tag{6.138}$$

At constant T, we have

$$p\,dV + V\,dp = 0. \tag{6.139}$$

Hence,

$$\left(\frac{\partial V}{\partial p}\right)_T = -\frac{V}{p}, \tag{6.140}$$

and the compressibility defined in Equation (6.131) becomes

$$\kappa_T = \frac{1}{p}. \tag{6.141}$$

Finally, the molar volume of an ideal gas is

$$v = \frac{V}{\nu} = \frac{R\,T}{p}. \tag{6.142}$$

Hence, Equations (6.134), (6.138), (6.141), and (6.142) yield

$$c_p - c_V = R, \tag{6.143}$$

which is identical to Equation (6.39).

6.14 Free Expansion of Gas

Consider a rigid container that is thermally insulated. The container is divided into two compartments separated by a valve that is initially closed. One compartment, of volume V_1, contains the gas under investigation. The other compartment is empty. The initial temperature of the system is T_1. The valve is now opened, and the gas is free to expand so as to fill the entire container, whose volume is V_2. What is the temperature, T_2, of the gas after the final equilibrium state has been reached?

Because the system consisting of the gas and the container is adiabatically insulated, no heat flows into the system; that is,

$$Q = 0. \tag{6.144}$$

Furthermore, the system does no work in the expansion process; that is,

$$W = 0. \tag{6.145}$$

It follows from the first law of thermodynamics that the total energy of the system is conserved; that is,

$$\Delta E = Q - W = 0. \tag{6.146}$$

Let us assume that the container itself has negligible heat capacity (which, it turns out, is not a particularly realistic assumption—see later), so that the internal energy of the container does not change. Under these circumstances, the energy change of the system is equivalent to that of the gas. The conservation of energy, (6.146), thus reduces to

$$E(T_2, V_2) = E(T_1, V_1), \tag{6.147}$$

where $E(T, V)$ is the gas's internal energy.

To predict the outcome of the experiment, it is only necessary to know the internal energy of the gas, $E(T, V)$, as a function of the temperature and the volume. If the initial parameters, T_1 and V_1, are known, as well as the final volume, V_2, then Equation (6.147) yields an equation that specifies the unknown final temperature, T_2.

For an ideal gas, the internal energy is independent of the volume; that is, $E = E(T)$. (See Section 6.2.) In this case, Equation (6.147) yields

$$E(T_2) = E(T_1), \tag{6.148}$$

which implies that $T_2 = T_1$. In other words, there is no temperature change in the free expansion of an ideal gas.

The change in temperature of an non-ideal gas that undergoes free expansion can be written

$$T_2 = T_1 + \int_{v_1}^{v_2} \eta\, dv, \tag{6.149}$$

where

$$\eta = \left(\frac{\partial T}{\partial v}\right)_E \tag{6.150}$$

is termed the *Joule coefficient*. Here, $v = V/\nu$ is the molar volume, where ν is the number of moles of gas in the container. Furthermore, $v_1 = V_1/\nu$ and $v_2 = V_2/\nu$.

According to the first law of thermodynamics,

$$dE = T\, dS - p\, dV. \tag{6.151}$$

However, if the internal energy is constant then we can write

$$0 = T\left[\left(\frac{\partial S}{\partial T}\right)_V dT + \left(\frac{\partial S}{\partial V}\right)_T dV\right] - p\, dV. \tag{6.152}$$

With the aid of Equations (6.109) and (6.120), this becomes

$$0 = C_V\, dT + \left[T\left(\frac{\partial p}{\partial T}\right)_V - p\right] dV. \tag{6.153}$$

Hence,

$$\left(\frac{\partial T}{\partial V}\right)_E = -\left[T\left(\frac{\partial p}{\partial T}\right)_V - p\right]\bigg/ C_V, \tag{6.154}$$

and the Joule coefficient can be written

$$\eta = -\left[T\left(\frac{\partial p}{\partial T}\right)_V - p\right]\bigg/ c_V = -\left(\frac{T\,\alpha_V}{\kappa_T} - p\right)\bigg/ c_V. \tag{6.155}$$

[See Equation (6.132).] Note that all terms on the right-hand side of the previous expression are easily measurable.

6.15 Van der Waals Gas

Consider a non-ideal gas whose equation of state takes the form

$$\left(p + \frac{a}{v^2}\right)(v - b) = R\,T, \tag{6.156}$$

where a and b are positive constants. (See Section 10.11.) Such a gas is known as a *van der Waals gas*.The previous approximate equation of state attempts to take into account the existence of long-range attractive forces between molecules in real gases, as well as the finite volume occupied by the molecules themselves. The

Table 6.2 Critical pressures and temperatures, and derived van der Waals parameters, for common gases. [Rumble *et al.* (2017)]

Gas	Symbol	p_c(bar)	T_c(K)	a(SI)	b(SI)
Helium	He	2.27	5.19	3.46×10^{-3}	2.38×10^{-5}
Hydrogen	H	12.9	32.9	2.45×10^{-2}	2.65×10^{-5}
Nitrogen	N_2	34.0	126.2	1.37×10^{-1}	3.86×10^{-5}
Oxygen	O_2	50.4	154.6	1.38×10^{-1}	3.19×10^{-5}

attractive forces gives rise to a slight compression of the gas (relative to an ideal gas)—the term a/v^2 represents this additional positive pressure. The parameter b is the volume occupied by a mole of gas molecules. Thus, $v \to b$ as $T \to 0$. Of course, in the limit $a \to 0$ and $b \to 0$, the van der Waals equation of state reduces to the ideal gas equation of state.

The van der Waals equation of state can be written

$$p = \frac{R\,T}{v-b} - \frac{a}{v^2}. \tag{6.157}$$

At fixed temperature, the previous equation yields $p(v)$ curves that exhibit a maximum and a minimum at two points where $(\partial p/\partial v)_T = 0$. At a particular temperature, the maximum and minimum coalesce into a single inflection point where $(\partial^2 p/\partial v^2)_T = 0$, in addition to $(\partial p/\partial v)_T = 0$. This point is called the *critical point*, and its temperature, pressure, and molar volume are denoted T_c, p_c, and v_c, respectively. It is readily demonstrated that

$$T_c = \frac{8\,a}{27\,R\,b}, \tag{6.158}$$

$$p_c = \frac{a}{27\,b^2}, \tag{6.159}$$

$$v_c = 3\,b. \tag{6.160}$$

(See Exercise 6.13.) The critical temperature, T_c, and the critical pressure, p_c, of a substance are easily determined experimentally, because they turn out to be the maximum temperature and pressure, respectively, at which distinct liquid and gas phases exist. (See Section 7.10.) This allows a determination of the constants a and b in the van der Waals equation of state. In fact,

$$a = \frac{27}{64}\,\frac{(R\,T_c)^2}{p_c}, \tag{6.161}$$

$$b = \frac{R\,T_c}{8\,p_c}. \tag{6.162}$$

Table 6.2 shows the experimentally measured critical pressures and temperatures, as well as the derived van der Waal parameters, for some common gases.

Let us calculate the Joule coefficient for a van der Waals gas. It follows from Equation (6.156) that

$$\left(\frac{\partial p}{\partial T}\right)_V (v - b) = R, \tag{6.163}$$

which implies that

$$T\left(\frac{\partial p}{\partial T}\right)_V = \frac{RT}{v - b} = p + \frac{a}{v^2}. \tag{6.164}$$

Substitution into Equation (6.155) gives

$$\eta = -\frac{a}{v^2\, c_V}. \tag{6.165}$$

Thus, the Joule coefficient for a van der Waals gas is negative. This implies that the temperature of the gas always decreases as it undergoes free expansion. Of course, this temperature decrease is a consequence of the work done in overcoming the inter-molecular attractive forces. Over a relatively small temperature range, $T_2 < T < T_1$, any possible temperature dependence of c_V is negligibly small. Thus, c_V can be regarded as essentially constant, and Equations (6.149) and (6.165) yield

$$T_2 - T_1 = -\frac{a}{c_V}\left(\frac{1}{v_1} - \frac{1}{v_2}\right). \tag{6.166}$$

For an expansion, where $v_2 > v_1$, this equation confirms that $T_2 < T_1$. In other words, the temperature of non-ideal gas that undergoes free expansion is reduced.

In principle, it appears that the free expansion of a gas could provide a method of cooling the gas to low temperatures. In practice, a substantial difficulty arises because of the appreciable heat capacity, C_c, of the container. Because the container's internal energy changes by an amount $C_c\,(T_2 - T_1)$, the molar heat capacity, c_V, in Equation (6.166), must be replaced by the total molar heat capacity, $c_V + C_c/\nu$. Given that the heat capacity of the container is generally much greater than that of the gas (i.e., $C_c \gg \nu\, c_V$), it follows that the actual temperature reduction is much smaller than that predicted by Equation (6.166).

6.16 Joule-Thompson Throttling

The difficulty associated with the presence of containing walls can be overcome by replacing the single-event free-expansion process just described (where this one event must supply the energy necessary to change the container temperature) with a continuous flow process (where the temperature of the walls can adjust itself initially, and remains unchanged after a steady-state has been achieved). We shall now discuss a specific example of such a process that was first suggested by Joule and Thompson.

Consider a pipe with thermally insulated walls. A porous plug in the pipe provides a constriction in the flow. A continuous stream of gas flows through the pipe. The constriction in the flow, due to the presence of the plug, results in a constant pressure difference being maintained across the plug. Let p_1 and T_1 be the pressure and temperature of the gas upstream of the plug, respectively, and let p_2 and T_2 be the downstream pressure and temperature, respectively. We expect the upstream pressure to exceed the downstream pressure (i.e., $p_1 > p_2$), otherwise gas would not be forced through the plug (in the upstream to downstream direction). But, how is the downstream temperature, T_2, related to the upstream temperature, T_1?

Consider a mass of gas, M, that flows from the upstream to the downstream side of the plug. In this so-called throttling process, the change in internal energy of the mass is

$$\Delta E = E_2 - E_1 = E(T_2, p_2) - E(T_1, p_1). \tag{6.167}$$

However, the mass also does work. Let V_1 and V_2 be the upstream and downstream volume of the mass, respectively. The mass does the work $p_2\,V_2$ in displacing gas on the downstream side of the plug, whereas the gas on the upstream side of the plug does the work $p_1\,V_1$ on the mass. Thus, the net work done by the mass is

$$W = p_2\,V_2 - p_1\,V_1. \tag{6.168}$$

No heat is absorbed by the mass of gas in the throttling process that we have just described. This is not just because the walls of the pipe are thermally insulated, so that no heat enters from the outside. More importantly, after a steady-state has been achieved, there is no temperature difference between the walls and the adjacent gas, so there is no heat flow from the walls to the gas. It follows that

$$Q = 0. \tag{6.169}$$

Application of the first law of thermodynamics to the mass of gas yields

$$\Delta E + W = Q = 0, \tag{6.170}$$

or

$$(E_2 - E_1) + (p_2\,V_2 - p_1\,V_1) = 0, \tag{6.171}$$

where use has been made of Equations (6.167) and (6.168). Now, the enthalpy of the gas mass is defined

$$H = E + p\,V. \tag{6.172}$$

(See Section 6.10.) Hence, we deduce that

$$H_2 = H_1, \tag{6.173}$$

or

$$H(T_2, p_2) = H(T_1, p_1). \tag{6.174}$$

In other words, the gas passes through the porous plug in such a manner that its enthalpy remains constant.

Suppose that $H(T, p)$ is a known function of T and p. In this case, given T_1 and p_1, as well as the downstream pressure, p_2, Equation (6.174) allows the downstream temperature, T_2, to be determined. In the case of an ideal gas,

$$H = E + p\,V = E(T) + \nu\,R\,T, \tag{6.175}$$

so that $H = H(T)$ is a function of the temperature only. Thus, Equation (6.174) reduces to

$$H(T_2) = H(T_1), \tag{6.176}$$

which implies that $T_2 = T_1$. Thus, the temperature of an ideal gas does not change in a throttling process.

For the case of a real gas, the parameter that controls whether the gas is heated or cooled in a throttling process is

$$\mu = \left(\frac{\partial T}{\partial p}\right)_H. \tag{6.177}$$

This parameter is known as the *Joule-Thompson coefficient*. Given that p decreases in a throttling process, a positive Joule-Thompson coefficient implies that the temperature also decreases, and vice versa. Let us derive a convenient expression for μ in term of readily measured experimental parameters. Starting from the fundamental thermodynamic relation [see Equation (6.5)]

$$dE = T\,dS - p\,dV, \tag{6.178}$$

we find that

$$dH = d(E + p\,V) = T\,dS + V\,dp. \tag{6.179}$$

For the case of throttling, $dH = 0$. Thus, we can write

$$0 = T\left[\left(\frac{\partial S}{\partial T}\right)_p dT + \left(\frac{\partial S}{\partial p}\right)_T dp\right] + V\,dp. \tag{6.180}$$

With the aid of Equations (6.119) and (6.121), this becomes

$$0 = C_p\,dT + \left[V - T\left(\frac{\partial V}{\partial T}\right)_p\right] dp. \tag{6.181}$$

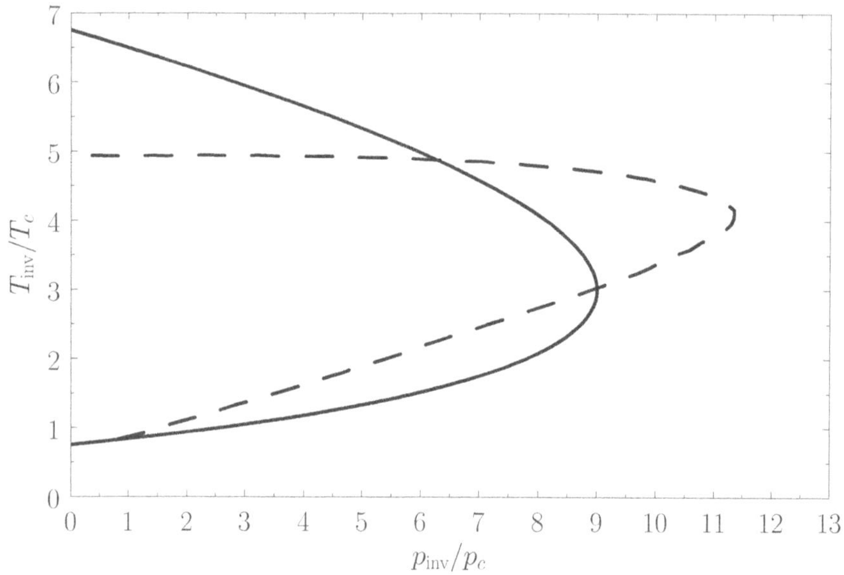

Figure 6.2 The solid curve shows the normalized inversion curve of a van der Waals gas. The dashed curve shows the normalized experimental inversion curve of nitrogen gas.

Hence,

$$\mu = \left(\frac{dT}{dp}\right)_H = \left[T\left(\frac{\partial V}{\partial T}\right)_p - V\right]\bigg/ C_p = \left[T\left(\frac{\partial v}{\partial T}\right)_p - v\right]\bigg/ c_p, \tag{6.182}$$

where $v = V/\nu$ is the molar volume.

Consider a van der Waals gas. The van der Waals equation of state, (6.156), can be written in the reduced form

$$p' = \frac{(8/3)\,T'}{v' - 1/3} - \frac{3}{v'^{\,2}}, \tag{6.183}$$

where $p' = p/p_c$, $T' = T/T_c$, and $v' = v/v_c$. [See Equations (6.158)–(6.160) and Exercise 6.13.] We can also write

$$\mu = \frac{v_c}{c_p}\,\mu', \tag{6.184}$$

where

$$\mu' = T'\left(\frac{\partial v'}{\partial T'}\right)_{p'} - v'. \tag{6.185}$$

Differentiation of Equation (6.183) yields

$$\left(\frac{\partial v'}{\partial T'}\right)_{p'} = \frac{v' - 1/3}{T' - (9/4)\,(v' - 1/3)^2/v'^{\,3}}. \tag{6.186}$$

Hence,

$$\mu = \frac{v_c}{3\,c_p}\left\{\frac{(27/4)\,[1-1/(3\,v')]^2 - T'}{T' - (27/4)\,[1-1/(3\,v')]^2\,[1/(3\,v')]}\right\}. \tag{6.187}$$

Thus, we deduce that $\mu > 0$, which is consistent with cooling of the gas in a throttling process, provided that $T < T_{\rm inv}$, where

$$T_{\rm inv} = \frac{27\,T_c}{4}\left(1-\frac{1}{3\,v'}\right)^2 = \frac{2\,a}{R\,b}\left(1-\frac{b}{v}\right)^2 \tag{6.188}$$

is known as the *inversion temperature*. Of course, if $T > T_{\rm inv}$ then $\mu < 0$, and the gas is heated by a throttling process. Let $p = p_{\rm inv}$ when $T = T_{\rm inv}$. Equations (6.183) and (6.188) give

$$\frac{T_{\rm inv}}{T_c} = \frac{27}{4}\left(1-\frac{1}{3\,v'}\right)^2, \tag{6.189}$$

$$\frac{p_{\rm inv}}{p_c} = \frac{18}{v'} - \frac{9}{v'^{\,2}}. \tag{6.190}$$

These equations define a curve in the *p*-*T* plane, known as the *inversion curve*, that is shown in Figure 6.2. (Also shown, for comparison, is the experimental inversion curve of nitrogen.) To the left of the curve, $\mu > 0$, and the gas is cooled by a throttling process. Conversely, to the right of the curve, $\mu < 0$, and the gas is heated by a throttling process. The maximum possible inversion temperature is $(T_{\rm inv})_{\rm max} = (27/4)\,T_c$. Thus, using the data in Table 6.2, we would predict that the maximum inversion temperatures of helium, hydrogen, nitrogen, and oxygen are 35 K, 224 K, 852 K, and 1044 K, respectively. In fact, the experimentally determined maximum inversion temperatures are 34 K, 202 K, 625 K, and 760 K, respectively. Thus, our estimate is fairly accurate in the case of helium, but becomes increasingly less accurate as the atomic weight of the gas increases. This is merely a deficiency in the van der Waals model.

The so-called Joule-Thompson throttling process, just described, can be used to cool nitrogen and oxygen to very low temperatures, starting from room temperature, because the maximum inversion temperatures of these two gases both exceed room temperature. On the other hand, an attempt to cool helium gas starting from room temperature would result in an increase, rather than a decrease, in the gas temperature (because room temperature exceeds the maximum inversion temperature). Thus, in order to use the Joule-Thompson process to cool helium gas to very low temperatures, it is first necessary to pre-cool the gas to a temperature that is less than 34 K. This can be achieved using liquid hydrogen.

6.17 Heat Engines

Thermodynamics was invented, almost by accident, in 1825, by a young French engineer called Sadi Carnot, who was investigating the theoretical limitations on the efficiency of steam engines. Although, nowadays, we are not particularly interested in steam engines, it is still highly instructive to review some of Carnot's arguments. We know, by observation, that it is possible to do mechanical work w upon a device M, and then to extract an equivalent amount of heat q, which goes to increase the internal energy of some heat reservoir. (Here, we use the small letters w and q to denote intrinsically positive amounts of work and heat, respectively.) An good example of this process is Joule's classic experiment by which he verified the first law of thermodynamics. A paddle wheel is spun in a liquid by a falling weight, and the work done by the weight on the wheel is converted into heat, and absorbed by the liquid. Carnot's question was the following. Is it possible to reverse this process, and build a device, called a *heat engine*, that extracts heat energy from a reservoir, and converts it into useful macroscopic work? For instance, is it possible to extract heat from the ocean, and use it to run an electric generator?

There are a few caveats to Carnot's question. First of all, the work should not be done at the expense of the heat engine itself, otherwise the conversion of heat into work could not continue indefinitely. We can ensure that this is the case by demanding that the heat engine perform some sort of cycle, by which it periodically returns to the same macrostate, but, in the meantime, has extracted heat from the reservoir, and done an equivalent amount of useful work. A cyclic process seems reasonable, because we know that both steam engines and internal combustion engines perform continuous cycles. The second caveat is that the work done by the heat engine should be such as to change a single parameter of some external device (e.g., by lifting a weight), without doing it at the expense of affecting the other degrees of freedom, or the entropy, of that device. For instance, if we are extracting heat from the ocean to generate electricity, we want to spin the shaft of the electrical generator without increasing the generator's entropy; that is, without causing the generator to heat up, or fall to bits.

Let us examine the feasibility of a heat engine using the laws of thermodynamics. Suppose that a heat engine, M, performs a single cycle. Because M has returned to its initial macrostate, its internal energy is unchanged, and the first law of thermodynamics tell us that the work done by the engine, w, must equal the heat extracted from the reservoir, q, so that

$$w = q. \tag{6.191}$$

The previous condition is certainly a necessary condition for a feasible heat engine. But, is it also a sufficient condition? In other words, does every device that satisfies this condition actually work? Let us think a little more carefully about what we are actually expecting a heat engine to do. We want to construct a device that will extract energy from a heat reservoir, where it is randomly distributed over very many degrees of freedom, and convert it into energy distributed over a single degree of freedom associated with some parameter of an external device. Once we have expressed the problem in these terms, it is fairly obvious that what we are really asking for is a spontaneous transition from a probable to an improbable state, which we know is forbidden by the second law of thermodynamics. So, unfortunately, we cannot run an electric generator off heat extracted from the ocean, because this is equivalent to expecting all of the molecules in the ocean, which are moving in random directions, to all suddenly all move in the same direction, so as to exert a force on some lever, say, that can then be converted into a torque on the generator shaft. We know, from our investigation of statistical thermodynamics, that such a process—although, in principle, possible—is fantastically improbable.

The improbability of the scenario just outlined is summed up in the second law of thermodynamics, which says that the total entropy of an isolated system can never spontaneously decrease, so

$$\Delta S \geq 0. \tag{6.192}$$

For the case of a heat engine, the isolated system consists of the engine, the reservoir from which it extracts heat, and the outside device upon which it does work. The engine itself returns periodically to the same state, so its entropy is clearly unchanged after each cycle. We have already specified that there is no change in the entropy of the external device upon which the work is done. On the other hand, the entropy change per cycle of the heat reservoir, which is held at absolute temperature T_1, say, is given by

$$\Delta S_{\text{reservoir}} = \oint \frac{đQ}{T_1} = -\frac{q}{T_1}, \tag{6.193}$$

where $đQ$ is the infinitesimal heat absorbed by the reservoir, and the integral is taken over a whole cycle of the heat engine. The integral can be converted into the expression $-q/T_1$ because the amount of heat extracted by the engine is assumed to be too small to modify the temperature of the reservoir (which is the definition of a heat reservoir), so that T_1 is a constant during the cycle. The second law of thermodynamics reduces to

$$\frac{-q}{T_1} \geq 0 \tag{6.194}$$

or, making use of the first law of thermodynamics,

$$\frac{q}{T_1} = \frac{w}{T_1} \leq 0. \tag{6.195}$$

Because we wish the work, w, done by the engine to be positive, the previous relation clearly cannot be satisfied, which proves that an engine that converts heat directly into work is thermodynamically impossible.

A perpetual motion device, which continuously executes a cycle without extracting heat from, or doing work on, its surroundings, is possible according to Equation (6.195). In fact, such a device corresponds to the equality sign in Equation (6.192), which implies that the device must be completely reversible. In reality, there is no such thing as a completely reversible engine. All engines, even the most efficient, have frictional losses that render them, at least, slightly irreversible. Thus, the equality sign in Equation (6.192) corresponds to an asymptotic limit that reality can closely approach, but never quite attain. It follows that a perpetual motion device is thermodynamically impossible. Nevertheless, the US patent office receives about 100 patent applications a year regarding perpetual motion devices. The British patent office, being slightly less open-minded that its American counterpart, refuses to entertain such applications on the basis that perpetual motion devices are forbidden by the second law of thermodynamics.

According to Equation (6.195), there is no thermodynamic objection to a heat engine that runs backwards, and converts work directly into heat. This is not surprising, because we know that this is essentially what frictional forces do. Clearly, we have, here, another example of a natural process that is fundamentally irreversible according to the second law of thermodynamics. In fact, the statement

> It is impossible to construct a perfect heat engine that converts heat directly into work

is called Kelvin's formulation of the second law.

We have demonstrated that a *perfect heat engine*, that converts heat directly into work, is impossible. But, there must be some way of obtaining useful work from heat energy, otherwise steam engines would not operate. The reason that our previous scheme did not work was that it decreased the entropy of a heat reservoir, held at some temperature T_1, by extracting an amount of heat q per cycle, without any compensating increase in the entropy of anything else, so the second law of thermodynamics was violated. How can we remedy this situation? We still want the heat engine itself to perform periodic cycles (so, by definition, its entropy cannot increase over a cycle), and we also do not want to increase the entropy of the external device upon which the work is done. Our only other option is to increase the entropy of some other body. In Carnot's analysis, this other body is a second heat reservoir held at temperature T_2. We can increase the entropy of the second reservoir by dumping some of the heat that we extracted from the first reservoir into it. Suppose that the heat per cycle that we extract from the first reservoir is q_1, and the heat per cycle that we reject into the second reservoir is q_2. Let the work done on the external device be w per cycle. The first law of thermodynamics implies that

$$q_1 = w + q_2. \tag{6.196}$$

Note that if $q_2 < q_1$ then positive (i.e., useful) work is done on the external device. The total entropy change per cycle is a consequence of the heat extracted from the first reservoir, and the heat dumped into the second, and must be positive (or zero), according to the second law of thermodynamics. So,

$$\Delta S = \frac{-q_1}{T_1} + \frac{q_2}{T_2} \geq 0. \tag{6.197}$$

We can combine the previous two equations to give

$$\frac{-q_1}{T_1} + \frac{q_1 - w}{T_2} \geq 0, \tag{6.198}$$

or

$$\frac{w}{T_2} \leq q_1 \left(\frac{1}{T_2} - \frac{1}{T_1} \right). \tag{6.199}$$

It is clear that the engine is only going to perform useful work (i.e., w is only going to be positive) if $T_2 < T_1$. So, the second reservoir must be colder than the first if the heat dumped into the former is to increase the entropy of the universe more than the heat extracted from the latter decreases it. It is useful to define the efficiency, η, of a heat engine. This is the ratio of the work done per cycle on the external device to the heat energy absorbed per cycle from the first reservoir. The efficiency of a perfect heat engine is unity, but we have already shown that such an engine is impossible. What is the efficiency of a realizable engine? It is clear, from the previous equation, that

$$\eta \equiv \frac{w}{q_1} \leq 1 - \frac{T_2}{T_1} = \frac{T_1 - T_2}{T_1}. \tag{6.200}$$

Note that the efficiency is always less than unity. A real engine must always reject some energy into the second heat reservoir, in order to satisfy the second law of thermodynamics, so less energy is available to do external work, and the efficiency of the engine is reduced. The equality sign in the previous expression corresponds to a completely reversible heat engine (i.e., one that is quasi-static). It is clear that real engines, which are always irreversible to some extent, are less efficient than reversible engines. Furthermore, all reversible engines that operate between the two temperatures T_1 and T_2 have the same efficiency,

$$\eta = \frac{T_1 - T_2}{T_1}, \tag{6.201}$$

irrespective of the manner in which they operate.

Let us consider how we might construct a reversible heat engine. Suppose that we have some gas in a cylinder equipped with a frictionless piston. The gas is not necessarily a perfect gas. Suppose that we also have two heat reservoirs held at temperatures T_1 and T_2 (where $T_1 > T_2$). These reservoirs might take the form of large water baths. Let us start off with the gas in thermal contact with the first reservoir. We now pull the piston out, very slowly, so that heat energy flows reversibly into the gas from the reservoir. Let us now thermally isolate the gas, and slowly pull out the piston some more. During this adiabatic process, the temperature of the gas falls (because there is no longer any heat flowing into it to compensate for the work it does on the piston). Let us continue this process until the temperature of the gas falls to T_2. We now place the gas in thermal contact with the second reservoir, and slowly push the piston in. During this isothermal process, heat flows out of the gas into the reservoir. We next thermally isolate the gas a second time, and slowly compress it some more. In this process, the temperature of the gas increases. We stop the compression when the temperature reaches T_1. If we carry out each step properly then we can return the gas to its initial state, and then repeat the cycle ad infinitum. We now have a set of reversible processes by which a quantity of heat is extracted from the first reservoir, and a quantity of heat is dumped into the second. We can best evaluate the work done by the system during each cycle by plotting out the locus of the gas in a p-V diagram. The locus takes the form of a closed curve. See Figure 6.3. The net work done per cycle is the "area" enclosed within this curve, because $đW = p\,dV$ [if p is plotted vertically, and V horizontally, then $p\,dV$ is clearly an element of area under the curve $p(V)$]. The engine we have just described is called a *Carnot engine*, and is the simplest conceivable device capable of converting heat energy into useful work.

For the specific case of an ideal gas, we can actually calculate the work done per cycle, and, thereby, verify Equation (6.201). Consider the isothermal expansion phase of the gas. For an ideal gas, the internal energy is a function of the temperature alone. The temperature does not change during isothermal expansion, so the internal energy remains constant, and the net heat absorbed by the gas must equal the work it does on the piston. Thus,

$$q_1 = \int_a^b p\,dV, \tag{6.202}$$

where the expansion takes the gas from state a to state b. Because $p\,V = \nu\,R\,T$, for an ideal gas, we have

$$q_1 = \int_a^b \nu\,R\,T_1\,\frac{dV}{V} = \nu\,R\,T_1\,\ln\left(\frac{V_b}{V_a}\right). \tag{6.203}$$

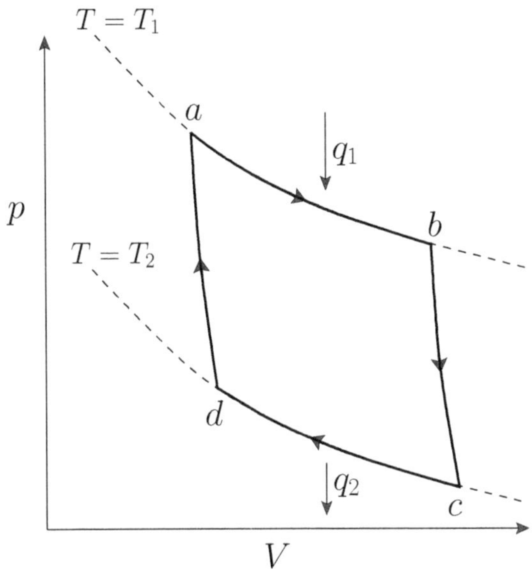

Figure 6.3 An ideal-gas Carnot engine.

Likewise, during the isothermal compression phase, in which the gas goes from state c to state d, the net heat rejected to the second reservoir is

$$q_2 = \nu R T_2 \ln\left(\frac{V_c}{V_d}\right). \tag{6.204}$$

Now, during adiabatic expansion, or compression,

$$T V^{\gamma-1} = \text{constant}. \tag{6.205}$$

It follows that, during the adiabatic expansion phase, which takes the gas from state b to state c,

$$T_1 V_b^{\gamma-1} = T_2 V_c^{\gamma-1}. \tag{6.206}$$

Likewise, during the adiabatic compression phase, which takes the gas from state d to state a,

$$T_1 V_a^{\gamma-1} = T_2 V_d^{\gamma-1}. \tag{6.207}$$

If we take the ratio of the previous two equations then we obtain

$$\frac{V_b}{V_a} = \frac{V_c}{V_d}. \tag{6.208}$$

Hence, the work done by the engine, which we can calculate using the first law of thermodynamics,

$$w = q_1 - q_2, \tag{6.209}$$

is

$$w = \nu R (T_1 - T_2) \ln\left(\frac{V_b}{V_a}\right). \tag{6.210}$$

Thus, the efficiency of the engine is

$$\eta = \frac{w}{q_1} = \frac{T_1 - T_2}{T_1} \tag{6.211}$$

which, not surprisingly, is identical to Equation (6.201).

The engine described previously is very idealized. Of course, real engines are far more complicated than this. Nevertheless, the maximum efficiency of an ideal heat engine places severe constraints on real engines. Conventional power stations have many different "front ends" (e.g., coal fired furnaces, oil fired furnaces, nuclear reactors), but their "back ends" are all very similar, and consist of a steam driven turbine connected to an electric generator. The front end heats water extracted from a local river (say), and turns it into steam, which is then used to drive the turbine, and, hence, to generate electricity. Finally, the steam is sent through a heat exchanger, so that it can heat up the incoming river water, which means that the incoming water does not have to be heated so much by the front end. At this stage, some heat is rejected to the environment, usually as clouds of steam escaping from the top of cooling towers. We can see that a power station possesses many of the same features as our idealized heat engine. There is a cycle that operates between two temperatures. The upper temperature is the temperature to which the steam is heated by the front end, and the lower temperature is the temperature of the environment into which heat is rejected. Suppose that the steam is only heated to 100°C (or 373 K), and the temperature of the environment is 15°C (or 288 K). It follows from Equation (6.200) that the maximum possible efficiency of the steam cycle is

$$\eta = \frac{373 - 288}{373} \simeq 0.23. \tag{6.212}$$

So, at least 77% of the heat energy generated by the front end goes straight up the cooling towers. Not surprisingly, commercial power stations do not operate with 100°C steam. The only way in which the thermodynamic efficiency of the steam cycle can be raised to an acceptable level is to use very hot steam (clearly, we cannot refrigerate the environment). Using 400°C steam, which is not uncommon, the maximum efficiency becomes

$$\eta = \frac{673 - 288}{673} \simeq 0.57, \tag{6.213}$$

which is more reasonable. In fact, the steam cycles of modern power stations are so well designed that they come surprisingly close to their maximum thermodynamic efficiencies.

6.18 Refrigerators

Let us now consider refrigerators. An idealized refrigerator is an engine that extracts heat from a cold heat reservoir (held at temperature T_2, say), and rejects it into a somewhat hotter heat reservoir, which is usually the environment (held at temperature T_1, say). To make this engine work, we always have to do some external work on the engine. For instance, the refrigerator in your home contains a small electric pump that does work on the freon in the cooling circuit. We can see that, in fact, a refrigerator is just a heat engine run in reverse. Hence, we can immediately carry over most of our heat engine analysis. Let q_2 be the heat absorbed per cycle from the colder reservoir, q_1 the heat rejected per cycle into the hotter reservoir, and w the external work done per cycle on the engine. The first law of thermodynamics implies that

$$w + q_2 = q_1. \tag{6.214}$$

The second law says that

$$\frac{q_1}{T_1} + \frac{-q_2}{T_2} \geq 0. \tag{6.215}$$

We can combine these two laws to give

$$\frac{w}{T_1} \geq q_2 \left(\frac{1}{T_2} - \frac{1}{T_1} \right). \tag{6.216}$$

The most sensible way of defining the efficiency of a refrigerator is as the ratio of the heat extracted per cycle from the cold reservoir to the work done per cycle on the engine. With this definition

$$\eta = \frac{T_2}{T_1 - T_2}. \tag{6.217}$$

We can see that this efficiency is, in general, greater than unity. In other words, for one joule of work done on the engine, or pump, more than one joule of energy is extracted from whatever it is we are cooling. Clearly, refrigerators are intrinsically very efficient devices. Domestic refrigerators cool their contents down to about 4°C (277K), and reject heat to the environment at about 15°C (288K). The maximum theoretical efficiency of such devices is

$$\eta = \frac{277}{288 - 277} = 25.2. \tag{6.218}$$

So, for every joule of electricity we put into a refrigerator, we can extract up to 25 joules of heat from its contents.

Exercises

6.1 Demonstrate that

$$\left(\frac{\partial x}{\partial y}\right)_z = 1\bigg/\left(\frac{\partial y}{\partial x}\right)_z.$$

This result is known as the *reciprocal rule* of partial differentiation.

6.2 Prove *Euler's chain rule*:

$$\left(\frac{\partial x}{\partial y}\right)_z\left(\frac{\partial y}{\partial z}\right)_x\left(\frac{\partial z}{\partial x}\right)_y = -1.$$

This result is also known as the *cyclic rule* of partial differentiation. Hence, deduce that

$$\left(\frac{\partial x}{\partial y}\right)_z = -\left(\frac{\partial z}{\partial y}\right)_x\bigg/\left(\frac{\partial z}{\partial x}\right)_y.$$

6.3 A cylindrical container 80 cm long is separated into two compartments by a thin piston, originally clamped in position 30 cm from the left end. The left compartment is filled with one mole of helium gas at a pressure of 5 atmospheres; the right compartment is filled with argon gas at 1 atmosphere of pressure. The gases may be considered ideal. The cylinder is submerged in 1 liter of water, and the entire system is initially at the uniform temperature of 25°C (with the previously mentioned pressures in the two compartments). The heat capacities of the cylinder and piston may be neglected. When the piston is unclamped, a new equilibrium situation is ultimately reached with the piston in a new position.

(a) What is the increase in temperature of the water?
(b) How far from the left end of the cylinder will the piston come to rest?
(c) What is the increase in total entropy of the system?

One atmosphere is equivalent to $10^5\,\mathrm{N\,m^{-2}}$. Absolute zero is −273°C. The specific heat of water is 4.18 joules/degree/gram. The density of water is $1\,\mathrm{g/cm^3}$. (From [Reif (1965)])

6.4 A vertical cylinder contains ν moles of an ideal gas, and is closed off by a piston of mass M and area A. The acceleration due to gravity is g. The molar specific heat c_V (at constant volume) of the gas is a constant independent of temperature. The heat capacities of the piston and cylinder are negligibly small, and any frictional forces between the piston and the cylinder walls can be neglected. The pressure of the atmosphere can also be neglected. The whole system is thermally insulated. Initially, the piston is clamped in position so that the gas has a volume V_0, and a temperature T_0. The piston is now released, and, after some oscillations, comes to rest in a final equilibrium situation corresponding to a larger volume of gas.

(a) Does the temperature of the gas increase, decrease, or remain the same?

(b) Does the entropy of the gas increase, decrease, or remain the same?

(c) Show that the final temperature of the gas is

$$T_1 = \left(\frac{c_V}{c_V + R}\right) T_0 + \left(\frac{M g/\nu A}{c_V + R}\right) V_0,$$

where R is the molar ideal gas constant.

(From [Reif (1965)])

6.5 The following describes a method used to measure the specific heat ratio, $\gamma \equiv c_p/c_V$, of a gas. The gas, assumed ideal, is confined within a vertical cylindrical container, and supports a freely-moving piston of mass m. The piston and cylinder both have the same cross-sectional area A. Atmospheric pressure is p_0, and when the piston is in equilibrium under the influence of gravity (acceleration g) and the gas pressure, the volume of the gas is V_0. The piston is now displaced slightly from its equilibrium position, and is found to oscillate about this position at the angular frequency ω. The oscillations of the piston are sufficiently slow that the gas always remains in internal equilibrium, but fast enough that the gas cannot exchange heat with its environment. The variations in gas pressure and volume are therefore adiabatic. Show that

$$\gamma = \frac{\omega^2 m V_0}{m g A + p_0 A^2}.$$

(From [Reif (1965)])

6.6 When sound passes through a fluid (liquid or gas), the period of vibration is short compared to the relaxation time necessary for a macroscopically small element of the fluid to exchange energy with the rest of the fluid through heat flow. Hence, compressions of such an element of volume can be considered adiabatic.

By analyzing one-dimensional compressions and rarefactions of the system of fluid contained in a slab of thickness dx, show that the pressure, $p(x, t)$, in the fluid depends on the position, x, and the time, t, so as to satisfy the wave equation

$$\frac{\partial^2 p}{\partial t^2} = u^2 \frac{\partial^2 p}{\partial x^2},$$

where the velocity of sound propagation, u, is a constant given by $u = (\rho\,\kappa_S)^{-1/2}$. Here ρ is the equilibrium mass density of the fluid, and κ_S is its *adiabatic compressibility*,

$$\kappa_S = -\frac{1}{V}\left(\frac{\partial V}{\partial p}\right)_S;$$

that is, its compressibility measured under conditions in which the fluid is thermally insulated. (From [Reif (1965)])

6.7 Demonstrate that

$$\kappa_S = \frac{c_V}{c_p}\,\kappa_T,$$

where κ_T is the isothermal compressibility. Hence, deduce that

$$\kappa_T - \kappa_S = \frac{v\,T\,\alpha_V^2}{c_p},$$

where v is the molar volume, and α_V the volume coefficient of expansion.

6.8 Refer to the results of the preceding two problems.

(a) Show that adiabatic compressibility, κ_S, of an ideal gas is

$$\kappa_S = \frac{1}{\gamma\,p},$$

where γ is the ratio of specific heats, and p the gas pressure.

(b) Show that the velocity of sound in an ideal gas is

$$u = \sqrt{\frac{\gamma\,p}{\rho}} = \sqrt{\frac{\gamma\,R\,T}{\mu}},$$

where ρ is the mass density, R the molar ideal gas constant, T the absolute temperature, and μ the molecular weight.

(c) How does the sound velocity depend on the gas temperature, T, at a fixed pressure? How does it depend on the gas pressure, p, at fixed temperature?

(d) Calculate the velocity of sound in nitrogen (N_2) gas at room temperature and pressure (i.e., 15°C at 1 bar). Take $\gamma = 1.4$.

6.9 Show that, in general,

$$\left(\frac{\partial\alpha_V}{\partial p}\right)_T + \left(\frac{\partial\kappa_T}{\partial T}\right)_p = 0,$$

where α_V is the volume coefficient of expansion, and κ_T the isothermal compressibility.

6.10 Show that

$$\left(\frac{\partial E}{\partial T}\right)_p = C_p - p\,V\,\alpha_V,$$

and

$$\left(\frac{\partial E}{\partial p}\right)_T = p\,V\,\kappa_T - (C_p - C_V)\,\frac{\kappa_T}{\alpha_V}.$$

6.11 Liquid mercury at 0°C (i.e., 273K) has a molar volume $v = 1.472 \times 10^{-5}\,\mathrm{m}^3$, a molar specific heat at constant pressure $c_p = 28.0\,\mathrm{J\,mol^{-1}}$, a volume coefficient of expansion $\alpha_V = 1.81 \times 10^{-4}\,\mathrm{K}^{-1}$, and an isothermal compressibility $\kappa_T = 3.88 \times 10^{-11}\,(\mathrm{N\,m^{-2}})^{-1}$. Find its molar specific heat at constant volume, and the ratio $\gamma = c_p/c_V$. (From [Reif (1965)])

6.12 Starting from the first Maxwell relation,

$$\left(\frac{\partial T}{\partial V}\right)_S = -\left(\frac{\partial p}{\partial S}\right)_V,$$

derive the other three by making use of the reciprocal and cyclic rules of partial differentiation (see Exercises 6.1 and 6.2), as well as the identity

$$\left(\frac{\partial x}{\partial y}\right)_f \left(\frac{\partial y}{\partial z}\right)_f \left(\frac{\partial z}{\partial x}\right)_f = 1.$$

6.13 Consider a van der Waals gas whose equation of state is

$$\left(p + \frac{a}{v^2}\right)(v - b) = R\,T.$$

The critical point is defined as the unique point at which $(\partial^2 p/\partial v^2)_T = (\partial p/\partial v)_T = 0$. (See Section 7.10.) Let p_c, v_c, and T_c be the temperature, molar volume, and temperature, respectively, at the critical point. Demonstrate that $p_c = a/(27\,b^2)$, $v_c = 3\,b$, and $T_c = 8\,a/(27\,R\,b)$. Hence, deduce that the van der Waals equation of state can be written in the reduced form

$$\left(p' + \frac{3}{v'^2}\right)\left(v' - \frac{1}{3}\right) = \frac{8\,T'}{3},$$

where $p' = p/p_c$, $v' = v/v_c$, and $T' = T/T_c$.

6.14 The behavior of a four-stroke gasoline engine can be approximated by the so-called *Otto cycle*, shown in Figure 6.4. The cycle is as follows:

$e \rightarrow a$: Isobaric (i.e., constant pressure) intake (at atmospheric pressure).

$a \rightarrow b$: Adiabatic compression (compression stroke).

$b \rightarrow c$: Isochoric (i.e., constant volume) increase of temperature during ignition. (Gas combustion is an irreversible process. Here, it is replaced by a reversible isochoric process in which heat is assumed to flow into the system from a reservoir.)

$c \rightarrow d$: Adiabatic expansion (power stroke).

$d \rightarrow a$: Isochoric decrease of temperature (exhaust value opened).

$a \rightarrow e$: Isobaric exhaust (at atmospheric pressure).

(a) Assume that the working substance is an ideal gas. Show that the efficiency of the cycle is

$$\eta = 1 - \left(\frac{T_d - T_a}{T_c - T_b}\right) = 1 - \frac{1}{r^{\gamma-1}},$$

where $r = V_1/V_2$ is the compression ratio of the engine.

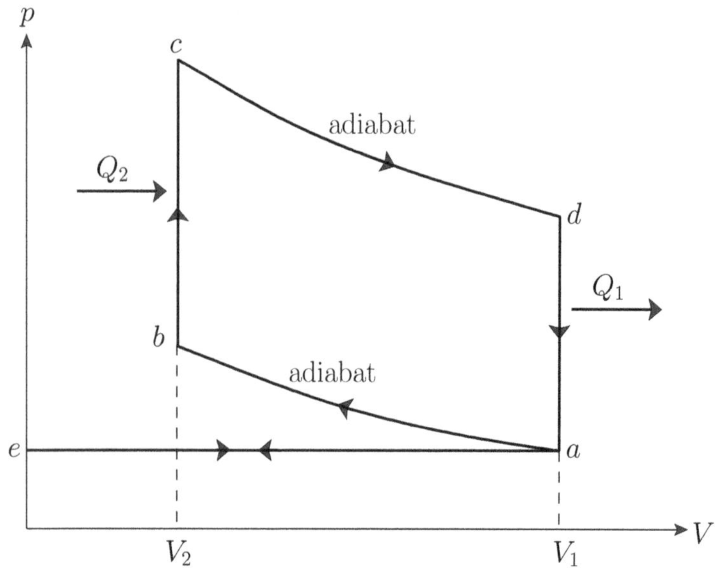

Figure 6.4 *p-V* diagram for the Otto cycle.

(b) Calculate η for the realistic values $r = 5$ and $\gamma = 1.5$.

Note: Because $\gamma > 1$, maximizing r maximizes the engine efficiency. The maximum practical value of r is about 7. For greater values, the rise in temperature during compression is large enough to cause ignition prior to the advent of the spark. This process is called *pre-ignition*, and is deleterious to the operation of the engine. Pre-ignition is not a problem for diesel engines (which depend on spontaneous ignition, rather than triggering ignition via a spark), so higher compression ratios are possible. This is partly the reason that diesel engines are inherently more efficient than gasoline engines.

Chapter 7

Multi-Phase Systems

7.1 Introduction

Up to now, we have dealt, almost exclusively, with systems consisting of a single phase (i.e., a single, spatially homogeneous, state of aggregation). In this chapter, we shall employ statistical thermodynamics to investigate the equilibria of multi-phase systems (e.g., ice and water). Most of our analysis turns out to be independent of the microscopic details of such systems, and, therefore, leads to fairly general results.

7.2 Equilibrium of Isolated System

Consider a thermally isolated system, A. According to Section 5.6, any spontaneously occurring process is such that the system's entropy tends to increase in time. In statistical terms, this means that the system evolves toward a situation of greater intrinsic probability. Thus, in any spontaneous process, the change in entropy satisfies

$$\Delta S \geq 0. \tag{7.1}$$

It follows that if a stable equilibrium state has been attained [i.e., one in which no further spontaneous processes (other than random fluctuations) can take place] then this state is such that S is maximized. In other words, it is the most probable state of the system, subject to the given constraints. Thus, we can make the following statement:

> For a thermally isolated system, the stable equilibrium state is such that
>
> $$S = \text{maximum}.$$

Now, in a thermally isolated system, the first law of thermodynamics implies that

$$0 = Q = W + \Delta E, \tag{7.2}$$

or

$$W = (-\Delta E)\,. \tag{7.3}$$

If the external parameters of the system (e.g., its volume) are kept fixed, so that no work is done (i.e., $W = 0$), then

$$E = \text{constant} \tag{7.4}$$

as S evolves toward its maximum value.

We can phrase the previous argument in more explicit statistical terms. Suppose that an isolated system is described by a parameter y (or by several such parameters), but that its total energy is constant. Let $\Omega(y)$ denote the number of microstates accessible to the system when this parameter lies between y and $y+\delta y$ (δy being some fixed small interval). The corresponding entropy of the system is $S = k \ln \Omega$. (See Section 5.6.) If the parameter y is free to vary then the principle of equal a priori probabilities asserts that, in an equilibrium state, the probability, $P(y)$, of finding the system with the parameter between y and $y + \delta y$ is given by

$$P(y) \propto \Omega(y) = \exp\left[\frac{S(y)}{k}\right]. \tag{7.5}$$

(See Section 3.3.) The previous expression shows explicitly that the most probable state is one in which S attains a maximum value, and also allows us to determine the relative probability of fluctuations about this state.

7.3 Equilibrium of Constant-Temperature System

A knowledge of the equilibrium conditions for an isolated system permits us to deduce similar conditions for other situations of physical interest. For instance, much experimental work is performed under conditions of constant temperature. Thus, it would be interesting to investigate the equilibrium of some system, A, in thermal contact with a heat reservoir, A', that is held at the constant absolute temperature T_0.

The combined system, $A^{(0)}$, consisting of the system A and the heat reservoir A', is an isolated system of the type discussed in the previous section. The entropy, $S^{(0)}$, of $A^{(0)}$ therefore satisfies condition (7.1). In other words, in any spontaneous process,

$$\Delta S^{(0)} \geq 0. \tag{7.6}$$

However, this condition can also be expressed in terms of quantities that only refer to the system A. In fact,

$$\Delta S^{(0)} = \Delta S + \Delta S', \tag{7.7}$$

where ΔS is the entropy change of A, and $\Delta S'$ that of the reservoir. But, if A absorbs heat Q from the reservoir A' during the process then A' absorbs heat $(-Q)$, and suffers a corresponding entropy change

$$\Delta S' = \frac{(-Q)}{T_0} \tag{7.8}$$

(because it remains in equilibrium at the constant temperature T_0). Furthermore, the first law of thermodynamics implies that

$$Q = \Delta E + W, \tag{7.9}$$

where ΔE is the change in internal energy of A, and W is the work done by A. Thus, Equation (7.7) can be written

$$\Delta S^{(0)} = \Delta S - \frac{Q}{T_0} = \frac{T_0\,\Delta S - (\Delta E + W)}{T_0} = \frac{\Delta\,(T_0\,S - E) - W}{T_0}, \tag{7.10}$$

or

$$\Delta S^{(0)} = \frac{-\Delta F_0 - W}{T_0}, \tag{7.11}$$

where use has been made of the fact that T_0 is a constant. Here,

$$F_0 \equiv E - T_0\,S, \tag{7.12}$$

reduces to the Helmholtz free energy, $F = E - T\,S$, of system A, when the latter has a temperature, T, equal to that of the heat reservoir, A'. Of course, in the general case in which A is not in equilibrium with A', the former system's temperature is not necessarily equal to T_0.

The fundamental condition (7.6) can be combined with Equation (7.11) to give

$$W \leq (-\Delta F_0) \tag{7.13}$$

(assuming that T_0 is positive, as is normally the case). This relation implies that the maximum work that can be done by a system in contact with a heat reservoir is $(-\Delta F_0)$. (Incidentally, this is the reason for the name "free energy" given to F.) The maximum work corresponds to the equality sign in the preceding equation, and is obtained when the process used is quasi-static (so that A is always in equilibrium with A', and $T = T_0$). Equation (7.13) should be compared to the rather different relation (7.3) that pertains to an isolated system.

If the external parameters of system A are held constant then $W = 0$, and Equation (7.13) yields the condition

$$\Delta F_0 \leq 0. \tag{7.14}$$

This equation is analogous to Equation (7.1) for an isolated system. It implies that if a system is in thermal contact with a heat reservoir then its Helmholtz free energy tends to decrease. Thus, we arrive at the following statement:

> If a system, whose external parameters are fixed, is in thermal contact with a heat reservoir then the stable equilibrium state is such that
>
> $$F_0 = \text{minimum}.$$

The preceding statement can again be phrased in more explicit statistical terms. Suppose that the external parameters of A are fixed, so that $W = 0$. Furthermore, let A be described by some parameter y. The thermodynamic functions of A—namely, S and E—have the definite values $S(y_1)$ and $E(y_1)$, respectively, when y has a given value $y = y_1$. If the parameter changes to some other value, y, then these functions change by the corresponding amounts $\Delta S = S(y) - S(y_1)$ and $\Delta E = E(y) - E(y_1) = Q$. The entropy of the heat reservoir, A', also changes because it absorbs heat. The corresponding change in the total entropy of $A^{(0)}$ is given by Equation (7.11) (with $W = 0$):

$$\Delta S^{(0)} = -\frac{\Delta F_0}{T_0}. \tag{7.15}$$

But, in an equilibrium state, the probability, $P(y)$, that the parameter lies between y and $y + \delta y$ is proportional to the number of states, $\Omega^{(0)}(y)$, accessible to the isolated total system, $A^{(0)}$, when the parameter lies in this range. Thus, by analogy with Equation (7.5), we have

$$P(y) \propto \Omega^{(0)}(y) = \exp\left[\frac{S^{(0)}(y)}{k}\right]. \tag{7.16}$$

However, from Equation (7.15),

$$S^{(0)}(y) = S^{(0)}(y_1) - \frac{\Delta F_0}{T_0} = S^{(0)}(y_1) - \left[\frac{F_0(y) - F_0(y_1)}{T_0}\right]. \tag{7.17}$$

Now, because y_1 is just some arbitrary constant, the corresponding constant terms can be absorbed into the constant of proportionality in Equation (7.16), which then becomes

$$P(y) \propto \exp\left[-\frac{F_0(y)}{k\,T_0}\right]. \tag{7.18}$$

This equation shows explicitly that the most probable state is one in which F_0 attains a minimum value, and also allows us to determine the relative probability of fluctuations about this state.

7.4 Equilibrium of Constant-Temperature Constant-Pressure System

Another case of physical interest is that where a system, A, is maintained under conditions of both constant temperature and constant pressure. This is a situation that occurs frequently in the laboratory, where we might carry out an experiment in a thermostat at atmospheric pressure (say). Therefore, let us suppose that A is in thermal contact with a heat reservoir, A', that is held at the constant temperature T_0, and at the constant pressure p_0. The system A can exchange heat with the reservoir, A', but the latter is so large that its temperature, T_0, remains constant. Likewise, the system A can change its volume, V, at the expense of that of the reservoir, A', doing work on the reservoir in the process. However, A' is so large that its pressure, p_0, remains unaffected by this relatively small volume change.

The analysis of the equilibrium conditions for system A follows similar lines to the that in the previous section. The entropy, $S^{(0)}$, of the combined system $A^{(0)} = A + A'$ satisfies the condition that in any spontaneous process

$$\Delta S^{(0)} = \Delta S + \Delta S' \geq 0. \tag{7.19}$$

If A absorbs heat Q from A' in this process then $\Delta S' = -Q/T_0$. Furthermore, the first law of thermodynamics yields

$$Q = \Delta E + p_0 \Delta V + W^*, \tag{7.20}$$

where $p_0 \Delta V$ is the work done by A against the constant pressure, p_0, of the reservoir, A', and where W^* denotes any other work done by A in the process. (For example, W^* might refer to electrical work.) Thus, we can write

$$\begin{aligned}\Delta S^{(0)} &= \Delta S - \frac{Q}{T_0} = \frac{T_0 \Delta S - (\Delta E + p_0 \Delta V + W^*)}{T_0}\\ &= \frac{\Delta(T_0 S - E - p_0 V) - W^*}{T_0},\end{aligned} \tag{7.21}$$

or

$$\Delta S^{(0)} = \frac{-\Delta G_0 - W^*}{T_0}. \tag{7.22}$$

Here, we have made use of the fact that T_0 and p_0 are constants. We have also introduced the definition

$$G_0 \equiv E - T_0 S + p_0 V, \tag{7.23}$$

where G_0 becomes the Gibbs free energy, $G = E - T S + p V$, of system A when the temperature and pressure of this system are both equal to those of the reservoir, A'.

The fundamental condition (7.19) can be combined with Equation (7.22) to give

$$W^* \leq (-\Delta G_0) \tag{7.24}$$

(assuming that T_0 is positive). This relation implies that the maximum work (other than the work done on the pressure reservoir) that can be done by system A is

$(-\Delta G_0)$. (Incidentally, this is the reason that G is also called a "free energy".) The maximum work corresponds to the equality sign in the preceding equation, and is obtained when the process used is quasi-static (so that A is always in equilibrium with A', and $T = T_0$, $p = p_0$).

If the external parameters of system A (except its volume) are held constant then $W^* = 0$, and Equation (7.24) yields the condition

$$\Delta G_0 \leq 0. \tag{7.25}$$

Thus, we arrive at the following statement:

> If a system is in contact with a reservoir at constant temperature and pressure then the stable equilibrium state is such that
>
> $$G_0 = \text{minimum}.$$

The preceding statement can again be phrased in more explicit statistical terms. Suppose that the external parameters of A (except its volume) are fixed, so that $W^* = 0$. Furthermore, let A be described by some parameter y. If y changes from some standard value y_1 then the corresponding change in the total entropy of $A^{(0)}$ is

$$\Delta S^{(0)} = -\frac{\Delta G_0}{T_0}. \tag{7.26}$$

But, in an equilibrium state, the probability, $P(y)$, that the parameter lies between y and $y + \delta y$ is given by

$$P(y) \propto \exp\left[\frac{S^{(0)}(y)}{k}\right]. \tag{7.27}$$

Now, from Equation (7.26),

$$S^{(0)}(y) = S^{(0)}(y_1) - \frac{\Delta G_0}{T_0} = S^{(0)}(y_1) - \left[\frac{G_0(y) - G_0(y_1)}{T_0}\right]. \tag{7.28}$$

However, because y_1 is just some arbitrary constant, the corresponding constant terms can be absorbed into the constant of proportionality in Equation (7.27), which then becomes

$$P(y) \propto \exp\left[-\frac{G_0(y)}{k\,T_0}\right]. \tag{7.29}$$

This equation shows explicitly that the most probable state is one in which G_0 attains a minimum value, and also allows us to determine the relative probability of fluctuations about this state.

7.5 Stability of Single-Phase Substance

Consider a system in a single phase (e.g., a liquid or a solid). Let us focus attention on a small, but macroscopic part, A, of this system. Here, A consists of a fixed number of particles, $N \gg 1$. The remainder of the system is, relatively, very large, and, therefore, acts like a reservoir held at some constant temperature, T_0, and constant pressure, p_0. According to Equation (7.25), the condition for stable equilibrium, applied to A, is

$$G_0 \equiv E - T_0\, S + p_0\, V = \text{minimum}. \tag{7.30}$$

Let the temperature, T, and the volume, V, be the two independent parameters that specify the macrostate of A. Consider, first, a situation where V is fixed, but T is allowed to vary. Suppose that the minimum of G_0 occurs for $T = \tilde{T}$, when $G_0 = G_{\rm min}$. Expanding G_0 about its minimum, and writing $\varDelta T = T - \tilde{T}$, we obtain

$$\varDelta_m G_0 \equiv G_0 - G_{\rm min} = \left(\frac{\partial G_0}{\partial T}\right)_V \varDelta T + \frac{1}{2}\left(\frac{\partial^2 G_0}{\partial T^2}\right)_V (\varDelta T)^2 + \cdots. \tag{7.31}$$

Here, all derivatives are evaluated at $T = \tilde{T}$. Because $G_0 = G_{\rm min}$ is a stationary point (i.e., a maximum or a minimum), $\varDelta_m G_0 = 0$ to first order in $\varDelta T$. In other words,

$$\left(\frac{\partial G_0}{\partial T}\right)_V = 0 \qquad \text{for } T = \tilde{T}. \tag{7.32}$$

Furthermore, because $G_0 = G_m$ is a minimum, rather than a maximum, $\varDelta_m G_0 \geq 0$ to second order in $\varDelta T$. In other words,

$$\left(\frac{\partial^2 G_0}{\partial T^2}\right)_V \geq 0 \qquad \text{for } T = \tilde{T}. \tag{7.33}$$

According to Equation (7.30), the condition, (7.32), that G_0 be stationary yields

$$\left(\frac{\partial G_0}{\partial T}\right)_V = \left(\frac{\partial E}{\partial T}\right)_V - T_0\left(\frac{\partial S}{\partial T}\right)_V = 0. \tag{7.34}$$

However, the fundamental thermodynamic relation

$$T\, dS = dE + p\, dV \tag{7.35}$$

implies that if V is fixed then

$$T\left(\frac{\partial S}{\partial T}\right)_V = \left(\frac{\partial E}{\partial T}\right)_V. \tag{7.36}$$

Thus, Equation (7.34) becomes

$$\left(\frac{\partial G_0}{\partial T}\right)_V = \left(1 - \frac{T_0}{T}\right)\left(\frac{\partial E}{\partial T}\right)_V = 0. \tag{7.37}$$

Recalling that the derivatives are evaluated at $T = \tilde{T}$, we get

$$\tilde{T} = T_0. \tag{7.38}$$

Thus, we arrive at the rather obvious conclusion that a necessary condition for equilibrium is that the temperature of the subsystem A be the same as that of the surrounding medium.

Equations (7.33) and (7.37) imply that

$$\left(\frac{\partial^2 G_0}{\partial T^2}\right)_V = \frac{T_0}{T^2}\left(\frac{\partial E}{\partial T}\right)_V + \left(1 - \frac{T_0}{T}\right)\left(\frac{\partial^2 E}{\partial T^2}\right)_V \geq 0. \tag{7.39}$$

Recalling that the derivatives are evaluated at $T = \tilde{T}$, and that $\tilde{T} = T_0$, the second term on the right-hand side of the previous equation vanishes, and we obtain

$$\left(\frac{\partial E}{\partial T}\right)_V \geq 0. \tag{7.40}$$

However, this derivative is simply the heat capacity, C_V, at constant volume. Thus, we deduce that a fundamental criterion for the stability of any phase is

$$C_V \geq 0. \tag{7.41}$$

The previous condition is a particular example of a very general rule, known as *Le Châtelier's principle*, which states that

> If a system is in a stable equilibrium then any spontaneous change of its parameters must give rise to processes that tend to restore the system to equilibrium.

To show how this principle applies to the present case, suppose that the temperature, T, of the subsystem A has increased above that of its surroundings, A', as a result of a spontaneous fluctuation. The process brought into play is the spontaneous flow of heat from system A, at the higher temperature, to its surroundings A', which brings about a decrease in the energy, E, of A (i.e., $\Delta E < 0$). By Le Châtelier's principle, this energy decrease must act to decrease the temperature of A (i.e., $\Delta T < 0$), so as to restore A to its equilibrium state (in which its temperature is the same as its surroundings). Hence, it follows that ΔE and ΔT have the same sign, so that $\Delta E/\Delta T > 0$, in agreement with Equation (7.40).

Suppose, now, that the temperature of subsystem A is fixed at $T = T_0$, but that its volume, V, is allowed to vary. We can write

$$\Delta_m G_0 = \left(\frac{\partial G_0}{\partial V}\right)_T \Delta V + \frac{1}{2}\left(\frac{\partial^2 G_0}{\partial V^2}\right)_T (\Delta V)^2 + \cdots, \tag{7.42}$$

where $\Delta V = V - \tilde{V}$, and the expansion is about the volume $V = \tilde{V}$ at which G_0 attains its minimum value. The condition that G_0 is stationary yields

$$\left(\frac{\partial G_0}{\partial V}\right)_T = 0. \tag{7.43}$$

The condition that the stationary point is a minimum gives

$$\left(\frac{\partial^2 G_0}{\partial V^2}\right)_T \geq 0. \tag{7.44}$$

Making use of Equation (7.30), we get

$$\left(\frac{\partial G_0}{\partial V}\right)_T = \left(\frac{\partial E}{\partial V}\right)_T - T_0\left(\frac{\partial S}{\partial V}\right)_T + p_0. \tag{7.45}$$

However, Equation (7.35) implies that

$$T\left(\frac{\partial S}{\partial V}\right)_T = \left(\frac{\partial E}{\partial V}\right)_T + p. \tag{7.46}$$

Hence,

$$\left(\frac{\partial G_0}{\partial V}\right)_T = T\left(\frac{\partial S}{\partial V}\right)_T - p - T_0\left(\frac{\partial S}{\partial V}\right)_T + p_0, \tag{7.47}$$

or

$$\left(\frac{\partial G_0}{\partial V}\right)_T = -p + p_0, \tag{7.48}$$

because $T = T_0$. Thus, the condition (7.43) yields

$$p = p_0. \tag{7.49}$$

This rather obvious result states that in equilibrium the pressure of subsystem A must be equal to that of the surrounding medium.

Equations (7.44) and (7.48) imply that

$$\left(\frac{\partial^2 G_0}{\partial V^2}\right)_T = -\left(\frac{\partial p}{\partial V}\right)_T \geq 0. \tag{7.50}$$

When expressed in terms of the isothermal compressibility,

$$\kappa_T = -\frac{1}{V}\left(\frac{\partial V}{\partial p}\right)_T, \tag{7.51}$$

the stability criterion (7.50) is equivalent to

$$\kappa_T \geq 0. \tag{7.52}$$

To show how the preceding stability condition is consistent with Le Chátelier's principle, suppose that the volume of subsystem A has increased by an amount $\Delta V > 0$, as a result of a spontaneous fluctuation. The pressure, p, of A must then decrease below that of its surroundings (i.e., $\Delta p < 0$) in order to ensure that the net force exerted on A by its surroundings is such as to reduce the volume of A back towards its equilibrium value. In other words, ΔV and Δp must have opposite signs, implying that $\kappa_T > 0$.

The preceding discussion allows us to characterize the spontaneous volume fluctuations of the small subsystem A. The most probable situation is that where V takes the value $\tilde{V}$ that minimizes G_0, so that $G_0(\tilde{V}) = G_{\rm min}$. Let $P(V)\,dV$ denote the probability that the volume of A lies between V and $V + dV$. According to Equation (7.29),

$$P(V)\,dV \propto \exp\left[-\frac{G_0(V)}{k\,T_0}\right]dV. \tag{7.53}$$

However, when $\Delta V = V - \tilde{V}$ is small, the expansion (7.42) is applicable. By virtue of Equations (7.43), (7.50), and (7.51), the expansion yields

$$G_0(V) = G_0(\tilde{V}) - \frac{1}{2}\left(\frac{\partial p}{\partial V}\right)_T (\Delta V)^2 = G_{\rm min} - \frac{(\Delta V)^2}{2\,\tilde{V}\,\kappa_T}. \tag{7.54}$$

Thus, Equation (7.53) becomes

$$P(V)\,dV = B\exp\left[-\frac{(V-\tilde{V})^2}{2\,k\,T_0\,\kappa_T\,\tilde{V}}\right]dV, \tag{7.55}$$

where we have absorbed $G_{\rm min}$ into the constant of proportionality, B. Of course, B can be determined via the requirement that the integral of $P(V)\,dV$ over all possible values of the volume, V, be equal to unity.

The probability distribution (7.55) is a Gaussian whose maximum lies at $V = \tilde{V}$. Thus, $\tilde{V}$ is equal to the mean volume, $\overline{V}$. Moreover, the standard deviation of the spontaneous fluctuations of V about its mean value is

$$\Delta^* V = \left(k\,T_0\,\kappa_T\,\overline{V}\right)^{1/2}. \tag{7.56}$$

(See Section 2.9.) The fluctuations in V imply corresponding fluctuations in the number of particles per unit volume, $n = N/V$, (and, hence, in the mean density) of the subsystem A. The fluctuations in n are centered on the mean value $\bar{n} = N/\overline{V}$. For relatively small fluctuations, $\Delta n \equiv n - \bar{n} \simeq -(N/\overline{V}^{\,2})\,\Delta V = -(\bar{n}/\overline{V})\,\Delta V$. Hence,

$$\frac{\Delta^* n}{\bar{n}} = \frac{\Delta^* V}{\overline{V}} = \left(\frac{k\,T_0\,\kappa_T}{\overline{V}}\right)^{1/2}. \tag{7.57}$$

Note that the relative magnitudes of the density and volume fluctuations are inversely proportional to the square-root of the volume, $\overline{V}$, under consideration.

An interesting situation arises when

$$\left(\frac{\partial p}{\partial V}\right)_T \to 0. \tag{7.58}$$

In this limit, $\kappa_T \to \infty$, and the density fluctuations become very large. [They do not, however, become infinite, because the neglect of third-order, and higher, terms in the expansion (7.42) is no longer justified.] The condition $(\partial p/\partial V)_T = 0$

defines the so-called *critical point* of the substance, at which the distinction between its liquid and gas phases disappears. (See Section 7.10.) The very large density fluctuations at the critical point lead to strong scattering of light. As a result, a substance that is ordinarily transparent will assume a milky-white appearance at its critical point (e.g., CO_2 when it approaches its critical point at 304 K and 73 atmospheres). This interesting phenomenon is called *critical-point opalescence*.

7.6 Equilibrium Between Phases

Consider a system that consists of two phases, which we shall denote by 1 and 2. For example, these might be solid and liquid, or liquid and gas. Suppose that the system is in equilibrium with a reservoir at the constant temperature T, and the constant pressure p, so that the system always has the temperature T and the mean pressure p. However, the system can exist in either one of its two phases, or some mixture of the two. Let us begin by finding the conditions that allow the two phases to coexist in equilibrium with one another.

In accordance with the discussion in Section 7.4, the equilibrium condition is that the Gibbs free energy, G, of the system is a minimum:

$$G = E - T\,S + p\,V = \text{minimum}. \tag{7.59}$$

Let ν_i be the number of moles of phase i present in the system, and let $g_i(T, p)$ be the Gibbs free energy per mole of phase i at the temperature T and the pressure p. It follows that

$$G = \nu_1\,g_1 + \nu_2\,g_2. \tag{7.60}$$

Furthermore, the conservation of matter implies that the total number of moles, ν, of the substance remains constant:

$$\nu_1 + \nu_2 = \nu = \text{constant}. \tag{7.61}$$

Thus, we can take ν_1 as the one independent parameter that is free to vary. In equilibrium, Equation (7.59) requires that G be stationary for small variations in ν_1. In other words,

$$dG = g_1\,d\nu_1 + g_2\,d\nu_2 = (g_1 - g_2)\,d\nu_1 = 0, \tag{7.62}$$

because $d\nu_2 = -d\nu_1$, as a consequence of Equation (7.61). Hence, a necessary condition for equilibrium between the two phases is

$$g_1 = g_2. \tag{7.63}$$

Clearly, when this condition is satisfied then the transfer of a mole of substance from one phase to another does not change the overall Gibbs free energy, G, of the system. Hence, G is stationary, as required. Incidentally, the condition that G is a minimum (rather than a maximum) is easily shown to reduce to the requirement that the heat capacities and isothermal compressibilities of both phases be positive, so that each phase is stable to temperature and volume fluctuations.

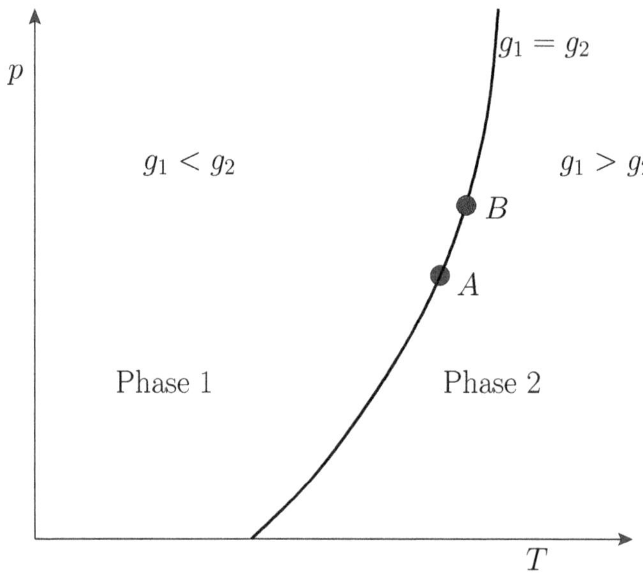

Figure 7.1 Pressure-temperature plot showing the domains of relative stability of two phases, and the phase-equilibrium line.

Now, for a given temperature and pressure, $g_1(T, p)$ is a well-defined function characteristic of phase 1. Likewise, $g_2(T, p)$ is a well-defined function characteristic of phase 2. If T and p are such that $g_1 < g_2$ then the minimum value of G in Equation (7.59) is achieved when all ν moles of the substance transform into phase 1, so that $G = \nu\, g_1$. In this case, phase 1 is the stable one. On the other hand, if $g_1 > g_2$ then the minimum value of G is achieved when all ν moles of the substance transform into phase 2, so that $G = \nu\, g_2$. In this case, phase 2 is the stable one. Finally, if $g_1 = g_2$ then the condition (7.59) is automatically satisfied, and any amount, ν_1, of phase 1 can coexist with the remaining amount, $\nu_2 = \nu - \nu_1$, of phase 2. The locus of the points in the T-p plane where $g_1 = g_2$ then represents a *phase-equilibrium line* along which the two phases can coexist in equilibrium. This line divides the T-p plane into two regions. The first corresponds to $g_1 < g_2$, so that phase 1 is stable. The second corresponds to $g_1 > g_2$, so that phase 2 is stable. See Figure 7.1.

7.7 Clausius-Clapeyron Equation

It is possible to characterize the phase-equilibrium line in terms of a differential equation. Referring to Figure 7.1, consider a point, such as A, that lies on the phase-equilibrium line, and corresponds to the temperature T and the pressure p. The condition (7.63) implies that

$$g_1(T, p) = g_2(T, p). \tag{7.64}$$

Now, consider a neighboring point, such as B, that also lies on the phase-equilibrium line, and corresponds to the temperature $T + dT$ and the pressure $p + dp$. The condition (7.63) yields

$$g_1(T + dT, p + dp) = g_2(T + dT, p + dp). \tag{7.65}$$

Taking the difference between the previous two equations, we obtain

$$dg_1 = dg_2, \tag{7.66}$$

where

$$dg_i = \left(\frac{\partial g_i}{\partial T}\right)_p dT + \left(\frac{\partial g_i}{\partial p}\right)_T dp \tag{7.67}$$

is the change in Gibbs free energy per mole of phase i in going from point A to point B.

The change, dg, for each phase can also be obtained from the fundamental thermodynamic relation

$$d\epsilon = T\,ds - p\,dv. \tag{7.68}$$

Here, ϵ refers to molar energy (i.e., energy per mole), s to molar entropy, and v to molar volume. Thus,

$$dg \equiv d(\epsilon - T\,s + p\,v) = -s\,dT + v\,dp. \tag{7.69}$$

Hence, Equation (7.66) implies that

$$-s_1\,dT + v_1\,dp = -s_2\,dT + v_2\,dp, \tag{7.70}$$

or

$$(s_2 - s_1)\,dT = (v_2 - v_2)\,dp, \tag{7.71}$$

which reduces to

$$\frac{dp}{dT} = \frac{\Delta s}{\Delta v}, \tag{7.72}$$

where $\Delta s \equiv s_2 - s_1$ and $\Delta v \equiv v_2 - v_1$. This result is known as the *Clausius-Clapeyron equation*.

Consider any point on the phase-equilibrium line at temperature T and pressure p. The Clausius-Clapeyron equation then relates the local slope of the line to the molar entropy change, Δs, and the molar volume change, Δv, of the substance in crossing the line at this point. Note, incidentally, that the quantities on the right-hand side of the Clausius-Clapeyron equation do not necessarily need to refer to one mole of the substance. In fact, both numerator and denominator can be multiplied by the same number of moles, leaving dp/dT unchanged.

Because there is an entropy change associated with a phase transformation, heat must also be absorbed during such a process. The *latent heat of transformation*, L_{12}, is defined as the heat absorbed when a given amount of phase 1 is transformed to phase 2. Because this process takes place at the constant temperature T, the corresponding entropy change is

$$\Delta S = S_2 - S_1 = \frac{L_{12}}{T}, \tag{7.73}$$

where L_{12} is the latent heat at this temperature. Thus, the Clausius-Clapeyron equation, (7.72), can also be written

$$\frac{dp}{dT} = \frac{\Delta S}{\Delta V} = \frac{L_{12}}{T\,\Delta V}. \tag{7.74}$$

7.8 Phase Diagrams

Simple substances are capable of existing in three different types of phase: namely, solid, liquid, and gas. (There may actually be several solid phases with different crystal structures.) In the T-p plane, the phase-equilibrium lines separating these phases typically appear as indicated in Figure 7.2. These lines separate solid from liquid, liquid from gas, and solid from gas. The gas phase is sometimes called the vapor phase. The transformation from solid to liquid is called *melting*, that from liquid to gas is called *vaporization*, and that from solid to gas is called *sublimation*. The three phase-equilibrium lines meet at one common point, A, known as the *triple point*. At this unique temperature and pressure, arbitrary amounts of all three phases can coexist in equilibrium. (This is the property that makes the triple point of water so suitable as a readily reproducible temperature standard.) At point B, the so-called *critical point*, the liquid-gas equilibrium line ends. The volume change, ΔV, between the liquid and gas phases is zero at this point. Beyond point B, there is no further phase transformation, because there exists a single fluid phase (i.e., the very dense gas has become indistinguishable from the liquid.)

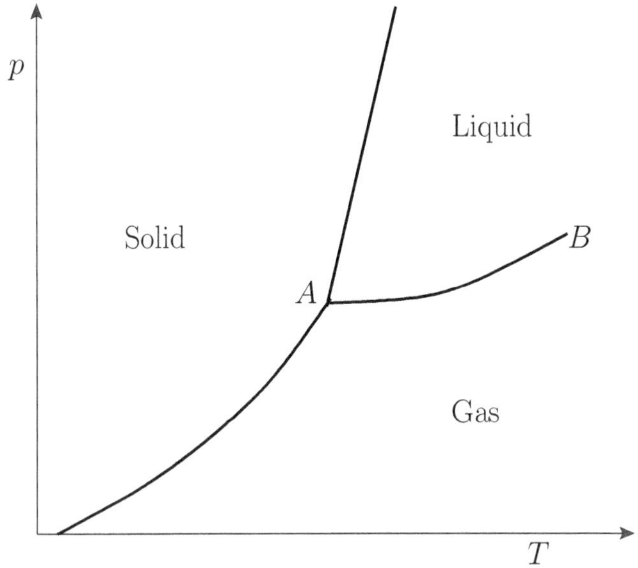

Figure 7.2 Phase diagram for a simple substance. Point A is the triple point, and point B is the critical point.

In going from solid (s) to liquid (l), the entropy (or degree of disorder) of the substance almost always increases. Thus, the corresponding latent heat, L_{sl}, is positive, and heat is absorbed in the transformation. In most cases, the solid expands upon melting, so $\Delta V > 0$. In this case, the Clausius-Clapeyron equation, (7.74), asserts that the slope of the solid-liquid equilibrium line (i.e., the melting curve) is positive, as shown in the diagram. However, there exist some substances, such as water, that contract upon melting, so that $\Delta V < 0$. For these substances, the slope of the melting curve is negative.

7.9 Vapor Pressure

The Clausius-Clapeyron equation can be used to derive an approximate expression for the pressure of the vapor in equilibrium with the liquid (or solid) at some temperature T. This pressure is called the *vapor pressure* of the liquid (or solid) at this temperature. According to Equation (7.74),

$$\frac{dp}{dT} = \frac{l}{T\,\Delta v}, \tag{7.75}$$

where $l = l_{12}$ is the latent heat per mole, and v the molar volume. Let 1 refer to the liquid (or solid) phase, and 2 to the vapor. It follows that

$$\Delta v = v_2 - v_1 \simeq v_2, \tag{7.76}$$

because the vapor is much less dense than the liquid, so that $v_2 \gg v_1$. Let us also assume that the vapor can be adequately treated as an ideal gas, so that its equation of state is written

$$p\, v_2 = R\, T. \tag{7.77}$$

Thus, $\Delta v \simeq R\,T/p$, and Equation (7.75) becomes

$$\frac{1}{p}\frac{dp}{dT} = \frac{1}{R\,T^2}. \tag{7.78}$$

Assuming that l is approximately temperature independent, we can integrate the previous equation to give

$$\ln p = -\frac{l}{R\,T} + \text{constant}, \tag{7.79}$$

which implies that

$$p = p_0 \exp\left(-\frac{l}{R\,T}\right), \tag{7.80}$$

where p_0 is some constant. This result shows that the vapor pressure, p, is a very rapidly increasing function of the temperature, T.

7.10 Phase Transformations in van der Waals Fluid

Consider a so-called *van der Waals fluid* whose equation of state takes the form

$$\left(p + \frac{a}{v^2}\right)(v - b) = R\,T, \tag{7.81}$$

where $a > 0$, $b > 0$ are constants, and v is the molar volume. (See Sections 6.15 and 10.11.) It is helpful to write this equation of state in the reduced form

$$p' = \frac{8}{3}\frac{T'}{v' - 1/3} - \frac{3}{v'^2}, \tag{7.82}$$

where $p' = p/p_c$, $T' = T'/T_c$, and $v' = v/v_c$. Here, the critical pressure, temperature and molar volume are $p_c = a/(27\,b^2)$, $T_c = 8\,a/(27\,R\,b)$, and $v_c = 3\,b$, respectively. (See Section 6.15.)

Figure 7.3 shows the isotherms of the van der Waals equation of state plotted in the v'-p' plane. It can be seen that, for temperatures that exceed the critical temperature (i.e., $T' > 1$), reducing the molar volume of the fluid causes its pressure to rise monotonically, eventually becoming infinite when $v' = 1/3$ (i.e., when

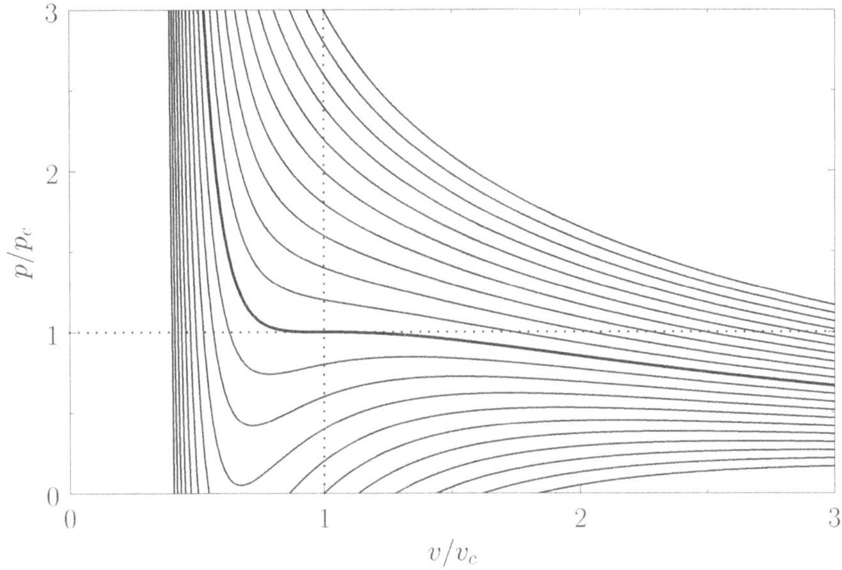

Figure 7.3 Isotherms of the van der Waals equation of state. The thick curve shows the critical isotherm, $T = T_c$, which passes through the so-called critical point, $v = v_c$, $p = p_c$, at which the isotherm has a point of inflection.

the molecules are closely packed). On the other hand, for temperatures less than the critical temperature (i.e., $T' < 1$), as v' is decreased the pressure rises, falls, and then rises again, which seems to imply that, for some molar volumes, compressing the fluid can cause its pressure to decrease. However, according to the analysis of Section 7.5, if a phase is such that compressing it causes its pressure to decrease (i.e., if its isothermal compressibility is negative) then the phase is unstable to density fluctuations. Thus, below the critical temperature, the stable states on a given isotherm are divided into two groups. The first group is characterized by relatively small molar volumes—these are liquid states. The second group is characterized by relatively large molar volumes—these are gas states. The liquid and gas states are separated by a set of unstable states. Above the critical temperature, the distinct liquid and gas states cease to exist—there is only a single stable fluid state. Consider the region of the v'-p' plane in which the states predicted by Equation (7.82) are unstable. It turns out that our analysis has broken down here because we assumed the existence of a single phase, whereas, in fact, in this region of the v'-p' plane, there exists a stable mixture of liquid and gas phases in equilibrium with one another. Let us investigate further.

At a given temperature and pressure, the true equilibrium state of a system is

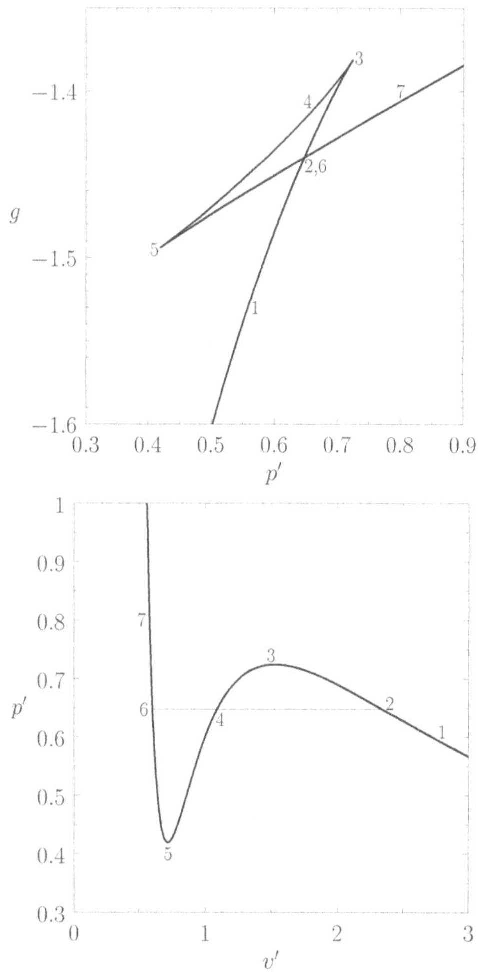

Figure 7.4 The reduced Gibbs free energy as a function of pressure for a van der Waals fluid at $T' = 0.9$. The corresponding isotherm is shown below. States in the range 2-3-4-5-6 are unstable.

that which minimizes the Gibbs free energy. (See Section 7.4.) For the case of a van der Waals fluid,

$$\frac{G}{\nu R T_c} = -T' \ln(v' - 1/3) + \frac{T'}{3\,v' - 1} - \frac{9}{4\,v'} + f(T'), \tag{7.83}$$

where $f(T')$ is an unimportant function of the reduced temperature. (See Exercise 10.9.) In fact, defining $g(v', T') = G/(\nu R T_c) - f(T')$, we can write

$$g = -T' \ln(v' - 1/3) + \frac{T'}{3\,v' - 1} - \frac{9}{4\,v'}. \tag{7.84}$$

Figure 7.4 shows g plotted as a function of p' for $T' = 0.9$. Also shown is the corresponding isotherm in the v'-p' plane. Although the van der Waals equation of state associates some pressures with more than one molar volume, the thermodynamically stable state is that with the lowest Gibbs free energy. Thus, the triangular loop in the graph of g versus p' (points 2-3-4-5-6) corresponds to unstable states. As the pressure is gradually increased, the system will go straight from point 2 to point 6, with an abrupt reduction in molar volume. Of course, this corresponds to a gas-liquid phase transition. At intermediate molar volumes, the thermodynamically stable state is a mixture of gaseous and liquid phases at the transition pressure, as illustrated by the straight-line in the v'-p' graph. The curved portion of the isotherm that is cut off by this straight-line correctly indicates what the allowed states would be if the fluid were homogeneous. However, these homogeneous states are unstable, because there is always another mixed state at the same pressure that possesses a lower Gibbs free energy.

The pressure at the phase transition can easily be determined from the graph of g versus p'. However, it is also possible to determine this pressure directly from the v'-p' isotherm. Note that the net change in G in going around the triangular loop 2-3-4-5-6 in Figure 7.4 is zero. Hence,

$$0 = \int_{\text{loop}} dG = \int_{\text{loop}} \left(\frac{\partial G}{\partial P}\right)_T = \int_{\text{loop}} V\,dP, \tag{7.85}$$

where use has been made of Equation (6.116). It follows that

$$\int_{\text{loop}} v'\,dp' = 0. \tag{7.86}$$

In other words, the straight-line in the bottom panel of Figure 7.4 must be drawn in such a manner that the area enclosed between it and the curve 6-5-4 is equal to that enclosed between it and the curve 4-3-2. Drawing the straight-line so as to enclose equal areas in this manner is known as the *Maxwell construction*, after James Clerk Maxwell.

Repeating the Maxwell construction for a variety of temperatures yields the phase diagram shown in Figure 7.5. [See J. Lekner, Am. J. Phys. **50**, 161 (1982) for a practical method of implementing the Maxwell construction analytically. See, also, Exercise 7.7.] For each temperature, there is a well-defined pressure, known as the *vapor pressure*, at which the liquid-gas transformation takes place. Plotting this pressure versus the temperature yields the phase diagram shown in Figure 7.6. Note that the liquid-gas phase boundary disappears at the critical temperature ($T = T_c$), because there is no phase transformation above this temperature. The point at which the phase boundary disappears is called the *critical point*. The corresponding temperature, pressure, and volume are known as the

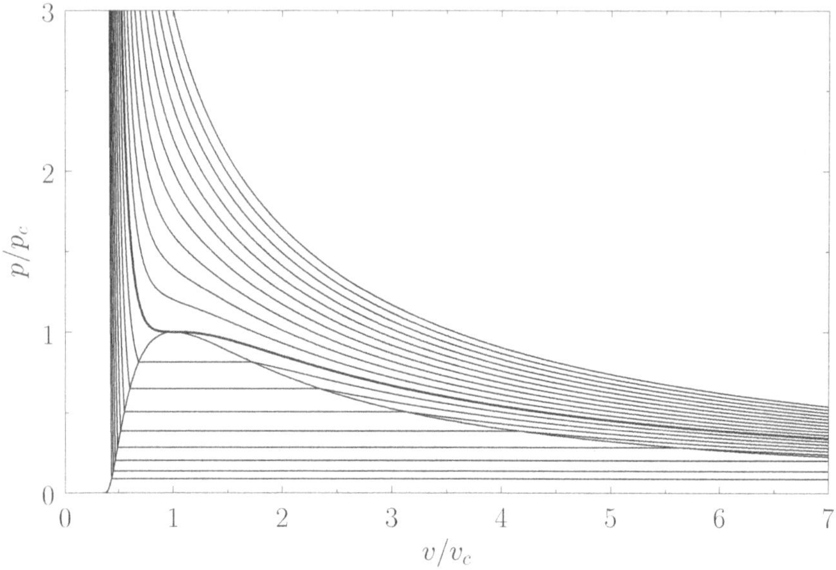

Figure 7.5 Complete v-p phase diagram predicted by the van der Waals model. The isotherms are for T/T_c ranging from 0.5 to 1.5 in increments of 0.05. The heavy curve shows the critical isotherm $T = T_c$. In the region in which the isotherms are horizontal, the stable state is a combination of liquid and gas phases.

critical temperature, pressure, and volume, respectively. At the critical point, the properties of the liquid and gas phases become identical.

Consider a phase transformation at some temperature $T < T_c$ in which the gaseous phase has the molar volume v_g, whereas the liquid phase has the molar volume $v_l < v_g$. Now, the molar entropy of a van der Waals gas can be written

$$s = R \ln(v' - 1/3) + g(T'), \tag{7.87}$$

where $g(T')$ is an unimportant function of the reduced temperature. (See Exercise 10.9.) Thus, the molar entropy increase associated with a liquid-gas phase transition is

$$s_g - s_l = R \ln\left(\frac{v_g' - 1/3}{v_l' - 1/3}\right), \tag{7.88}$$

where $v_g' = v_g/v_c$ and $v_l' = v_l/v_c$. It follows from Equation (7.73) that the molar latent heat of vaporization of a van der Waals fluid takes the form

$$\frac{l_{lg}}{R\,T_c} = T' \ln\left(\frac{v_g' - 1/3}{v_l' - 1/3}\right). \tag{7.89}$$

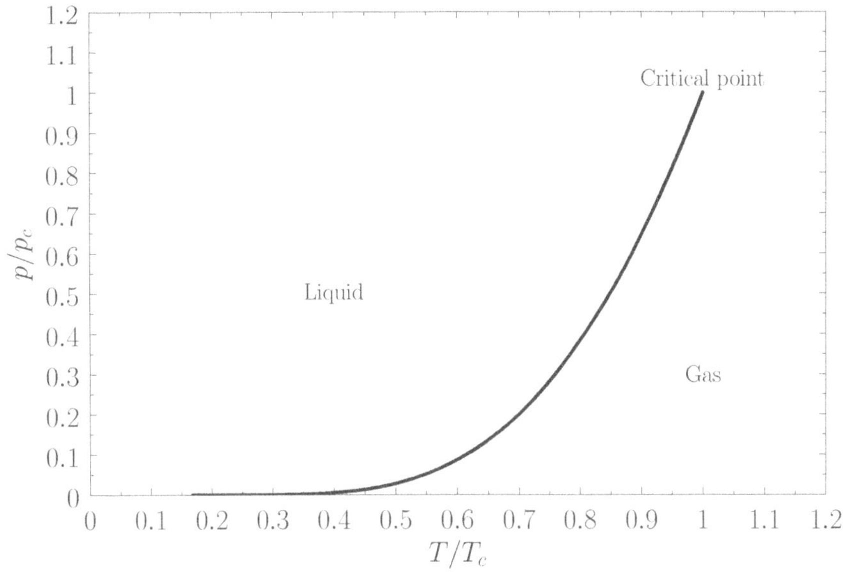

Figure 7.6 Complete T-p phase diagram predicted by the van der Waals model.

Figure 7.7 shows the latent heat of vaporization of a van der Waals fluid plotted as a function of the temperature. It can be seen that the latent heat drops abruptly to zero as the temperature exceeds the critical temperature. Note the great similarity between this figure and Figure 8.10, which illustrates the variation of the spontaneous magnetization of a ferromagnet with temperature, according to mean-field theory. This similarity is not surprising, because it turns out that the simplifying assumptions that were made during the derivation of the van der Waals equation of state in Section 10.11 amount to a form of mean-field theory.

Exercises

7.1 Consider a relatively small region of a homogeneous substance, containing $N \gg 1$ particles, that is in equilibrium with the remainder of the substance. The region is characterized by a volume, V, and a temperature, T.

(a) Show that the properly normalized probability that the volume lies between V and $V + dV$, and the temperature lies between T and $T + dT$, is

$$P(V,T)\,dV\,dT = \frac{1}{2\pi\,\Delta^* V\,\Delta^* T}\exp\left[-\frac{(V-\overline{V})^2}{2\,(\Delta^* V)^2}\right]\exp\left[-\frac{(T-\overline{T})^2}{2\,(\Delta^* T)^2}\right]dV\,dT,$$

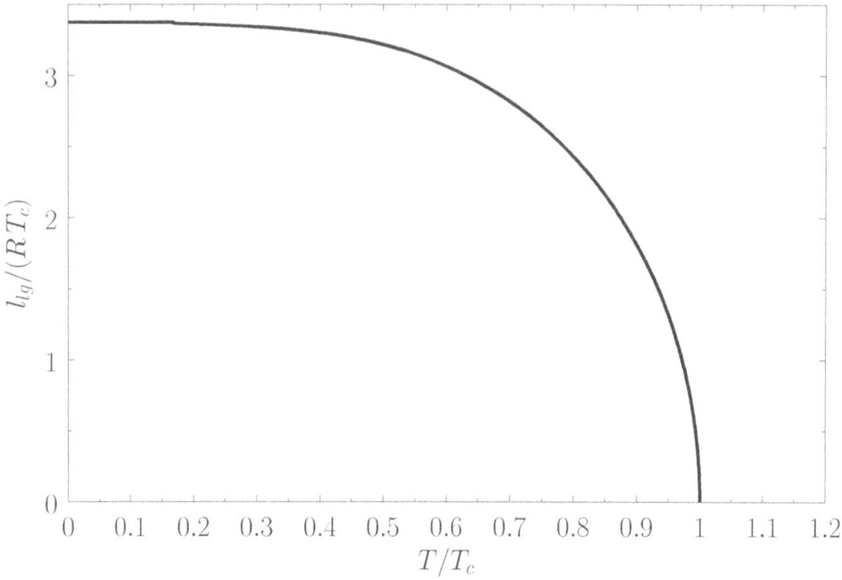

Figure 7.7 Molar latent heat of vaporization of a van der Waals fluid as a function of the temperature.

where

$$\Delta^* V = \left(k\,\overline{T}\,\kappa_T\,\overline{V}\right)^{1/2},$$

$$\Delta^* T = \left(\frac{k\,\overline{T}^{\,2}}{C_V}\right)^{1/2}.$$

Here, the mean temperature, $\overline{T}$, is the same as that of the remainder of the substance, whereas the mean volume, $\overline{V}$, is such that when $T = \overline{T}$ the pressure of the region matches that of the surrounding substance.

(b) Hence, deduce that

$$\langle (V - \overline{V})\,(T - \overline{T})\rangle = 0.$$

In other words, the volume and temperature fluctuations are uncorrelated.

(c) Show that for the case of a monatomic ideal gas

$$\frac{\Delta^* V}{\overline{V}} = \left(\frac{1}{N}\right)^{1/2},$$

$$\frac{\Delta^* T}{\overline{T}} = \left(\frac{2}{3\,N}\right)^{1/2},$$

$$\frac{\Delta^* p}{p} = \left(\frac{5}{3N}\right)^{1/2},$$

where p is the pressure of the region. Furthermore, demonstrate that

$$\frac{\langle (p-)(V-\overline{V})\rangle}{\bar{p}\,\overline{V}} = -\frac{1}{N},$$

$$\frac{\langle (p-\bar{p})(T-\overline{T})\rangle}{\bar{p}\,\overline{T}} = \frac{2}{3N}.$$

7.2 A substance of molecular weight μ has its triple point at the absolute temperature T_0 and pressure p_0. At this point, the mass densities of the solid and liquid phases are ρ_s and ρ_l, respectively, while the vapor phase can be approximated as a dilute ideal gas. If, at the triple point, the slope of the melting curve is $(dp/dT)_m$, and that of the liquid vaporization curve is $(dp/dT)_v$, show that the slope of the sublimation curve can be written

$$\left(\frac{dp}{dT}\right)_s = \frac{\mu\, p_0}{R\,T_0}\left(\frac{\rho_s - \rho_l}{\rho_s\,\rho_l}\right)\left(\frac{dp}{dT}\right)_m + \left(\frac{dp}{dT}\right)_v.$$

7.3 The vapor pressure of solid ammonia (in millimeters of mercury) is given by $\ln p = 23.02 - 3754/T$, and that of liquid ammonia by $\ln p = 19.48 - 3063/T$. Here, T is in degrees kelvin.

(a) Deduce that the triple point of ammonia occurs at 195.2 K.

(b) Show that, at the triple point, the latent heats of sublimation, vaporization, and melting of ammonia are 31.22 kJ/mol, 25.47 kJ/mol, and 5.75 kJ/mol, respectively.

7.4 Water boils when its vapor pressure is equal to that of the atmosphere. The boiling point of pure water at ground level is 100° C. Moreover, the latent heat of vaporization at this temperature is 40.66 kJ/mol. Show that the boiling point of water decreases approximately linearly with increasing altitude such that

$$T_b(z) \simeq 100 - \frac{z}{295.0},$$

where z measures altitude above ground level (in meters), and $T_b(z)$ is the boiling point in degrees centigrade at this altitude. In other words, the boiling point of pure water is depressed by about 3.4° C per kilometer increase in altitude.

7.5 The *relative humidity* of air is defined as the ratio of the partial pressure of water vapor to the equilibrium vapor pressure at the ambient temperature. The *dew point* is defined as the temperature at which the relative humidity becomes 100%. Show that if the relative humidity of air at (absolute) temperature T is $h(T)$ then the dew point is given by

$$T_d \simeq T - \frac{R\,T^2}{l}\,\ln\left[\frac{1}{h(T)}\right].$$

Here, R is the ideal gas constant, and l the molar latent heat of vaporization of water. The molar latent heat of water at 25°C is 43.99 kJ/mol. Suppose that air at 25°C has a relative humidity of 90%. Estimate the dew point. Suppose that the relative humidity is 40%. Estimate the dew point.

7.6 When a rising air mass in the atmosphere becomes saturated (i.e., attains 100% relative humidity), condensing water droplets give up energy, thereby slowing the adiabatic cooling process.

(a) Use the first law of thermodynamics to show that, as condensation forms during adiabatic expansion, the temperature of the air mass changes by

$$dT = \frac{2}{7}\,\frac{T}{p}\,dp - \frac{2}{7}\,\frac{l}{R}\,\frac{d\nu_w}{d\nu},$$

where ν_w is the number of moles of water vapor, ν the total number of moles, and l the molar latent heat of vaporization of water. Here, we have assumed that $\gamma = 7/5$ for air.

(b) Assuming that the air is always saturated during this process, show that

$$d\left(\frac{\nu_w}{\nu}\right) = \frac{p_v(T)}{p}\,\frac{l}{R\,T}\,\frac{dT}{T} - \frac{p_v}{p}\,\frac{dp}{p},$$

where $p_v(T)$ is the vapor pressure of water at temperature T, and p is the pressure of the atmosphere.

(c) Use the equation of hydrostatic equilibrium of the atmosphere,

$$\frac{dp}{p} = -\frac{\mu\,g}{R\,T}\,dz,$$

where z represents altitude, μ is the molecular weight of air, and g the acceleration due to gravity, to obtain the following expression so-called *wet adiabatic lapse-rate* of the atmosphere:

$$\frac{dT}{dz} = -\left(\frac{2}{7}\,\frac{\mu\,g}{R}\right)\frac{1 + (p_v/p)\,(l/R\,T)}{1 + (2/7)\,(p_v/p)\,(l/R\,T)^2}.$$

Of course, $dT/dz = -(2/7)\,(\mu\,g/R)$ is the dry adiabatic lapse-rate derived in Section 6.8.

(d) At 25°C, the vapor pressure of water is 0.0317 bar, and the molar latent heat of vaporization is 43.99 kJ/mol. At 0°C, the vapor pressure of water is 0.00611 bar, and the molar latent heat of vaporization is 51.07 kJ/mol. What is the ratio of the wet adiabatic lapse-rate to the dry adiabatic lapse-rate at these two temperatures?

7.7 Consider a phase transition in a van der Waals fluid whose reduced equation of state is

$$p' = \frac{8}{3}\,\frac{T'}{v' - 1/3} - \frac{3}{v'^2}.$$

Let T' and p'_v be the constant temperature and reduced vapor pressure at the transition, respectively, and let v'_l and v'_g be the reduced molar volumes of the liquid and gas phases, respectively. The fact that the phase transition takes place at constant temperature and pressure implies that

$$p'_v = \frac{8}{3}\,\frac{T'}{v'_l - 1/3} - \frac{3}{v'^2_l} = \frac{8}{3}\,\frac{T'}{v'_g - 1/3} - \frac{3}{v'^2_g}.$$

Moreover, the Maxwell construction yields

$$\int_l^g v'\,dp' = 0.$$

(a) Demonstrate that the Maxwell construction implies that

$$p'_v\,(v'_g - v'_l) = \frac{8}{3}\,T'\,\ln\left(\frac{v'_g - 1/3}{v'_l - 1/3}\right) + 3\left(\frac{1}{v'_g} - \frac{1}{v'_l}\right).$$

(b) Eliminate T' and p'_v from the previous equations to obtain the transcendental equation

$$\ln\left(\frac{v'_g - 1/3}{v'_l - 1/3}\right) = \left(\frac{v'_g - v'_l}{v'_g + v'_l}\right)\left(\frac{v'_g}{v'_g - 1/3} + \frac{v'_l}{v'_l - 1/3}\right)$$

that relates the molar volumes of the liquid and gas phases.

(c) Writing $x_+ = 1/(3\,v'_l - 1)$ and $x_- = 1/(3\,v'_g - 1)$, show that the previous equation reduces to

$$\ln\left(\frac{x_+}{x_-}\right) = \frac{(x_+ - x_-)\,(x_+ + x_- + 2)}{(x_+ + x_- + 2\,x_+\,x_-)}.$$

(d) By setting both sides of the previous equation equal to y, show that it can be solved parametrically to give

$$x_+ = \mathrm{e}^{+y/2}\,f(y/2),$$
$$x_- = \mathrm{e}^{-y/2}\,f(y/2),$$

where

$$f(x) = \frac{x\cosh x - \sinh x}{\sinh x\,\cosh x - x}.$$

(e) Furthermore, demonstrate that

$$T' = \frac{27}{4}\,\frac{f\,(c+f)}{g^2},$$

$$p_v' = \frac{27\,f^2\,(1-f^2)}{g^2},$$

$$v_g' - v_l' = \frac{2\,s}{3\,f},$$

$$\frac{s_g - s_l}{R} = y,$$

where

$$f = f(y/2),$$

$$c = \cosh(y/2),$$

$$s = \sinh(y/2),$$

$$g = 1 + 2\,c\,f + f^2.$$

Here, s_g and s_l denote the molar entropies of the gas and liquid phases, respectively.

(f) Finally, by considering the limits $y \gg 1$ and $y \ll 1$, show that

$$p_v' = 27\,\exp\left(-\frac{27}{8\,T'}\right),$$

$$\frac{l_{lg}}{R\,T_c} = \frac{27}{4},$$

in the limit $T' \ll 1$, and

$$v_g' - v_l' = 4\,(1-T')^{1/2},$$

$$\frac{l_{lg}}{R\,T_c} = 6\,(1-T')^{1/2},$$

in the limit $T' \to 1$. Here, l_{lg} is the molar latent heat of vaporization.

Chapter 8

Applications of Statistical Thermodynamics

8.1 Introduction

In our study of classical thermodynamics, we concentrated on the application of statistical physics to macroscopic systems. Somewhat paradoxically, statistical arguments did not figure very prominently in this investigation. The resolution of this paradox is, of course, that macroscopic systems contain a very large number of particles, and their statistical fluctuations are, therefore, negligible. Let us now apply statistical physics to microscopic systems, such as atoms and molecules. In this study, the underlying statistical nature of thermodynamics will become far more apparent.

8.2 Canonical Probability Distribution

We have gained some understanding of the macroscopic properties of the air in a classroom (say). For instance, we know something about its internal energy and specific heat capacity. How can we obtain information about the statistical properties of the molecules that make up this air? Consider a specific molecule. It constantly collides with its immediate neighbor molecules, and occasionally bounces off the walls of the room. These interactions "inform" it about the macroscopic state of the air, such as its temperature, pressure, and volume. The statistical distribution of the molecule over its own particular microstates must be consistent with this macrostate. In other words, if we have a large group of such molecules with similar statistical distributions then they must be equivalent to air with the appropriate macroscopic properties. So, it ought to be possible to calculate the probability distribution of the molecule over its microstates from a knowledge of these macroscopic properties.

We can think of the interaction of a molecule with the air in a classroom as analogous to the interaction of a small system, A, in thermal contact with a heat reservoir, A'. The air acts like a heat reservoir because its energy fluctuations due to interactions with the molecule are far too small to affect any of its macroscopic parameters. Let us determine the probability, P_r, of finding system A in one particular microstate, r, of energy E_r, when it is thermal equilibrium with the heat reservoir, A'.

As usual, we assume fairly weak interaction between A and A', so that the energies of these two systems are additive. The energy of A is not known at this stage. In fact, only the total energy of the combined system, $A^{(0)} = A + A'$, is known. Suppose that the total energy lies in the range $E^{(0)}$ to $E^{(0)}+\delta E$. The overall energy is constant in time, because $A^{(0)}$ is assumed to be an isolated system, so

$$E_r + E' = E^{(0)}, \tag{8.1}$$

where E' denotes the energy of the reservoir A'. Let $\Omega'(E')$ be the number of microstates accessible to the reservoir when its energy lies in the range E' to $E' + \delta E$. Clearly, if system A has an energy E_r then the reservoir A' must have an energy close to $E' = E^{(0)} - E_r$. Hence, because A is in one definite state (i.e., state r), and the total number of states accessible to A' is $\Omega'(E^{(0)}-E_r)$, it follows that the total number of states accessible to the combined system is simply $\Omega'(E^{(0)} - E_r)$. The principle of equal a priori probabilities tells us the probability of occurrence of a particular situation is proportional to the number of accessible microstates. Thus,

$$P_r = C'\,\Omega'(E^{(0)} - E_r), \tag{8.2}$$

where C' is a constant of proportionality that is independent of r. This constant can be determined by the normalization condition

$$\sum_r P_r = 1, \tag{8.3}$$

where the sum is over all possible states of system A, irrespective of their energy.

Let us now make use of the fact that system A is far smaller than system A'. It follows that $E_r \ll E^{(0)}$, so the slowly-varying logarithm of P_r can be Taylor expanded about $E' = E^{(0)}$. Thus,

$$\ln P_r = \ln C' + \ln \Omega'(E^{(0)}) - \left[\frac{\partial \ln \Omega'}{\partial E'}\right]_0 E_r + \cdots . \tag{8.4}$$

Note that we must expand $\ln P_r$, rather than P_r itself, because the latter function varies so rapidly with energy that the radius of convergence of its Taylor series is too small for the series to be of any practical use. The higher-order terms in Equation (8.4) can be safely neglected, because $E_r \ll E^{(0)}$. Now, the derivative

$$\left[\frac{\partial \ln \Omega'}{\partial E'}\right]_0 \equiv \beta \tag{8.5}$$

is evaluated at the fixed energy $E' = E^{(0)}$, and is, thus, a constant, independent of the energy, E_r, of A. In fact, we know, from Chapter 5, that this derivative is just the temperature parameter $\beta = (k\,T)^{-1}$ characterizing the heat reservoir A'. Here, T is the absolute temperature of the reservoir. Hence, Equation (8.4) becomes

$$\ln P_r = \ln C' + \ln \Omega'(E^{(0)}) - \beta\, E_r, \tag{8.6}$$

giving

$$P_r = C \exp(-\beta\, E_r), \tag{8.7}$$

where C is a constant independent of r. The parameter C is determined by the normalization condition, which gives

$$C^{-1} = \sum_r \exp(-\beta\, E_r), \tag{8.8}$$

so that the distribution becomes

$$P_r = \frac{\exp(-\beta\, E_r)}{\sum_r \exp(-\beta\, E_r)}. \tag{8.9}$$

This distribution is known as the *canonical probability distribution* (it is also sometimes called the Boltzmann probability distribution), and is an extremely important result in statistical physics.

The canonical distribution often causes confusion. People who are familiar with the principle of equal a priori probabilities, which says that all microstates are equally probable, are understandably surprised when they come across the canonical distribution, which says that high energy microstates are markedly less probable then low energy states. However, there is no need for any confusion. The principle of equal a priori probabilities applies to the whole system, whereas the canonical distribution only applies to a small part of the system. The two results are perfectly consistent. If the small system is in a microstate with a comparatively high energy, E_r, then the remainder of the system (i.e., the reservoir) has a slightly lower energy, E', than usual (because the overall energy is fixed). The number of accessible microstates of the reservoir is a very strongly increasing function of its energy. It follows that if the small system has a high energy then significantly less states than usual are accessible to the reservoir, so the number of microstates accessible to the overall system is reduced, and, hence, the configuration is comparatively unlikely. The strong increase in the number of accessible microstates of the reservoir with increasing E' gives rise to the strong (i.e., exponential) decrease in the likelihood of a state r of the small system with increasing E_r. The exponential factor $\exp(-\beta\, E_r)$ is called the *Boltzmann factor*.

The canonical distribution gives the probability of finding the small system A in one particular state r of energy E_r. The probability $P(E)$ that A has an energy in the small range between E and $E + \delta E$ is just the sum of all the probabilities of the states that lie in this range. However, because each of these states has approximately the same Boltzmann factor, this sum can be written

$$P(E) = C\,\Omega(E)\,\exp(-\beta\,E), \tag{8.10}$$

where $\Omega(E)$ is the number of microstates of A whose energies lie in the appropriate range. Suppose that system A is itself a large system, but still very much smaller than system A'. For a large system, we expect $\Omega(E)$ to be a very rapidly increasing function of energy, so the probability $P(E)$ is the product of a rapidly increasing function of E, and another rapidly decreasing function (i.e., the Boltzmann factor). This gives a sharp maximum of $P(E)$ at some particular value of the energy. As system A becomes larger, this maximum becomes sharper. Eventually, the maximum becomes so sharp that the energy of system A is almost bound to lie at the most probable energy. As usual, the most probable energy is evaluated by looking for the maximum of $\ln P$, so

$$\frac{\partial \ln P}{\partial E} = \frac{\partial \ln \Omega}{\partial E} - \beta = 0, \tag{8.11}$$

giving

$$\frac{\partial \ln \Omega}{\partial E} = \beta. \tag{8.12}$$

Of course, this corresponds to the situation in which the temperature of A is the same as that of the reservoir. This is a result that we have seen before. (See Chapter 5.) Note, however, that the canonical distribution is applicable no matter how small system A is, so it is a far more general result than any that we have previously obtained.

8.3 Spin-1/2 Paramagnetism

The simplest microscopic system that we can analyze using the canonical probability distribution is one that has only two possible states. (There would clearly be little point in analyzing a system with only one possible state.) Most elements, and some compounds, are *paramagnetic*. In other words, their constituent atoms, or molecules, possess a permanent magnetic moment due to the presence of one or more unpaired electrons. Consider a substance whose constituent particles contain only one unpaired electron (with zero orbital angular momentum). Such particles have spin $1/2$ [i.e., their spin angular momentum is $(1/2)\,\hbar$], and consequently possess an intrinsic magnetic moment, μ. According to quantum mechanics, the magnetic moment of a spin-$1/2$ particle can point either parallel or antiparallel to an external magnetic field, $\mathbf{B}$. Let us determine the mean magnetic moment (parallel to **B**), $\overline{\mu_{\parallel}}$, of the constituent particles of the substance when its absolute temperature is T. We shall assume, for the sake of simplicity, that each atom (or molecule) only interacts weakly with its neighboring atoms. This enables us to focus attention on a single atom, and to treat the remaining atoms as a heat bath at temperature T.

Our atom can be in one of two possible states. Namely, the (+) state in which its spin points up (i.e., parallel to **B**), or the (−) state in which its spin points down (i.e., antiparallel to **B**). In the (+) state, the atomic magnetic moment is parallel to the magnetic field, so that $\mu_{\parallel} = \mu$. The magnetic energy of the atom is $\epsilon_+ = -\mu\, B$. In the (−) state, the atomic magnetic moment is antiparallel to the magnetic field, so that $\mu_{\parallel} = -\mu$. The magnetic energy of the atom is $\epsilon_- = \mu\, B$.

According to the canonical distribution, the probability of finding the atom in the (+) state is

$$P_+ = C \exp(-\beta\,\epsilon_+) = C \exp(\beta\,\mu\,B), \tag{8.13}$$

where C is a constant, and $\beta = (k\,T)^{-1}$. Likewise, the probability of finding the atom in the (−) state is

$$P_- = C \exp(-\beta\,\epsilon_-) = C \exp(-\beta\,\mu\,B). \tag{8.14}$$

Clearly, the most probable state is the state with the lower energy [i.e., the (+) state]. Thus, the mean magnetic moment points in the direction of the magnetic field (i.e., the atomic spin is more likely to point parallel to the field than antiparallel).

It is apparent that the critical parameter in a paramagnetic system is

$$y \equiv \beta\,\mu\,B = \frac{\mu\,B}{k\,T}. \tag{8.15}$$

This parameter measures the ratio of the typical magnetic energy of the atom, $\mu\,B$, to its typical thermal energy, $k\,T$. If the thermal energy greatly exceeds the magnetic energy then $y \ll 1$, and the probability that the atomic moment points parallel to the magnetic field is about the same as the probability that it points antiparallel. In this situation, we expect the mean atomic moment to be small, so that $\overline{\mu_{\parallel}} \simeq 0$. On the other hand, if the magnetic energy greatly exceeds the thermal energy then $y \gg 1$, and the atomic moment is far more likely to be directed parallel to the magnetic field than antiparallel. In this situation, we expect $\overline{\mu_{\parallel}} \simeq \mu$.

Let us calculate the mean atomic moment, $\overline{\mu_{\parallel}}$. The usual definition of a mean value gives

$$\overline{\mu_{\parallel}} = \frac{P_+\,\mu + P_-\,(-\mu)}{P_+ + P_-} = \mu \left[\frac{\exp(\beta\,\mu\,B) - \exp(-\beta\,\mu\,B)}{\exp(\beta\,\mu\,B) + \exp(-\beta\,\mu\,B)}\right]. \tag{8.16}$$

This can also be written

$$\overline{\mu_{\parallel}} = \mu \tanh\left(\frac{\mu B}{k T}\right), \tag{8.17}$$

where the hyperbolic tangent is defined

$$\tanh y \equiv \frac{\exp(y) - \exp(-y)}{\exp(y) + \exp(-y)}. \tag{8.18}$$

For small arguments, $y \ll 1$,

$$\tanh y \simeq y - \frac{y^3}{3} + \cdots, \tag{8.19}$$

whereas for large arguments, $y \gg 1$,

$$\tanh y \simeq 1. \tag{8.20}$$

It follows that at comparatively high temperatures, $k T \gg \mu B$,

$$\overline{\mu_{\parallel}} \simeq \frac{\mu^2 B}{k T}, \tag{8.21}$$

whereas at comparatively low temperatures, $k T \ll \mu B$,

$$\overline{\mu_{\parallel}} \simeq \mu. \tag{8.22}$$

Suppose that the substance contains N_0 atoms (or molecules) per unit volume. The *magnetization* is defined as the mean magnetic moment per unit volume, and is given by

$$\overline{M_{\parallel}} = N_0 \, \overline{\mu_{\parallel}}. \tag{8.23}$$

At high temperatures, $k T \gg \mu B$, the mean magnetic moment, and, hence, the magnetization, is proportional to the applied magnetic field, so we can write

$$\overline{M_{\parallel}} \simeq \chi \frac{B}{\mu_0}, \tag{8.24}$$

where χ is a dimensionless constant of proportionality known as the *magnetic susceptibility*, and μ_0 the magnetic permeability of free space. It is clear that the magnetic susceptibility of a spin-1/2 paramagnetic substance takes the form

$$\chi = \frac{N_0 \, \mu_0 \, \mu^2}{k T}. \tag{8.25}$$

The fact that $\chi \propto T^{-1}$ is known as *Curie's law*, because it was discovered experimentally by Pierre Curie at the end of the nineteenth century. At low temperatures, $k T \ll \mu B$,

$$\overline{M_{\parallel}} \to N_0 \, \mu, \tag{8.26}$$

so the magnetization becomes independent of the applied field. This corresponds to the maximum possible magnetization, in which all atomic moments are aligned parallel to the field. The breakdown of the $\overline{M_{\parallel}} \propto B$ law at low temperatures (or high magnetic fields) is known as *saturation*.

8.4 System with Specified Mean Energy

Consider an isolated system, A, that consists of a fixed number of particles contained in a given volume. Suppose that the only available information about the system's energy is its mean value, $\overline{E}$. This might be the case, for example, if the system is brought to some final macrostate as a result of interaction with other macroscopic systems. Here, the measurement of the macroscopic work done on the system, or the heat that it absorbs, during this process does not tell us about the energy of each system in the corresponding ensemble, but, instead, only provides information about the mean energy of the final macrostate of A.

A system, A, with specified mean energy, $\overline{E}$, is also described by a canonical distribution. For, if such a system were placed in thermal contact with a heat reservoir held at some "temperature," $\beta = 1/(k\,T)$, then the mean energy of the system would be determined. Thus, an appropriate choice of β would guarantee that the mean energy of the system had the specific value $\overline{E}$.

We can give a more direct proof of the previous assertion. Let E_r be the energy of system A in some state r. Suppose that the statistical ensemble consists of a very large number, a, of such systems, a_r of which are in state r. Thus, the information available to us is that

$$\frac{1}{a}\sum_s a_s\,E_s = \overline{E}. \tag{8.27}$$

Hence, it follows that

$$\sum_s a_s\,E_s = a\,\overline{E} = \text{constant}. \tag{8.28}$$

This implies that the situation is equivalent to one in which a fixed amount of energy, $a\,\overline{E}$, is to be distributed over all the systems in the ensemble, such that each system is equally likely to be in any one state. If a system in the ensemble is in state r then the remaining $a-1$ systems must then have a combined energy $E' = a\,\overline{E} - E_r$. These $a-1$ systems are distributed over some very large number, $\Phi(E')$, of accessible states. In other words, if the one system under consideration is in state r then the remaining $a-1$ systems are distributed with equal probability in any of the $\Phi(a\,\overline{E} - E_r)$ states accessible to them. Because $E_r \ll a\,\overline{E}$, the mathematical problem here is exactly the same as that considered in Section 8.2, where we dealt with a system in thermal contact with a heat bath. In this case, the role of the heat bath is played by the totality of all the other systems in the ensemble. Accordingly, we again get the canonical distribution,

$$P_r \propto \exp(-\beta\,E_r). \tag{8.29}$$

The parameter $\beta \equiv (\partial \ln \Phi/\partial E')$ does not here have any immediate physical significance in terms of the temperature of a real heat bath. Instead, it is to be determined by the condition that the mean energy calculated using the distribution has the specified value $\overline{E}$. In other words,

$$\frac{\sum_r \exp(-\beta\, E_r)\, E_r}{\sum_r \exp(-\beta\, E_r)} = \overline{E}. \tag{8.30}$$

In conclusion, if one is dealing with a system in contact with a heat bath held at "temperature" $\beta = 1/(k\,T)$ then the canonical distribution, (8.29), is valid, and the mean energy, $\overline{E}$, can be calculated from Equation (8.30), using the known value of β. On the other hand, if one is dealing with a system of specified mean energy, $\overline{E}$, then the canonical distribution, (8.29), is again valid, but the parameter β is to be calculated using Equation (8.30) and the known value of $\overline{E}$.

8.5 Calculation of Mean Values

Consider a system in contact with a heat reservoir, or with a specified mean energy. The systems in the representative ensemble are distributed over their accessible states in accordance with the canonical distribution. Thus, the probability of occurrence of some state r with energy E_r is given by

$$P_r = \frac{\exp(-\beta\, E_r)}{\sum_r \exp(-\beta\, E_r)}. \tag{8.31}$$

The mean energy is written

$$\overline{E} = \frac{\sum_r \exp(-\beta\, E_r)\, E_r}{\sum_r \exp(-\beta\, E_r)}, \tag{8.32}$$

where the sum is taken over all states of the system, irrespective of their energy. Note that

$$\sum_r \exp(-\beta\, E_r)\, E_r = -\sum_r \frac{\partial}{\partial\beta} \exp(-\beta\, E_r) = -\frac{\partial Z}{\partial\beta}, \tag{8.33}$$

where

$$Z = \sum_r \exp(-\beta\, E_r). \tag{8.34}$$

It follows that

$$\overline{E} = -\frac{1}{Z}\frac{\partial Z}{\partial\beta} = -\frac{\partial \ln Z}{\partial\beta}. \tag{8.35}$$

The quantity Z, which is defined as the sum of the Boltzmann factor over all states, irrespective of their energy, is called the *partition function*. (Incidentally, the partition function is represented by the symbol Z because in German it is known as the "zustandssumme," which means the "sum over states.") We have just demonstrated that it is fairly easy to work out the mean energy of a system using its partition function. In fact, as we shall discover, it is straightforward to calculate virtually any piece of statistical information from the partition function.

Let us evaluate the variance of the energy. We know that

$$\overline{(\Delta E)^2} = \overline{E^2} - \overline{E}^2. \tag{8.36}$$

(See Chapter 2.) Now, according to the canonical distribution,

$$\overline{E^2} = \frac{\sum_r \exp(-\beta\, E_r)\, E_r^2}{\sum_r \exp(-\beta\, E_r)}. \tag{8.37}$$

However,

$$\begin{aligned}\sum_r \exp(-\beta\, E_r)\, E_r^2 &= -\frac{\partial}{\partial\beta}\left[\sum_r \exp(-\beta\, E_r)\, E_r\right] \\ &= \left(-\frac{\partial}{\partial\beta}\right)^2 \left[\sum_r \exp(-\beta\, E_r)\right].\end{aligned} \tag{8.38}$$

Hence,

$$\overline{E^2} = \frac{1}{Z}\frac{\partial^2 Z}{\partial\beta^2}. \tag{8.39}$$

We can also write

$$\overline{E^2} = \frac{\partial}{\partial\beta}\left(\frac{1}{Z}\frac{\partial Z}{\partial\beta}\right) + \frac{1}{Z^2}\left(\frac{\partial Z}{\partial\beta}\right)^2 = -\frac{\partial\overline{E}}{\partial\beta} + \overline{E}^2, \tag{8.40}$$

where use has been made of Equation (8.35). It follows from Equation (8.36) that

$$\overline{(\Delta E)^2} = -\frac{\partial\overline{E}}{\partial\beta} = \frac{\partial^2 \ln Z}{\partial\beta^2}. \tag{8.41}$$

Thus, the variance of the energy can be worked out from the partition function almost as easily as the mean energy. Because, by definition, a variance can never be negative, it follows that $\partial\overline{E}/\partial\beta \leq 0$, or, equivalently, $\partial\overline{E}/\partial T \geq 0$. Hence, the mean energy of a system governed by the canonical distribution always increases with increasing temperature.

Suppose that the system is characterized by a single external parameter x (such as its volume). The generalization to the case where there are several external parameters is straightforward. Consider a quasi-static change of the external parameter from x to $x + dx$. In this process, the energy of the system in state r changes by

$$\delta E_r = \frac{\partial E_r}{\partial x}\, dx. \tag{8.42}$$

The macroscopic work $đW$ done by the system due to this parameter change is

$$đW = \frac{\sum_r \exp(-\beta\, E_r)\,(-\partial E_r/\partial x\; dx)}{\sum_r \exp(-\beta\, E_r)}. \tag{8.43}$$

In other words, the work done is minus the average change in internal energy of the system, where the average is calculated using the canonical distribution. We can write

$$\sum_r \exp(-\beta\, E_r)\,\frac{\partial E_r}{\partial x} = -\frac{1}{\beta}\frac{\partial}{\partial x}\left[\sum_r \exp(-\beta\, E_r)\right] = -\frac{1}{\beta}\frac{\partial Z}{\partial x}, \tag{8.44}$$

which gives

$$đW = \frac{1}{\beta Z}\frac{\partial Z}{\partial x}\,dx = \frac{1}{\beta}\frac{\partial \ln Z}{\partial x}\,dx. \tag{8.45}$$

We also have the following general expression for the work done by the system

$$đW = \overline{X}\,dx, \tag{8.46}$$

where

$$\overline{X} = -\overline{\frac{\partial E_r}{\partial x}} \tag{8.47}$$

is the mean generalized force conjugate to x. (See Chapter 4.) It follows that

$$\overline{X} = \frac{1}{\beta}\frac{\partial \ln Z}{\partial x}. \tag{8.48}$$

Suppose that the external parameter is the volume, so $x = V$. It follows that

$$đW = \overline{p}\,dV = \frac{1}{\beta}\frac{\partial \ln Z}{\partial V}\,dV \tag{8.49}$$

and

$$\overline{p} = \frac{1}{\beta}\frac{\partial \ln Z}{\partial V}, \tag{8.50}$$

where p is the pressure. Because the partition function is a function of β and V (the energies E_r depend on V), it follows that the previous equation relates the mean pressure, $\overline{p}$, to T (via $\beta = 1/k\,T$) and V. In other words, the previous expression is the equation of state. Hence, we can work out the pressure, and even the equation of state, using the partition function.

8.6 Partition Function

It is clear that all important macroscopic quantities associated with a system can be expressed in terms of its partition function, Z. Let us investigate how the partition function is related to thermodynamical quantities. Recall that Z is a function of both β and x (where x is the single external parameter). Hence, $Z = Z(\beta, x)$, and we can write

$$d\ln Z = \frac{\partial \ln Z}{\partial x}\,dx + \frac{\partial \ln Z}{\partial \beta}\,d\beta. \tag{8.51}$$

Consider a quasi-static change by which x and β change so slowly that the system stays close to equilibrium, and, thus, remains distributed according to the canonical distribution. It follows from Equations (8.35) and (8.45) that

$$d\ln Z = \beta\,đW - \overline{E}\,d\beta. \tag{8.52}$$

The last term can be rewritten

$$d\ln Z = \beta\,đW - d\left(\overline{E}\,\beta\right) + \beta\,d\overline{E}, \tag{8.53}$$

giving

$$d\left(\ln Z + \beta\,\overline{E}\right) = \beta\left(đW + d\overline{E}\right) \equiv \beta\,đQ. \tag{8.54}$$

The previous equation shows that although the heat absorbed by the system, $đQ$, is not an exact differential, it becomes one when multiplied by the temperature parameter, β. This is essentially the second law of thermodynamics. In fact, we know that

$$dS = \frac{đQ}{T}. \tag{8.55}$$

Hence,

$$S \equiv k\left(\ln Z + \beta\,\overline{E}\right). \tag{8.56}$$

This expression enables us to calculate the entropy of a system from its partition function.

Suppose that we are dealing with a system $A^{(0)}$ consisting of two systems A and A' that only interact weakly with one another. Let each state of A be denoted by an index r, and have a corresponding energy E_r. Likewise, let each state of A' be denoted by an index s, and have a corresponding energy E'_s. A state of the combined system $A^{(0)}$ is then denoted by two indices r and s. Because A and A' only interact weakly, their energies are additive, and the energy of state rs is

$$E^{(0)}_{rs} = E_r + E'_s. \tag{8.57}$$

By definition, the partition function of $A^{(0)}$ takes the form

$$Z^{(0)} = \sum_{r,s} \exp\left[-\beta\, E_{rs}^{(0)}\right] = \sum_{r,s} \exp(-\beta\,[E_r + E_s']) = \sum_{r,s} \exp(-\beta\, E_r)\exp(-\beta\, E_s')$$

$$= \left[\sum_r \exp(-\beta\, E_r)\right]\left[\sum_s \exp(-\beta\, E_s')\right]. \tag{8.58}$$

Hence,

$$Z^{(0)} = Z\, Z', \tag{8.59}$$

giving

$$\ln Z^{(0)} = \ln Z + \ln Z', \tag{8.60}$$

where Z and Z' are the partition functions of A and A', respectively. It follows from Equation (8.35) that the mean energies of $A^{(0)}$, A, and A' are related by

$$\overline{E}^{(0)} = \overline{E} + \overline{E}'. \tag{8.61}$$

It also follows from Equation (8.56) that the respective entropies of these systems are related via

$$S^{(0)} = S + S'. \tag{8.62}$$

Hence, the partition function tells us that the extensive (see Section 8.8) thermodynamic functions of two weakly-interacting systems are simply additive.

It is clear that we can perform statistical thermodynamical calculations using the partition function, Z, instead of the more direct approach in which we use the density of states, Ω. The former approach is advantageous because the partition function is an unrestricted sum of Boltzmann factors taken over all accessible states, irrespective of their energy, whereas the density of states is a restricted sum taken over all states whose energies lie in some narrow range. In general, it is far easier to perform an unrestricted sum than a restricted sum. Thus, it is invariably more straightforward to derive statistical thermodynamical results using Z rather than Ω, although Ω has a far more direct physical significance than Z.

8.7 Ideal Monatomic Gas

Let us now practice calculating thermodynamic relations using the partition function by considering an example with which we are already quite familiar: namely, an ideal monatomic gas. Consider a gas consisting of N identical monatomic molecules of mass m, enclosed in a container of volume V. Let us denote the position and momentum vectors of the ith molecule by $\mathbf{r}_i$ and $\mathbf{p}_i$, respectively.

Because the gas is ideal, there are no interatomic forces, and the total energy is simply the sum of the individual kinetic energies of the molecules:

$$E = \sum_{i=1,N} \frac{p_i^2}{2\,m}, \tag{8.63}$$

where $p_i^2 = \mathbf{p}_i\cdot\mathbf{p}_i$.

Let us treat the problem classically. In this approach, we divide up phase-space into cells of equal volume h_0^f. Here, f is the number of degrees of freedom, and h_0 is a small constant with dimensions of angular momentum that parameterizes the precision to which the positions and momenta of molecules are determined. (See Section 3.2.) Each cell in phase-space corresponds to a different state. The partition function is the sum of the Boltzmann factor $\exp(-\beta\,E_r)$ taken over all possible states, where E_r is the energy of state r. Classically, we can approximate the summation over cells in phase-space as an integration over all phase-space. Thus,

$$Z = \int\cdots\int \exp(-\beta\,E)\,\frac{d^3\mathbf{r}_1\cdots d^3\mathbf{r}_N\,d^3\mathbf{p}_1\cdots d^3\mathbf{p}_N}{h_0^{3N}}, \tag{8.64}$$

where $3N$ is the number of degrees of freedom of a monatomic gas containing N molecules. Making use of Equation (8.63), the previous expression reduces to

$$Z = \frac{V^N}{h_0^{3N}}\int\cdots\int \exp\left(-\frac{\beta}{2\,m}\,p_1^2\right)d^3\mathbf{p}_1\cdots\exp\left(-\frac{\beta}{2\,m}\,p_N^2\right)d^3\mathbf{p}_N. \tag{8.65}$$

Note that the integral over the coordinates of a given molecule simply yields the volume of the container, V, because the energy, E, is independent of the locations of the molecules in an ideal gas. There are N such integrals, so we obtain the factor V^N in the previous expression. Note, also, that each of the integrals over the molecular momenta in Equation (8.65) are identical: they differ only by irrelevant dummy variables of integration. It follows that the partition function Z of the gas is made up of the product of N identical factors: that is,

$$Z = \zeta^N, \tag{8.66}$$

where

$$\zeta = \frac{V}{h_0^3}\int \exp\left(-\frac{\beta}{2\,m}\,p^2\right)d^3\mathbf{p} \tag{8.67}$$

is the partition function for a single molecule. Of course, this result is obvious, because we have already shown that the partition function for a system made up of a number of weakly interacting subsystems is just the product of the partition functions of the subsystems. (See Section 8.6.)

The integral in Equation (8.67) is easily evaluated:

$$\int \exp\left(-\frac{\beta}{2m}\,p^2\right)d^3\mathbf{p} = \int_{-\infty}^{\infty} \exp\left(-\frac{\beta}{2m}\,p_x^2\right)dp_x \int_{-\infty}^{\infty} \exp\left(-\frac{\beta}{2m}\,p_y^2\right)dp_y$$
$$\times \int_{-\infty}^{\infty} \exp\left(-\frac{\beta}{2m}\,p_z^2\right)dp_z$$
$$= \left(\sqrt{\frac{2\pi\,m}{\beta}}\right)^3, \tag{8.68}$$

where use has been made of Equation (2.79). Thus,

$$\zeta = V\left(\frac{2\pi\,m}{h_0^2\,\beta}\right)^{3/2}, \tag{8.69}$$

and

$$\ln Z = N\ln\zeta = N\left[\ln V - \frac{3}{2}\ln\beta + \frac{3}{2}\ln\left(\frac{2\pi\,m}{h_0^2}\right)\right]. \tag{8.70}$$

The expression for the mean pressure, (8.50), yields

$$\overline{p} = \frac{1}{\beta}\frac{\partial\ln Z}{\partial V} = \frac{1}{\beta}\frac{N}{V}, \tag{8.71}$$

which reduces to the ideal gas equation of state

$$\overline{p}\,V = N\,k\,T = \nu\,R\,T, \tag{8.72}$$

where use has been made of $N = \nu\,N_A$ and $R = N_A\,k$. According to Equation (8.35), the mean energy of the gas is given by

$$\overline{E} = -\frac{\partial\ln Z}{\partial\beta} = \frac{3}{2}\frac{N}{\beta} = \frac{3}{2}\,\nu\,R\,T. \tag{8.73}$$

Note that the internal energy is a function of temperature alone, with no dependence on volume. The molar heat capacity at constant volume of the gas is given by

$$c_V = \frac{1}{\nu}\left(\frac{\partial\overline{E}}{\partial T}\right)_V = \frac{3}{2}\,R, \tag{8.74}$$

so the mean energy can be written

$$\overline{E} = \nu\,c_V\,T. \tag{8.75}$$

We have seen all of the previous results before. Let us now use the partition function to calculate a new result. The entropy of the gas can be calculated quite simply from the expression

$$S = k\left(\ln Z + \beta\,\overline{E}\right). \tag{8.76}$$

Thus,

$$S = \nu R\left[\ln V - \frac{3}{2}\ln\beta + \frac{3}{2}\ln\left(\frac{2\pi\,m}{h_0^2}\right) + \frac{3}{2}\right], \tag{8.77}$$

or

$$S = \nu R\left(\ln V + \frac{3}{2}\ln T + \sigma\right), \tag{8.78}$$

where

$$\sigma = \frac{3}{2}\ln\left(\frac{2\pi\,m\,k}{h_0^2}\right) + \frac{3}{2}. \tag{8.79}$$

The previous expression for the entropy of an ideal gas is certainly a new result. Unfortunately, it is also quite obviously incorrect.

8.8 Gibbs Paradox

First of all, let us be clear why Equation (8.78) is incorrect.

We can see that $S \rightarrow -\infty$ as $T \rightarrow 0$, which contradicts the third law of thermodynamics. However, this is not a problem. Equation (8.78) was derived using classical physics, which breaks down at low temperatures. Thus, we would not expect this equation to give a sensible answer close to the absolute zero of temperature.

Equation (8.78) is wrong because it implies that the entropy does not behave properly as an extensive quantity. Thermodynamic quantities can be divided into two groups; *extensive* and *intensive*. Extensive quantities increase by a factor α when the size of the system under consideration is increased by the same factor. Intensive quantities stay the same. Energy and volume are typical extensive quantities. Pressure and temperature are typical intensive quantities. Entropy is very definitely an extensive quantity. We have previously shown that the entropies of two weakly-interacting systems are additive. [See Equation (8.62).] Thus, if we double the size of a system then we expect the entropy to double as well. Suppose that we have a system of volume V, containing ν moles of ideal gas at temperature T. Doubling the size of the system is like joining two identical systems together to form a new system of volume $2\,V$, containing $2\,\nu$ moles of gas at temperature T. Let

$$S = \nu R\left(\ln V + \frac{3}{2}\ln T + \sigma\right) \tag{8.80}$$

denote the entropy of the original system, and let

$$S' = 2\,\nu R\left[\ln(2\,V) + \frac{3}{2}\ln T + \sigma\right] \tag{8.81}$$

denote the entropy of the double-sized system. Clearly, if entropy is an extensive quantity then we should have

$$S' = 2\,S. \tag{8.82}$$

But, in fact, we find that

$$S' - 2\,S = 2\,\nu R\,\ln 2. \tag{8.83}$$

So, the entropy of the double-sized system is more than double the entropy of the original system.

Where does this extra entropy come from? Let us consider a little more carefully how we might go about doubling the size of our system. Suppose that we put another identical system adjacent to it, and separate the two systems by a partition. Let us now suddenly remove the partition. If entropy is a properly extensive quantity then the entropy of the overall system should be the same before and after the partition is removed. It is certainly the case that the energy (another extensive quantity) of the overall system stays the same. However, according to Equation (8.83), the overall entropy of the system increases by $2\,\nu R\,\ln 2$ after the partition is removed. Suppose, now, that the second system is identical to the first system in all respects, except that its molecules are in some way slightly different to the molecules in the first system, so that the two sets of molecules are distinguishable. In this case, we would certainly expect an overall increase in entropy when the partition is removed. Before the partition is removed, it separates type 1 molecules from type 2 molecules. After the partition is removed, molecules of both types become jumbled together. This is clearly an irreversible process. We cannot imagine the molecules spontaneously sorting themselves out again. The increase in entropy associated with this jumbling is called *entropy of mixing*, and is easily calculated. We know that the number of accessible states of an ideal gas varies with volume like $\Omega \propto V^N$. The volume accessible to type 1 molecules clearly doubles after the partition is removed, as does the volume accessible to type 2 molecules. Using the fundamental formula $S = k\,\ln\Omega$, the increase in entropy due to mixing is given by

$$S = 2\,k\ln\left(\frac{\Omega_f}{\Omega_i}\right) = 2\,N\,k\ln\left(\frac{V_f}{V_i}\right) = 2\,\nu R\,\ln 2. \tag{8.84}$$

It is clear that the additional entropy, $2\,\nu R\,\ln 2$, that appears when we double the size of an ideal gas system, by joining together two identical systems, is entropy of mixing of the molecules contained in the original systems. But, if the molecules in these two systems are indistinguishable then why should there be any entropy of mixing? Obviously, there is no entropy of mixing in this case. At this point, we can begin to understand what has gone wrong in our calculation.

We have calculated the partition function assuming that all of the molecules in our system have the same mass and temperature, but we have not explicitly taken into account the fact that we consider the molecules to be indistinguishable. In other words, we have been treating the molecules in our ideal gas as if each carried a little license plate, or a social security number, so that we could always tell one from another. In quantum mechanics, which is what we really should be using to study microscopic phenomena, the essential indistinguishability of atoms and molecules is hard-wired into the theory at a very low level. Our problem is that we have been taking the classical approach a little too seriously. It is plainly nonsense to pretend that we can distinguish molecules in a statistical problem where we cannot closely follow the motions of individual particles. A paradox arises if we try to treat molecules as if they were distinguishable. This is called the *Gibbs paradox*, after the American physicist Josiah Gibbs, who first discussed it. The resolution of the Gibbs paradox is quite simple: treat all molecules of the same species as if they were indistinguishable.

In our previous calculation of the ideal gas partition function, we inadvertently treated each of the N molecules in the gas as distinguishable. Because of this, we over-counted the number of states of the system. Because the $N!$ possible permutations of the molecules amongst themselves do not lead to physically different situations, and, therefore, cannot be counted as separate states, the number of actual states of the system is a factor $N!$ less than what we initially thought. We can easily correct our partition function by simply dividing by this factor, so that

$$Z = \frac{\zeta^N}{N!}. \tag{8.85}$$

This gives

$$\ln Z = N \ln \zeta - \ln N!, \tag{8.86}$$

or

$$\ln Z = N \ln \zeta - N \ln N + N, \tag{8.87}$$

using Stirling's approximation. Note that our new version of $\ln Z$ differs from our previous version by an additive term involving the number of particles in the system. This explains why our calculations of the mean pressure and mean energy, which depend on partial derivatives of $\ln Z$ with respect to the volume and the temperature parameter β, respectively, came out all right. However, our expression for the entropy S is modified by this additive term. The new expression is

$$S = \nu R \left[\ln V - \frac{3}{2} \ln \beta + \frac{3}{2} \ln \left(\frac{2\pi m k}{h_0^2} \right) + \frac{3}{2} \right] + k(-N \ln N + N). \tag{8.88}$$

This gives

$$S = \nu R \left[\ln\left(\frac{V}{N}\right) + \frac{3}{2} \ln T + \sigma_0 \right] \tag{8.89}$$

where

$$\sigma_0 = \frac{3}{2} \ln\left(\frac{2\pi m k}{h_0^2}\right) + \frac{5}{2}. \tag{8.90}$$

It is clear that the entropy behaves properly as an extensive quantity in Equation (8.89). In other words, it increases by a factor α when ν, V, and N each increase by the same factor.

8.9 General Paramagnetism

In Section 8.3, we considered the special case of paramagnetism in which all the atoms that make up the substance under investigation possess spin 1/2. Let us now discuss the general case.

Consider a system consisting of N non-interacting atoms in a substance held at absolute temperature T, and placed in an external magnetic field, $\mathbf{B} = B_z \mathbf{e}_z$, that points in the z-direction. The magnetic energy of a given atom is written

$$\epsilon = -\boldsymbol{\mu} \cdot \mathbf{B}, \tag{8.91}$$

where $\boldsymbol{\mu}$ is the atom's magnetic moment. Now, the magnetic moment of an atom is proportional to its total electronic angular momentum, $\hbar \mathbf{J}$. In fact,

$$\boldsymbol{\mu} = g \mu_B \mathbf{J}, \tag{8.92}$$

where

$$\mu_B = \frac{e \hbar}{2 m_e} \tag{8.93}$$

is a standard unit of magnetic moment known as the *Bohr magneton*. Here, e is the magnitude of the electron charge, and m_e the electron mass. Moreover, g is a dimensionless number of order unity that is called the *g-factor* of the atom.

Combining Equations (8.91) and (8.92), we obtain

$$\epsilon = -g \mu_B \mathbf{J} \cdot \mathbf{B} = -g \mu_B B_z J_z. \tag{8.94}$$

Now, according to standard quantum mechanics,

$$J_z = m, \tag{8.95}$$

where m is an integer that can take on all values between $-J$ and $+J$. Here, $|\mathbf{J}|^2 = J(J+1)$, where J is a positive number that can either take integer or half-integer values. Thus, there are $2J + 1$ allowed values of m, corresponding to different possible projections of the angular momentum vector along the z-direction. It follows from Equation (8.94) that the possible magnetic energies of an atom are

$$\epsilon_m = -g\,\mu_B\,B_z\,m. \tag{8.96}$$

For example, if $J = 1/2$ (corresponding to a spin-1/2 system) then there are only two possible energies, corresponding to $m = \pm 1/2$. This was the situation considered in Section 8.3. (To be more exact, a spin-1/2 system corresponds to $J = 1/2$ and $g = 2$.)

The probability, P_m, that an atom is in a state labeled m is

$$P_m \propto \exp(-\beta\,\epsilon_m) = \exp(\beta\,g\,\mu_B\,B_z\,m), \tag{8.97}$$

where $\beta = 1/(k\,T)$. According to Equation (8.92), the corresponding z-component of the magnetic moment is

$$\mu_z = g\,\mu_B\,m. \tag{8.98}$$

Hence, the mean z-component of the magnetic moment is

$$\overline{\mu_z} = \frac{\sum_{m=-J,+J} \exp(\beta\,g\,\mu_B\,B_z\,m)\,g\,\mu_B\,m}{\sum_{m=-J,+J} \exp(\beta\,g\,\mu_B\,B_z\,m)}. \tag{8.99}$$

The numerator in the previous expression is conveniently expressed as a derivative with respect to the external parameter B_z:

$$\sum_{m=-J,+J} \exp(\beta\,g\,\mu_B\,B_z\,m)\,g\,\mu_B\,m = \frac{1}{\beta}\frac{\partial Z_a}{\partial B_z}, \tag{8.100}$$

where

$$Z_a = \sum_{m=-J,+J} \exp(\beta\,g\,\mu_B\,B_z\,m) \tag{8.101}$$

is the partition function of a single atom. Thus, Equation (8.99) becomes

$$\overline{\mu_z} = \frac{1}{Z_a}\frac{1}{\beta}\frac{\partial Z_a}{\partial B_z} = \frac{1}{\beta}\frac{\partial \ln Z_a}{\partial B_z}. \tag{8.102}$$

In order to calculate Z_a, it is convenient to define

$$\eta = \beta\,g\,\mu_B\,B_z = \frac{g\,\mu_B\,B_z}{k\,T}, \tag{8.103}$$

which is a dimensionless parameter that measures the ratio of the magnetic energy, $g\,\mu_B\,B_z$, that acts to align the atomic magnetic moments parallel to the external magnetic field, to the thermal energy, $k\,T$, that acts to keep the magnetic moment randomly orientated. Thus, Equation (8.101) becomes

$$Z_a = \sum_{m=-J,+J} \exp(m\,\eta) = \mathrm{e}^{-\eta\,J} \sum_{k=0,2J} [\exp(\eta)]^k. \tag{8.104}$$

The previous geometric series can be summed to give

$$Z_a = \mathrm{e}^{-\eta\,J} \left\{ \frac{1 - [\exp(\eta)]^{2J+1}}{1 - \exp(\eta)} \right\}. \tag{8.105}$$

(See Exercise 8.1.) Multiplying the numerator and denominator by $\exp(-\eta/2)$, we obtain

$$Z_a = \frac{\exp[-\eta\,(J + 1/2)] - \exp[\,\eta\,(J + 1/2)]}{\exp(-\eta/2) - \exp(\,\eta/2)}, \tag{8.106}$$

or

$$Z_a = \frac{\sinh[(J + 1/2)\,\eta]}{\sinh(\eta/2)}, \tag{8.107}$$

where the hyperbolic sine is defined

$$\sinh z \equiv \frac{\exp(\,y) - \exp(-y)}{2}. \tag{8.108}$$

Thus,

$$\ln Z_a = \ln\,\sinh[(J + 1/2)\,\eta] - \ln\sinh(\eta/2). \tag{8.109}$$

According to Equations (8.102) and (8.103),

$$\overline{\mu_z} = \frac{1}{\beta}\frac{\partial \ln Z_a}{\partial B_z} = \frac{1}{\beta}\frac{\partial \ln Z_a}{\partial \eta}\frac{\partial \eta}{\partial B_z} = g\,\mu_B\,\frac{\partial \ln Z_a}{\partial \eta}. \tag{8.110}$$

Hence,

$$\overline{\mu_z} = g\,\mu_B \left\{ \frac{(J + 1/2)\,\cosh[(J + 1/2)\,\eta]}{\sinh[(J + 1/2)\,\eta]} - \frac{(1/2)\,\cosh(\eta/2)}{\sinh(\eta/2)} \right\}, \tag{8.111}$$

or

$$\overline{\mu_z} = g\,\mu_B\,J\,\mathcal{B}_J(\eta), \tag{8.112}$$

where

$$\mathcal{B}_J(\eta) = \frac{1}{J}\left\{ \left(J + \frac{1}{2}\right) \coth\left[\left(J + \frac{1}{2}\right)\eta\right] - \frac{1}{2}\,\coth\left(\frac{\eta}{2}\right) \right\} \tag{8.113}$$

is known as the *Brillouin function*.

The hyperbolic cotangent is defined

$$\coth z \equiv \frac{\cosh z}{\sinh z} = \frac{\exp(\,z) + \exp(-z)}{\exp(\,z) - \exp(-z)}. \tag{8.114}$$

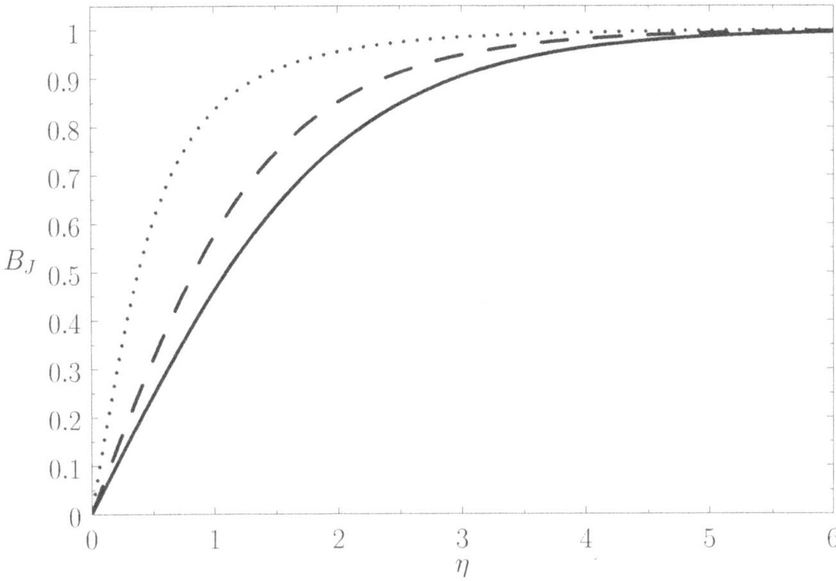

Figure 8.1 The solid, dashed, and dotted curves show the Brillouin function, $\mathcal{B}_J(\eta)$, for $J = 1/2$, $J = 1$, and $J = 7/2$, respectively.

For $z \gg 1$, we have

$$\coth z \simeq 1. \tag{8.115}$$

On the other hand, for $z \ll 1$,

$$\coth z = \frac{1}{z} + \frac{1}{3}z + O\left(z^3\right). \tag{8.116}$$

Thus, it follows that

$$\mathcal{B}_J(\eta) = \frac{1}{J}\left[\left(J + \frac{1}{2}\right) - \frac{1}{2}\right] = 1 \tag{8.117}$$

when $\eta \gg 1$. On the other hand, when $\eta \ll 1$, we get

$$\begin{aligned}\mathcal{B}_J(\eta) &= \frac{1}{J}\left\{\left(J + \frac{1}{2}\right)\left[\frac{1}{(J + 1/2)\eta} + \frac{1}{3}\left(J + \frac{1}{2}\right)\eta\right] - \frac{1}{2}\left(\frac{2}{\eta} + \frac{\eta}{6}\right)\right\} \\ &= \frac{1}{J}\left[\frac{1}{3}\left(J + \frac{1}{2}\right)^2 \eta - \frac{1}{12}\eta\right] = \frac{\eta}{3J}\left(J^2 + J + \frac{1}{4} - \frac{1}{4}\right),\end{aligned} \tag{8.118}$$

which reduces to

$$\mathcal{B}_J(\eta) = \left(\frac{J + 1}{3}\right)\eta. \tag{8.119}$$

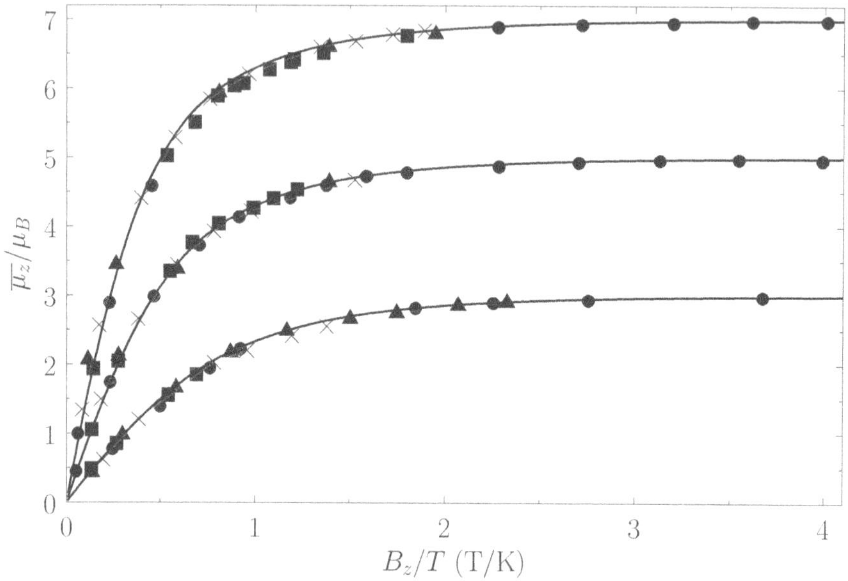

Figure 8.2 The magnetization versus B_z/T curves for (lower curve) chromium potassium alum ($J = 3/2$, $g = 2$), (middle curve) iron ammonium alum ($J = 5/2$, $g = 2$), and (top curve) gadolinium sulphate ($J = 7/2$, $g = 2$). The solid lines are the theoretical predictions, whereas the data points are experimental measurements. The circles, triangles, crosses, and squares correspond to $T = 1.30$K, 2.00K, 3.00K, and 4.21K, respectively. Data from [Henry (1952)].

Figure 8.1 shows how $\mathcal{B}_J(\eta)$ depends on η for various values of J.

If there are N_0 atoms per unit volume then the mean magnetic moment per unit volume (or magnetization) becomes

$$\overline{M_z} = N_0\,\overline{\mu_z} = N_0\,g\,\mu_B\,J\,\mathcal{B}_J(\eta), \tag{8.120}$$

where use has been made of Equation (8.112). If $\eta \ll 1$ then Equation (8.119) implies that $\overline{M_z} \propto \eta \propto B_z/T$. This relation can be written

$$\overline{M_z} \simeq \chi\,\frac{B_z}{\mu_0}, \tag{8.121}$$

where the dimensionless constant of proportionality,

$$\chi = \frac{N_0\,\mu_0\,\mu_B^2\,g^2\,J\,(J+1)}{3\,k\,T}, \tag{8.122}$$

is known as the magnetic susceptibility. Thus, $\chi \propto T^{-1}$—a result that is called Curie's law. (See Section 8.3.) If $\eta \gg 1$ then

$$\overline{M_z} \to N_0\,g\,\mu_B\,J. \tag{8.123}$$

This corresponds to magnetic saturation—a situation in which each atom has the maximum component of the magnetic moment parallel to the magnetic field, $g\,\mu_B\,J$, that it can possibly have.

Figure 8.2 compares the experimental and theoretical magnetization versus field-strength curves for three different paramagnetic substances containing atoms of total angular momentum 3/2, 5/2, and 7/2 particles, showing excellent agreement between theory and experiment. Note that, in all cases, the magnetization is proportional to the magnetic field-strength at small field-strengths, but saturates at some constant value as the field-strength increases.

The previous analysis completely neglects any interaction between the spins of neighboring atoms or molecules. It turns out that this is a fairly good approximation for paramagnetic substances. However, for *ferromagnetic* substances, in which the spins of neighboring atoms interact very strongly, this approximation breaks down completely. (See Section 8.17.) Thus, the previous analysis does not apply to ferromagnetic substances.

8.10 Equipartition Theorem

The internal energy of a monatomic ideal gas containing N particles is $(3/2)\,N\,k\,T$. This means that each particle possess, on average, $(3/2)\,k\,T$ units of energy. Monatomic particles have only three translational degrees of freedom, corresponding to their motion in three dimensions. They possess no internal rotational or vibrational degrees of freedom. Thus, the mean energy per degree of freedom in a monatomic ideal gas is $(1/2)\,k\,T$. In fact, this is a special case of a rather general result. Let us now try to prove this.

Suppose that the energy of a system is determined by f generalized coordinates, q_k, and f corresponding generalized momenta, p_k, so that

$$E = E(q_1, \cdots, q_f, p_1, \cdots, p_f). \tag{8.124}$$

Suppose further that:

(1) The total energy splits additively into the form

$$E = \epsilon_i(p_i) + E'(q_1, \cdots, p_f), \tag{8.125}$$

where ϵ_i involves only one variable, p_i, and the remaining part, E', does not depend on p_i.

(2) The function ϵ_i is quadratic in p_i, so that

$$\epsilon_i(p_i) = b\,p_i^2, \tag{8.126}$$

where b is a constant.

The most common situation in which the previous assumptions are valid is where p_i is a momentum. This is because the kinetic energy is usually a quadratic function of each momentum component, whereas the potential energy does not involve the momenta at all. However, if a coordinate, q_i, were to satisfy assumptions 1 and 2 then the theorem that we are about to establish would hold just as well.

What is the mean value of ϵ_i in thermal equilibrium if conditions 1 and 2 are satisfied? If the system is in equilibrium at absolute temperature $T \equiv (k\beta)^{-1}$ then it is distributed according to the canonical probability distribution. In the classical approximation, the mean value of ϵ_i is expressed in terms of integrals over all phase-space:

$$\overline{\epsilon_i} = \frac{\int_{-\infty}^{\infty} \exp[-\beta\, E(q_1,\cdots,p_f)]\,\epsilon_i\,dq_1\cdots dp_f}{\int_{-\infty}^{\infty} \exp[-\beta\, E(q_1,\cdots,p_f)]\,dq_1\cdots dp_f}. \tag{8.127}$$

Condition 1 gives

$$\begin{aligned}\overline{\epsilon_i} &= \frac{\int_{-\infty}^{\infty} \exp[-\beta\,(\epsilon_i + E')]\,\epsilon_i\,dq_1\cdots dp_f}{\int_{-\infty}^{\infty} \exp[-\beta\,(\epsilon_i + E')]\,dq_1\cdots dp_f} \\ &= \frac{\int_{-\infty}^{\infty} \exp(-\beta\,\epsilon_i)\,\epsilon_i\,dp_i\,\int_{-\infty}^{\infty} \exp(-\beta\,E')\,dq_1\cdots dp_f}{\int_{-\infty}^{\infty} \exp(-\beta\,\epsilon_i)\,dp_i\,\int_{-\infty}^{\infty} \exp(-\beta\,E')\,dq_1\cdots dp_f},\end{aligned} \tag{8.128}$$

where use has been made of the multiplicative property of the exponential function, and where the final integrals in both the numerator and denominator extend over all variables, q_k and p_k, except for p_i. These integrals are equal and, thus, cancel. Hence,

$$\overline{\epsilon_i} = \frac{\int_{-\infty}^{\infty} \exp(-\beta\,\epsilon_i)\,\epsilon_i\,dp_i}{\int_{-\infty}^{\infty} \exp(-\beta\,\epsilon_i)\,dp_i}. \tag{8.129}$$

This expression can be simplified further because

$$\int_{-\infty}^{\infty} \exp(-\beta\,\epsilon_i)\,\epsilon_i\,dp_i \equiv -\frac{\partial}{\partial\beta}\left[\int_{-\infty}^{\infty} \exp(-\beta\,\epsilon_i)\,dp_i\right], \tag{8.130}$$

so

$$\overline{\epsilon_i} = -\frac{\partial}{\partial\beta}\ln\left[\int_{-\infty}^{\infty} \exp(-\beta\,\epsilon_i)\,dp_i\right]. \tag{8.131}$$

According to condition 2,

$$\int_{-\infty}^{\infty} \exp(-\beta\,\epsilon_i)\,dp_i = \int_{-\infty}^{\infty} \exp\left(-\beta\,b\,p_i^2\right)dp_i = \frac{1}{\sqrt{\beta}}\int_{-\infty}^{\infty} \exp\left(-b\,y^2\right)dy, \tag{8.132}$$

where $y = \sqrt{\beta}\,p_i$. Thus,

$$\ln\int_{-\infty}^{\infty} \exp(-\beta\,\epsilon_i)\,dp_i = -\frac{1}{2}\ln\beta + \ln\int_{-\infty}^{\infty} \exp\left(-b\,y^2\right)dy. \tag{8.133}$$

Note that the integral on the right-hand side is independent of β. It follows from Equation (8.131) that

$$\overline{\epsilon}_i = -\frac{\partial}{\partial\beta}\left(-\frac{1}{2}\ln\beta\right) = \frac{1}{2\beta}, \tag{8.134}$$

giving

$$\overline{\epsilon}_i = \frac{1}{2}\,k\,T. \tag{8.135}$$

This is the famous *equipartition theorem* of classical physics. It states that the mean value of every independent quadratic term in the energy is equal to $(1/2)\,k\,T$. If all terms in the energy are quadratic then the mean energy is spread equally over all degrees of freedom. (Hence, the name "equipartition.")

8.11 Harmonic Oscillators

Our proof of the equipartition theorem depends crucially on the classical approximation. To see how quantum effects modify this result, let us examine a particularly simple system that we know how to analyze using both classical and quantum physics: namely, a simple harmonic oscillator. Consider a one-dimensional harmonic oscillator in equilibrium with a heat reservoir held at absolute temperature T. The energy of the oscillator is given by

$$E = \frac{p^2}{2\,m} + \frac{1}{2}\,\kappa\,x^2, \tag{8.136}$$

where the first term on the right-hand side is the kinetic energy, involving the momentum, p, and the mass, m, and the second term is the potential energy, involving the displacement, x, and the force constant, κ. Each of these terms is quadratic in the respective variable. So, in the classical approximation, the equipartition theorem yields:

$$\frac{\overline{p^2}}{2\,m} = \frac{1}{2}\,k\,T, \tag{8.137}$$

$$\frac{1}{2}\,\kappa\,\overline{x^2} = \frac{1}{2}\,k\,T. \tag{8.138}$$

That is, the mean kinetic energy of the oscillator is equal to the mean potential energy, which equals $(1/2)\,k\,T$. It follows that the mean total energy is

$$\overline{E} = \frac{1}{2}\,k\,T + \frac{1}{2}\,k\,T = k\,T. \tag{8.139}$$

According to quantum mechanics, the energy levels of a harmonic oscillator are equally spaced, and satisfy

$$E_n = \left(n + \frac{1}{2}\right)\hbar\,\omega, \tag{8.140}$$

where n is a non-negative integer, and

$$\omega = \sqrt{\frac{\kappa}{m}}. \tag{8.141}$$

(See Section C.11.) The partition function for such an oscillator is given by

$$Z = \sum_{n=0,\infty} \exp(-\beta\,E_n) = \exp\left(-\frac{1}{2}\,\beta\,\hbar\,\omega\right) \sum_{n=0,\infty} \exp(-n\,\beta\,\hbar\,\omega). \tag{8.142}$$

Now,

$$\sum_{n=0,\infty} \exp(-n\,\beta\,\hbar\,\omega) = 1 + \exp(-\beta\,\hbar\,\omega) + [\exp(-\beta\,\hbar\,\omega)]^2 + \cdots \tag{8.143}$$

is simply the sum of an infinite geometric series, and can be evaluated immediately to give

$$\sum_{n=0,\infty} \exp(-n\,\beta\,\hbar\,\omega) = \frac{1}{1 - \exp(-\beta\,\hbar\,\omega)}. \tag{8.144}$$

(See Exercise 8.1.) Thus, the partition function takes the form

$$Z = \frac{\exp[-(1/2)\,\beta\,\hbar\,\omega]}{1 - \exp(-\beta\,\hbar\,\omega)}, \tag{8.145}$$

and

$$\ln Z = -\frac{1}{2}\,\beta\,\hbar\,\omega - \ln\left[1 - \exp(-\beta\,\hbar\,\omega)\right]. \tag{8.146}$$

The mean energy of the oscillator is given by [see Equation (8.35)]

$$\overline{E} = -\frac{\partial}{\partial\beta}\ln Z = -\left[-\frac{1}{2}\,\hbar\,\omega - \frac{\exp(-\beta\,\hbar\,\omega)\,\hbar\,\omega}{1 - \exp(-\beta\,\hbar\,\omega)}\right], \tag{8.147}$$

or

$$\overline{E} = \hbar\,\omega\left[\frac{1}{2} + \frac{1}{\exp(\beta\,\hbar\,\omega) - 1}\right]. \tag{8.148}$$

Consider the limit

$$\beta\,\hbar\,\omega = \frac{\hbar\,\omega}{k\,T} \ll 1, \tag{8.149}$$

in which the thermal energy, $k\,T$, is large compared to the separation, $\hbar\,\omega$, between successive energy levels. In this limit,

$$\exp(\beta\,\hbar\,\omega) \simeq 1 + \beta\,\hbar\,\omega, \tag{8.150}$$

so

$$\overline{E} \simeq \hbar\,\omega\left[\frac{1}{2} + \frac{1}{\beta\,\hbar\,\omega}\right] \simeq \hbar\,\omega\left[\frac{1}{\beta\,\hbar\,\omega}\right], \tag{8.151}$$

giving

$$\overline{E} \simeq \frac{1}{\beta} = k\,T. \tag{8.152}$$

Thus, the classical result, (8.139), holds whenever the thermal energy greatly exceeds the typical spacing between quantum energy levels.

Consider the limit

$$\beta\,\hbar\,\omega = \frac{\hbar\,\omega}{k\,T} \gg 1, \tag{8.153}$$

in which the thermal energy is small compared to the separation between the energy levels. In this limit,

$$\exp(\beta\,\hbar\,\omega) \gg 1, \tag{8.154}$$

and so

$$\overline{E} \simeq \hbar\,\omega\left[\frac{1}{2} + \exp(-\beta\,\hbar\,\omega)\right] \simeq \frac{1}{2}\,\hbar\,\omega. \tag{8.155}$$

Thus, if the thermal energy is much less than the spacing between quantum states then the mean energy approaches that of the ground-state (the so-called *zero-point energy*). Clearly, the equipartition theorem is only valid in the former limit, where $k\,T \gg \hbar\,\omega$, and the oscillator possess sufficient thermal energy to explore many of its possible quantum states.

8.12 Specific Heats

We have discussed the internal energies and entropies of substances (mostly ideal gases) at some length. Unfortunately, these quantities cannot be directly measured. Instead, they must be inferred from other information. The thermodynamic property of substances that is the easiest to measure is, of course, the heat capacity, or specific heat. In fact, once the variation of the specific heat with temperature is known, both the internal energy and entropy can be easily reconstructed via

$$E(T, V) = \nu \int_0^T c_V(T, V)\,dT + E(0, V), \tag{8.156}$$

$$S(T, V) = \nu \int_0^T \frac{c_V(T, V)}{T}\,dT. \tag{8.157}$$

Here, use has been made of $dS = đQ/T$, and the third law of thermodynamics. Clearly, the optimum way of verifying the results of statistical thermodynamics is to compare the theoretically predicted heat capacities with the experimentally measured values.

Classical physics, in the guise of the equipartition theorem, says that each independent degree of freedom associated with a quadratic term in the energy possesses an average energy $(1/2)\,k\,T$ in thermal equilibrium at temperature T. Consider a substance made up of N molecules. Every molecular degree of freedom contributes $(1/2)\,N\,k\,T$, or $(1/2)\,\nu\,R\,T$, to the mean energy of the substance (with the tacit proviso that each degree of freedom is associated with a quadratic term in the energy). Thus, the contribution to the molar heat capacity at constant volume (we wish to avoid the complications associated with any external work done on the substance) is

$$\frac{1}{\nu}\left(\frac{\partial\overline{E}}{\partial T}\right)_V = \frac{1}{\nu}\frac{\partial[(1/2)\,\nu\,R\,T]}{\partial T} = \frac{1}{2}\,R, \tag{8.158}$$

per molecular degree of freedom. The total classical heat capacity is therefore

$$c_V = \frac{g}{2}\,R, \tag{8.159}$$

where g is the number of molecular degrees of freedom. Because large complicated molecules clearly have very many more degrees of freedom than small simple molecules, the previous formula predicts that the molar heat capacities of substances made up of the former type of molecules should greatly exceed those of substances made up of the latter. In fact, the experimental heat capacities of substances containing complicated molecules are generally greater than those of substances containing simple molecules, but by nowhere near the large factor predicted in Equation (8.159). This equation also implies that heat capacities are temperature independent. In fact, this is not the case for most substances. Experimental heat capacities generally increase with increasing temperature. These two experimental facts pose severe problems for classical physics. Incidentally, these problems were fully appreciated as far back as 1850. Stories that physicists at the end of the nineteenth century thought that classical physics explained absolutely everything are largely apocryphal.

The equipartition theorem (and the whole classical approximation) is only valid when the typical thermal energy, $k\,T$, greatly exceeds the spacing between quantum energy levels. Suppose that the temperature is sufficiently low that this condition is not satisfied for one particular molecular degree of freedom. In fact, suppose that $k\,T$ is much less than the spacing between the energy levels. According to Section 8.11, in this situation, the degree of freedom only contributes the ground-state energy, E_0 (say) to the mean energy of the molecule. Now, the ground-state energy can be a quite complicated function of the internal properties of the molecule, but is certainly not a function of the temperature, because this is a collective property of all molecules. It follows that the contribution to the molar

Table 8.1 Effective "temperatures" of various different types of electromagnetic radiation.

Radiation type	Frequency (hz)	$T_{\rm rad}$(K)
Radio	$< 10^9$	< 0.05
Microwave	$10^9 - 10^{11}$	$0.05 - 5$
Infrared	$10^{11} - 10^{14}$	$5 - 5000$
Visible	5×10^{14}	2×10^4
Ultraviolet	$10^{15} - 10^{17}$	$5\times 10^4 - 5\times 10^6$
X-ray	$10^{17} - 10^{20}$	$5\times 10^6 - 5\times 10^9$
γ-ray	$> 10^{20}$	$> 5\times 10^9$

heat capacity is

$$\frac{1}{\nu}\left(\frac{\partial[N\,E_0]}{\partial T}\right)_V = 0. \tag{8.160}$$

Thus, if $k\,T$ is much less than the spacing between the energy levels then the degree of freedom contributes nothing at all to the molar heat capacity. We say that this particular degree of freedom is "frozen out." Clearly, at very low temperatures, just about all degrees of freedom are frozen out. As the temperature is gradually increased, degrees of freedom successively kick in, and eventually contribute their full $(1/2)\,R$ to the molar heat capacity, as $k\,T$ approaches, and then greatly exceeds, the spacing between their quantum energy levels. We can use these simple ideas to explain the behaviors of most experimental heat capacities.

To make further progress, we need to estimate the typical spacing between the quantum energy levels associated with various degrees of freedom. We can do this by observing the frequency of the electromagnetic radiation emitted and absorbed during transitions between these energy levels. If the typical spacing between energy levels is ΔE then transitions between the various levels are associated with photons of frequency ν, where $h\,\nu = \Delta E$. (Here, h is Planck's constant.) We can define an *effective temperature* of the radiation via $h\,\nu = k\,T_{\rm rad}$. If $T \gg T_{\rm rad}$ then $k\,T \gg \Delta E$, and the degree of freedom makes its full contribution to the heat capacity. On the other hand, if $T \ll T_{\rm rad}$ then $k\,T \ll \Delta E$, and the degree of freedom is frozen out. Table 8.1 lists the "temperatures" of various different types of radiation. It is clear that degrees of freedom that give rise to emission or absorption of radio or microwave radiation contribute their full $(1/2)\,R$ to the molar heat capacity at room temperature. On the other hand, degrees of freedom that give rise to emission or absorption in the visible, ultraviolet, X-ray, or γ-ray regions of the electromagnetic spectrum are frozen out at room temperature. Degrees of freedom that emit or absorb infrared radiation are on the border line.

8.13 Specific Heats of Gases

Let us now investigate the specific heats of gases. Consider, first of all, translational degrees of freedom. Every molecule in a gas is free to move in three dimensions. If one particular molecule has mass m and momentum $\mathbf{p} = m\,\mathbf{v}$ then its kinetic energy of translation is

$$K = \frac{1}{2\,m}\left(p_x^{\,2} + p_y^{\,2} + p_z^{\,2}\right). \tag{8.161}$$

The kinetic energy of other molecules does not involve the momentum, $\mathbf{p}$, of this particular molecule. Moreover, the potential energy of interaction between molecules depends only on their position coordinates, and is, thus, independent of $\mathbf{p}$. Any internal rotational, vibrational, electronic, or nuclear degrees of freedom of the molecule also do not involve $\mathbf{p}$. Hence, the essential conditions of the equipartition theorem are satisfied. (At least, in the classical approximation.) Because Equation (8.161) contains three independent quadratic terms, there are clearly three degrees of freedom associated with translation (one for each dimension of space), so the translational contribution to the molar heat capacity of gases is

$$(c_V)_{\rm translation} = \frac{3}{2}\,R. \tag{8.162}$$

Suppose that our gas is contained in a cubic enclosure of dimensions L. According to Schrödinger's equation, the quantized translational energy levels of an individual molecule are given by

$$E = \frac{\hbar^2\,\pi^2}{2\,m\,L^2}\left(n_1^{\,2} + n_2^{\,2} + n_3^{\,2}\right), \tag{8.163}$$

where n_1, n_2, and n_3 are positive integer quantum numbers. (See Appendix C.) Clearly, the spacing between the energy levels can be made arbitrarily small by increasing the size of the enclosure. This implies that translational degrees of freedom can be treated classically, so that Equation (8.162) is always valid. (Except very close to absolute zero.) We conclude that all gases possess a minimum molar heat capacity of $(3/2)\,R$ due to the translational degrees of freedom of their constituent molecules.

The electronic degrees of freedom of gas molecules (i.e., the possible configurations of electrons orbiting the atomic nuclei) typically give rise to absorption and emission in the ultraviolet or visible regions of the spectrum. It follows from Table 8.1 that electronic degrees of freedom are frozen out at room temperature. Similarly, nuclear degrees of freedom (i.e., the possible configurations of protons and neutrons in the atomic nuclei) are frozen out because they are associated with absorption and emission in the X-ray and γ-ray regions of the electromagnetic spectrum. In fact, the only additional degrees of freedom that we need worry about for gases are rotational and vibrational degrees of freedom. These typically give rise to absorption lines in the infrared region of the spectrum.

The rotational kinetic energy of a molecule tumbling in space can be written

$$K = \frac{1}{2}\frac{L_x^2}{I_x} + \frac{1}{2}\frac{L_y^2}{I_y} + \frac{1}{2}\frac{L_z^2}{I_z}, \tag{8.164}$$

where the x-, y-, and z-axes are the so called *principal axes of rotation* of the molecule (these are mutually perpendicular), L_x, L_y, and L_z are the angular momenta about these axes, and I_x, I_y, and I_z are the moments of inertia about these axes. No other degrees of freedom depend on the angular momenta. Because the kinetic energy of rotation is the sum of three quadratic terms, the rotational contribution to the molar heat capacity of gases is

$$(c_V)_{\rm rotation} = \frac{3}{2}\,R, \tag{8.165}$$

according to the equipartition theorem. Note that the typical magnitude of a molecular moment of inertia is $m\,d^2$, where m is the molecular mass, and d is the typical interatomic spacing in the molecule. A special case arises if the molecule is linear (e.g., if the molecule is diatomic). In this case, one of the principal axes lies along the line of centers of the atoms. The moment of inertia about this axis is of order $m\,a^2$, where a is a typical nuclear dimension (remember that nearly all of the mass of an atom resides in the nucleus). Because $a \sim 10^{-5}\,d$, it follows that the moment of inertia about the line of centers is minuscule compared to the moments of inertia about the other two principle axes. In quantum mechanics, angular momentum is quantized in units of $\hbar$. The energy levels of a rigid rotator spinning about a principal axis are written

$$E = \frac{\hbar^2}{2\,I}\,J\,(J+1), \tag{8.166}$$

where I is the moment of inertia, and J is a non-negative integer. (See Section C.12.) Note the inverse dependence of the spacing between energy levels on the moment of inertia. It is clear that, for the case of a linear molecule, the rotational degree of freedom associated with spinning along the line of centers of the atoms is frozen out at room temperature, given the very small moment of inertia along this axis, and, hence, the very widely spaced rotational energy levels.

Classically, the vibrational degrees of freedom of a molecule are studied by standard normal mode analysis of the molecular structure. Each normal mode behaves like an independent harmonic oscillator, and, therefore, contributes R to the molar specific heat of the gas [$(1/2)\,R$ from the kinetic energy of vibration, and

$(1/2)\,R$ from the potential energy of vibration]. A molecule containing n atoms has $n-1$ normal modes of vibration. For instance, a diatomic molecule has just one normal mode (corresponding to periodic stretching of the bond between the two atoms). Thus, the classical contribution to the specific heat from vibrational degrees of freedom is

$$(c_V)_{\rm vibration} = (n-1)\,R. \tag{8.167}$$

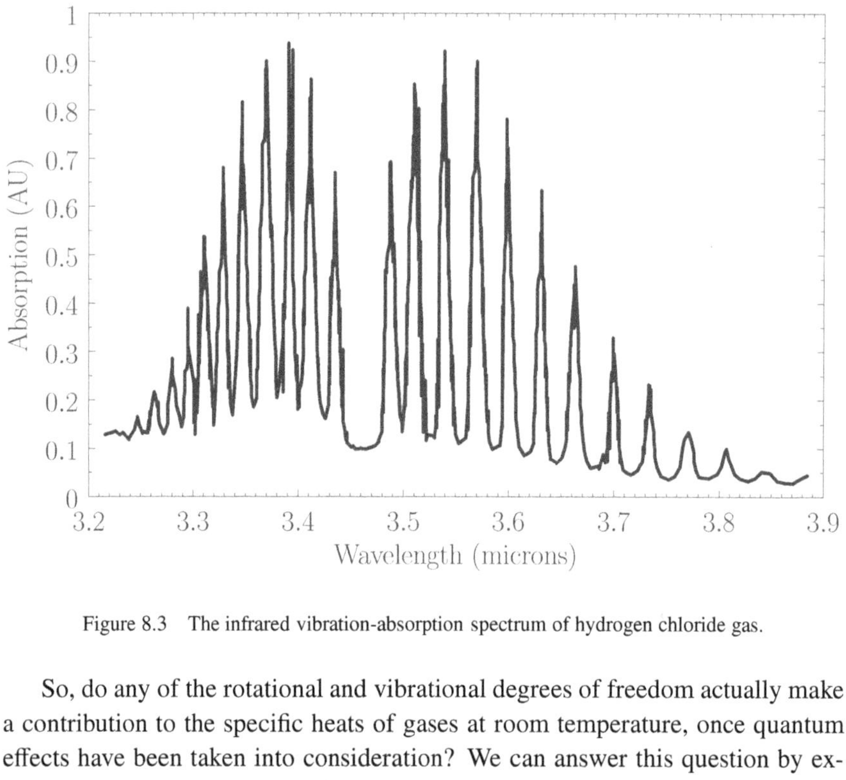

Figure 8.3 The infrared vibration-absorption spectrum of hydrogen chloride gas.

So, do any of the rotational and vibrational degrees of freedom actually make a contribution to the specific heats of gases at room temperature, once quantum effects have been taken into consideration? We can answer this question by examining just one piece of data. Figure 8.3 shows the infrared absorption spectrum of hydrogen chloride gas. The absorption lines correspond to simultaneous transitions between different vibrational and rotational energy levels. Hence, this is usually called a *vibration-rotation spectrum.* The missing line at about 3.47 microns corresponds to a pure vibrational transition from the ground-state to the first excited state. (Pure vibrational transitions are forbidden—hydrogen chloride molecules always have to simultaneously change their rotational energy level if they are to couple effectively to electromagnetic radiation.) The longer wavelength absorption lines correspond to vibrational transitions in which there is a simultaneous decrease in the rotational energy level. Likewise, the shorter wavelength absorption lines correspond to vibrational transitions in which there is a simultaneous increase in the rotational energy level. It is clear that the rotational energy levels are more closely spaced than the vibrational energy levels. The pure vibrational transition gives rise to absorption at about 3.47 microns, which corresponds to infrared radiation of frequency 8.5×10^{13} hertz with an associated radiation "temperature" of 4,100 K. We conclude that the vibrational degrees of freedom of hydrogen chloride, or any other small molecule, are frozen out at room temperature. The rotational transitions split the vibrational lines by about 0.2 microns. This implies that pure rotational transitions would be associated with infrared radiation of frequency 5×10^{12} hertz and corresponding radiation "temperature" 240 K. We conclude that the rotational degrees of freedom of hydrogen chloride, or any other small molecule, are not frozen out at room temperature, and probably contribute the classical $(1/2)\,R$ to the molar specific heat. There is one proviso, however. Linear molecules (like hydrogen chloride) effectively only have two rotational degrees of freedom (instead of the usual three), because of the very small moment of inertia of such molecules along the line of centers of the atoms.

We are now in a position to make some predictions regarding the specific heats of various gases. Monatomic molecules only possess three translational degrees of freedom, so monatomic gases should have a molar heat capacity $(3/2)\,R = 12.47$ joules/degree/mole. Moreover, the ratio of specific heats $\gamma = c_p/c_V = (c_V + R)/c_V$ should be $5/3 = 1.667$. It can be seen from Table 6.1 that both of these predictions are borne out pretty well for helium and argon. Diatomic molecules possess three translational degrees of freedom, and two rotational degrees of freedom. (All other degrees of freedom are frozen out at room temperature.) Thus, diatomic gases should have a molar heat capacity $(5/2)\,R = 20.8$ joules/degree/mole. Moreover, the ratio of specific heats should be $7/5 = 1.4$. It can be seen from Table 6.1 that these are reasonably accurate predictions for nitrogen and oxygen. The freezing out of vibrational degrees of freedom becomes gradually less effective as molecules become heavier and more complex. This is partly because such molecules are generally less stable, so the force constant, κ, is reduced, and partly because the molecular mass is increased. Both these effects reduce the frequency of vibration of the molecular normal modes [see Equation (8.141)], and, hence, the spacing between vibrational energy levels [see Equation (8.140)]. This accounts for the obviously non-classical [i.e., not a multiple of $(1/2)\,R$] specific heats of carbon dioxide and ethane in Table 6.1. In both molecules, vibrational degrees of freedom contribute to the molar specific heat. (But not the full R, because the temperature is not sufficiently high.)

Figure 8.4 shows the variation of the molar heat capacity at constant volume

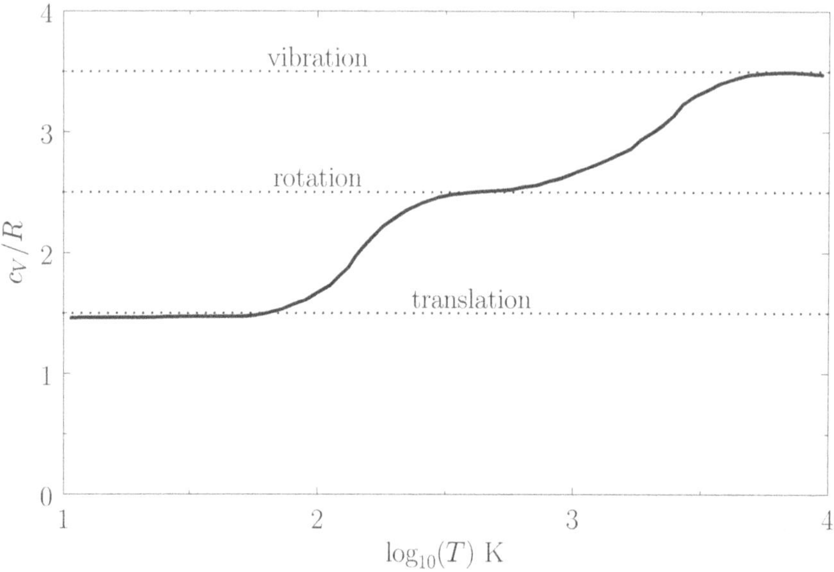

Figure 8.4 The molar heat capacity at constant volume of gaseous molecular hydrogen versus temperature.

of gaseous molecular hydrogen (i.e., H_2) with temperature. The expected contribution from the translational degrees of freedom is $(3/2)\,R$ (there are three translational degrees of freedom per molecule). The expected contribution at high temperatures from the rotational degrees of freedom is R (there are effectively two rotational degrees of freedom per molecule). Finally, the expected contribution at high temperatures from the vibrational degrees of freedom is R (there is one vibrational degree of freedom per molecule). It can be seen that, as the temperature rises, the rotational, and then the vibrational, degrees of freedom eventually make their full classical contributions to the heat capacity.

8.14 Specific Heats of Solids

Consider a simple solid containing N atoms. Now, atoms in solids cannot translate (unlike those in gases), but are free to vibrate about their equilibrium positions. Such vibrations are termed *lattice vibrations*, and can be thought of as sound waves propagating through the crystal lattice. Each atom is specified by three independent position coordinates, and three conjugate momentum coordinates. Let us only consider small-amplitude vibrations. In this case, we can expand the potential energy of interaction between the atoms to give an expression that is quadratic in the atomic displacements from their equilibrium positions. It is always possible to perform a *normal mode analysis* of the oscillations. In effect, we can find $3N$ independent modes of oscillation of the solid. Each mode has its own particular oscillation frequency, and its own particular pattern of atomic displacements. Any general oscillation can be written as a linear combination of these *normal modes*. Let q_i be the (appropriately normalized) amplitude of the ith normal mode, and p_i the momentum conjugate to this coordinate. In *normal-mode coordinates*, the total energy of the lattice vibrations takes the particularly simple form

$$E = \frac{1}{2} \sum_{i=1,3N} \left(p_i^2 + \omega_i^2 q_i^2\right), \tag{8.168}$$

where ω_i is the (angular) oscillation frequency of the ith normal mode. It is clear that, when expressed in normal-mode coordinates, the linearized lattice vibrations are equivalent to $3N$ independent harmonic oscillators. (Of course, each oscillator corresponds to a different normal mode.)

The typical value of ω_i is the (angular) frequency of a sound wave propagating through the lattice. Sound wave frequencies are far lower than the typical vibration frequencies of gaseous molecules. In the latter case, the mass involved in the vibration is simply that of the molecule, whereas in the former case the mass involved is that of very many atoms (because lattice vibrations are non-localized). The strength of interatomic bonds in gaseous molecules is similar to those in solids, so we can use the estimate $\omega \simeq \sqrt{\kappa/m}$ (κ is the force constant that measures the strength of interatomic bonds, and m is the mass involved in the oscillation) as proof that the typical frequencies of lattice vibrations are very much less than the vibration frequencies of simple molecules. It follows, from $\Delta E = \hbar\,\omega$, that the quantum energy levels of lattice vibrations are far more closely spaced than the vibrational energy levels of gaseous molecules. Thus, it is likely (and is, indeed, the case) that lattice vibrations are not frozen out at room temperature, but, instead, make their full classical contribution to the molar specific heat of the solid.

If the lattice vibrations behave classically then, according to the equipartition theorem, each normal mode of oscillation has an associated mean energy $k\,T$, in equilibrium at temperature T [$(1/2)\,k\,T$ resides in the kinetic energy of the oscillation, and $(1/2)\,k\,T$ resides in the potential energy]. Thus, the mean internal energy per mole of the solid is

$$\overline{E} = 3\,N\,k\,T = 3\,\nu\,R\,T. \tag{8.169}$$

It follows that the molar heat capacity at constant volume is

$$c_V = \frac{1}{\nu}\left(\frac{\partial\overline{E}}{\partial T}\right)_V = 3\,R \tag{8.170}$$

Table 8.2 Values of c_p (joules/mole/degree) for some solids at $T = 298°$ K. From [Reif (1965)].

Solid	c_p	Solid	c_p
Copper	24.5	Aluminium	24.4
Silver	25.5	Tin (white)	26.4
Lead	26.4	Sulphur (rhombic)	22.4
Zinc	25.4	Carbon (diamond)	6.1

for solids. This gives a value of 24.9 joules/mole/degree. In fact, at room temperature, most solids (in particular, metals) have heat capacities that lie remarkably close to this value. This fact was discovered experimentally by Dulong and Petite at the beginning of the nineteenth century, and was used to make some of the first crude estimates of the molecular weights of solids. (If we know the molar heat capacity of a substance then we can easily work out how much of it corresponds to one mole, and by weighing this amount, and then dividing the result by Avogadro's number, we can then obtain an estimate of the molecular weight.) Table 8.2 lists the experimental molar heat capacities, c_p, at constant pressure for various solids. The heat capacity at constant volume is somewhat less than the constant pressure value, but not by much, because solids are fairly incompressible. It can be seen that *Dulong and Petite's law* (i.e., that all solids have a molar heat capacities close to 24.9 joules/mole/degree) holds pretty well for metals. However, the law fails badly for diamond. This is not surprising. As is well known, diamond is an extremely hard substance, so its interatomic bonds must be very strong, suggesting that the force constant, κ, is large. Diamond is also a fairly low-density substance, so the mass, m, involved in lattice vibrations is comparatively small. Both these facts suggest that the typical lattice vibration frequency of diamond ($\omega \simeq \sqrt{\kappa/m}$) is high. In fact, the spacing between the different vibrational energy levels (which scales like $\hbar\,\omega$) is sufficiently large in diamond for the vibrational degrees of freedom to be largely frozen out at room temperature. This accounts for the anomalously low heat capacity of diamond in Table 8.2.

Dulong and Petite's law is essentially a high-temperature limit. The molar heat capacity cannot remain a constant as the temperature approaches absolute zero, because, by Equation (8.157), this would imply $S \to \infty$, which violates the third law of thermodynamics. We can make a crude model of the behavior of c_V at low temperatures by assuming that all of the normal modes oscillate at the same frequency, ω (say). This approximation was first employed by Albert Einstein in a paper published in 1907. According to Equation (8.168), the solid acts like a set of $3N$ independent oscillators which, making use of Einstein's approximation, all vibrate at the same frequency. We can use the quantum mechanical result (8.148) for a single oscillator to write the mean energy of the solid in the form

$$\overline{E} = 3\,N\,\hbar\,\omega\left[\frac{1}{2} + \frac{1}{\exp(\beta\,\hbar\,\omega) - 1}\right]. \tag{8.171}$$

The molar heat capacity is defined

$$c_V = \frac{1}{\nu}\left(\frac{\partial\overline{E}}{\partial T}\right)_V = \frac{1}{\nu}\left(\frac{\partial\overline{E}}{\partial\beta}\right)_V\frac{\partial\beta}{\partial T} = -\frac{1}{\nu\,k\,T^2}\left(\frac{\partial\overline{E}}{\partial\beta}\right)_V, \tag{8.172}$$

giving

$$c_V = -\frac{3\,N_A\,\hbar\,\omega}{k\,T^2}\left\{-\frac{\exp(\beta\,\hbar\,\omega)\,\hbar\,\omega}{[\exp(\beta\,\hbar\,\omega) - 1]^2}\right\}, \tag{8.173}$$

which reduces to

$$c_V = 3\,R\left(\frac{\theta_E}{T}\right)^2\frac{\exp(\theta_E/T)}{[\exp(\theta_E/T) - 1]^2}. \tag{8.174}$$

Here,

$$\theta_E = \frac{\hbar\,\omega}{k} \tag{8.175}$$

is termed the *Einstein temperature*. If the temperature is sufficiently high that $T \gg \theta_E$ then $k\,T \gg \hbar\,\omega$, and the previous expression reduces to $c_V = 3\,R$, after expansion of the exponential functions. Thus, the law of Dulong and Petite is recovered for temperatures significantly in excess of the Einstein temperature. On the other hand, if the temperature is sufficiently low that $T \ll \theta_E$ then the exponential factors appearing in Equation (8.174) become very much larger than unity, giving

$$c_V \simeq 3\,R\left(\frac{\theta_E}{T}\right)^2\exp\left(-\frac{\theta_E}{T}\right). \tag{8.176}$$

So, in this simple model, the specific heat approaches zero exponentially as $T \to 0$.

In reality, the specific heats of solids do not approach zero quite as quickly as suggested by Einstein's model when $T \to 0$. The experimentally observed low-temperature behavior is more like $c_V \propto T^3$. (See Figure 8.6.) The reason for this discrepancy is the crude approximation that all normal modes have the same frequency. In fact, long-wavelength (i.e., low-frequency) modes have lower frequencies than short-wavelength (i.e., high-freqeuncy) modes, so the former are much harder to freeze out than the latter (because the spacing between quantum energy levels, $\hbar\,\omega$, is smaller in the former case). The molar heat capacity does not decrease with temperature as rapidly as suggested by Einstein's model because these long-wavelength modes are able to make a significant contribution to the heat capacity, even at very low temperatures. A more realistic model of lattice vibrations was developed by the Dutch physicist Peter Debye in 1912. In the Debye model, the frequencies of the normal modes of vibration are estimated by treating the solid as an isotropic continuous medium. This approach is reasonable because the only modes that really matter at low temperatures are the long-wavelength modes: more explicitly, those whose wavelengths greatly exceed the interatomic spacing. It is plausible that these modes are not particularly sensitive to the discrete nature of the solid. In other words, the fact that the solid is made up of atoms, rather than being continuous.

Consider a sound wave propagating through an isotropic continuous medium. The disturbance varies with position vector $\mathbf{r}$ and time t like $\exp[-{\rm i}\,(\mathbf{k}\cdot\mathbf{r}-\omega\,t)]$, where the wavevector, $\mathbf{k}$, and the frequency of oscillation, ω, satisfy the dispersion relation for sound waves in an isotropic medium:

$$\omega = k\,c_s. \tag{8.177}$$

Here, c_s is the speed of sound in the medium. Suppose, for the sake of argument, that the medium is periodic in the x-, y-, and z-directions with periodicity lengths L_x, L_y, and L_z, respectively. In order to maintain periodicity, we need

$$k_x\,(x+L_x) = k_x\,x + 2\pi\,n_x, \tag{8.178}$$

where n_x is an integer. There are analogous constraints on k_y and k_z. It follows that, in a periodic medium, the components of the wavevector are quantized, and can only take the values

$$k_x = \frac{2\pi}{L_x}\,n_x, \tag{8.179}$$

$$k_y = \frac{2\pi}{L_y}\,n_y, \tag{8.180}$$

$$k_z = \frac{2\pi}{L_z}\,n_z, \tag{8.181}$$

where n_x, n_y, and n_z are all integers. It is assumed that L_x, L_y, and L_z are macroscopic lengths, so the allowed values of the components of the wavevector are very closely spaced. For given values of k_y and k_z, the number of allowed values of k_x that lie in the range k_x to $k_x + dk_x$ is given by

$$\Delta n_x = \frac{L_x}{2\pi}\,dk_x. \tag{8.182}$$

It follows that the number of allowed values of $\mathbf{k}$ (i.e., the number of allowed modes) when k_x lies in the range k_x to $k_x + dk_x$, k_y lies in the range k_y to $k_y + dk_y$, and k_z lies in the range k_z to $k_z + dk_z$, is

$$\rho\,d^3\mathbf{k} = \left(\frac{L_x}{2\pi}\,dk_x\right)\left(\frac{L_y}{2\pi}\,dk_y\right)\left(\frac{L_z}{2\pi}\,dk_z\right) = \frac{V}{(2\pi)^3}\,dk_x\,dk_y\,dk_z, \tag{8.183}$$

where $V = L_x L_y L_z$ is the periodicity volume, and $d^3\mathbf{k} \equiv dk_x\, dk_y\, dk_z$. The quantity ρ is called the density of modes. Note that this density is independent of $\mathbf{k}$, and proportional to the periodicity volume. Thus, the density of modes per unit volume is a constant, independent of the magnitude, or shape, of the periodicity volume. The density of modes per unit volume when the magnitude of $\mathbf{k}$ lies in the range k to $k + dk$ is given by multiplying the density of modes per unit volume by the "volume" in $\mathbf{k}$-space of the spherical shell lying between radii k and $k + dk$. Thus,

$$\rho_k\, dk = \frac{4\pi\, k^2\, dk}{(2\pi)^3} = \frac{k^2}{2\pi^2}\, dk. \tag{8.184}$$

Consider an isotropic continuous medium of volume V. According to the previous relation, the number of normal modes whose frequencies lie between ω and $\omega + d\omega$ (which is equivalent to the number of modes whose k values lie in the range ω/c_s to $\omega/c_s + d\omega/c_s$) is

$$\sigma_c(\omega)\, d\omega = 3\, V\, \rho_k(k)\, \frac{dk}{d\omega}\, d\omega = \frac{3\, V}{2\pi^2\, c_s^3}\, \omega^2\, d\omega. \tag{8.185}$$

The factor of 3 comes from the three possible polarizations of sound waves in solids. For every allowed wavenumber (or frequency), there are two independent torsional modes, where the displacement is perpendicular to the direction of propagation, and one longitudinal mode, where the displacement is parallel to the direction of propagation. Torsion waves are vaguely analogous to electromagnetic waves (these also have two independent polarizations). The longitudinal mode is very similar to the compressional sound wave in gases. Of course, torsion waves can not propagate in gases, because gases have no resistance to deformation without change of volume.

The Debye approach consists in approximating the actual density of normal modes, $\sigma(\omega)$, by the density in a continuous medium, $\sigma_c(\omega)$, not only at low frequencies (long wavelengths) where these should be nearly the same, but also at higher frequencies where they may differ substantially. Suppose that we are dealing with a solid consisting of N atoms. We know that there are only $3N$ independent normal modes. It follows that we must cut off the density of states above some critical frequency, ω_D (say), otherwise we will have too many modes. Thus, in the Debye approximation, the density of normal modes takes the form

$$\sigma_D(\omega) = \begin{cases} \sigma_c(\omega) & \omega < \omega_D \\ 0 & \omega > \omega_D \end{cases}. \tag{8.186}$$

Here, ω_D is termed the *Debye frequency*, and is chosen such that the total number of normal modes is $3N$:

$$\int_0^\infty \sigma_D(\omega)\, d\omega = \int_0^{\omega_D} \sigma_c(\omega)\, d\omega = 3N. \tag{8.187}$$

Substituting Equation (8.185) into the previous formula yields

$$\frac{3V}{2\pi^2 c_s^3}\int_0^{\omega_D}\omega^2\,d\omega = \frac{V}{2\pi^2 c_s^3}\,\omega_D^3 = 3N. \tag{8.188}$$

This implies that

$$\omega_D = c_s\left(6\pi^2\,\frac{N}{V}\right)^{1/3}. \tag{8.189}$$

Thus, the Debye frequency depends only on the sound velocity in the solid, and the number of atoms per unit volume. The wavelength corresponding to the Debye frequency is $2\pi\,c_s/\omega_D$, which is clearly on the order of the interatomic spacing, $a \simeq (V/N)^{1/3}$. It follows that the cut-off of normal modes whose frequencies exceed the Debye frequency is equivalent to a cut-off of normal modes whose wavelengths are less than the interatomic spacing. Of course, it makes physical sense that such modes should be absent.

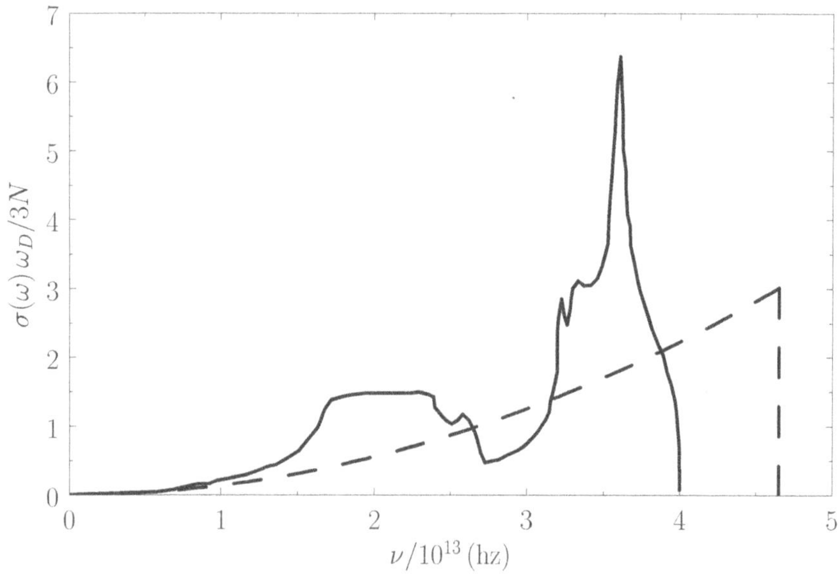

Figure 8.5 The true density of normal modes in diamond (solid curve) compared with the density of normal modes predicted by Debye theory (dashed curve). Here, $\nu = \omega/2\pi$.

Figure 8.5 compares the actual density of normal modes in diamond with the density predicted by Debye theory. Not surprisingly, there is not a particularly strong resemblance between these two curves, because Debye theory is highly idealized. Nevertheless, both curves exhibit sharp cut-offs at high frequencies, and coincide at low frequencies. Furthermore, the areas under both curves are the same. As we shall see, this is sufficient to allow Debye theory to correctly account for the temperature variation of the specific heat of solids at low temperatures.

We can use the quantum-mechanical expression for the mean energy of a single oscillator, Equation (8.148), to calculate the mean energy of lattice vibrations in the Debye approximation. We obtain

$$\overline{E} = \int_0^\infty \sigma_D(\omega)\,\hbar\,\omega\left[\frac{1}{2} + \frac{1}{\exp(\beta\,\hbar\,\omega) - 1}\right]d\omega. \tag{8.190}$$

According to Equation (8.172), the molar heat capacity takes the form

$$c_V = \frac{1}{\nu\,k\,T^2}\int_0^\infty \sigma_D(\omega)\,\hbar\,\omega\left\{\frac{\exp(\beta\,\hbar\,\omega)\,\hbar\,\omega}{[\exp(\beta\,\hbar\,\omega) - 1]^2}\right\}d\omega. \tag{8.191}$$

Making use of Equations (8.185) and (8.186), we find that

$$c_V = \frac{k}{\nu}\int_0^{\omega_D} \frac{\exp(\beta\,\hbar\,\omega)\,(\beta\,\hbar\,\omega)^2}{[\exp(\beta\,\hbar\,\omega) - 1]^2}\,\frac{3\,V}{2\pi^2\,c_s^3}\,\omega^2\,d\omega, \tag{8.192}$$

giving

$$c_V = \frac{3\,V\,k}{2\pi^2\,\nu\,(c_s\,\beta\,\hbar)^3}\int_0^{\beta\,\hbar\,\omega_D} \frac{\exp x}{(\exp x - 1)^2}\,x^4\,dx, \tag{8.193}$$

in terms of the dimensionless variable $x = \beta\,\hbar\,\omega$. According to Equation (8.189), the volume can be written

$$V = 6\pi^2\,N\left(\frac{c_s}{\omega_D}\right)^3, \tag{8.194}$$

so the heat capacity reduces to

$$c_V = 3\,R\,f_D(\beta\,\hbar\,\omega_D) = 3\,R\,f_D\left(\frac{\theta_D}{T}\right), \tag{8.195}$$

where the *Debye function* is defined

$$f_D(y) \equiv \frac{3}{y^3}\int_0^y \frac{\exp x}{(\exp x - 1)^2}\,x^4\,dx. \tag{8.196}$$

We have also defined the *Debye temperature*, θ_D, as

$$k\,\theta_D = \hbar\,\omega_D. \tag{8.197}$$

Consider the asymptotic limit in which $T \gg \theta_D$. For small y, we can approximate $\exp x$ as $1 + x$ in the integrand of Equation (8.196), so that

$$f_D(y) \to \frac{3}{y^3}\int_0^y x^2\,dx = 1. \tag{8.198}$$

Table 8.3 Comparison of Debye temperatures (in degrees kelvin) obtained from the low temperature behavior of the heat capacity with those calculated from the sound speed. From [Kittel (1956)].

Solid	θ_D (low temperature)	θ_D (sound speed)
NaCl	308	320
KCl	230	246
Ag	225	216
Zn	308	305

Thus, if the temperature greatly exceeds the Debye temperature then we recover the law of Dulong and Petite that $c_V = 3\,R$. Consider, now, the asymptotic limit in which $T \ll \theta_D$. For large y,

$$\int_0^y \frac{\exp x}{(\exp x - 1)^2}\, x^4\, dx \simeq \int_0^\infty \frac{\exp x}{(\exp x - 1)^2}\, x^4\, dx = \frac{4\pi^4}{15}. \tag{8.199}$$

The latter integration is standard (if rather obscure), and can be looked up in any (large) reference book on integration. Thus, in the low-temperature limit,

$$f_D(y) \rightarrow \frac{4\pi^4}{5}\frac{1}{y^3}. \tag{8.200}$$

This yields

$$c_V \simeq \frac{12\pi^4}{5}\, R \left(\frac{T}{\theta_D}\right)^3 \tag{8.201}$$

in the limit $T \ll \theta_D$, Note that c_V varies with temperature like T^3.

The fact that c_V goes like T^3 at low temperatures is quite well verified experimentally, although it is sometimes necessary to go to temperatures as low as $0.02\,\theta_D$ to obtain this asymptotic behavior. Theoretically, θ_D should be calculable from Equation (8.189) in terms of the sound speed in the solid, and the molar volume. Table 8.3 shows a comparison of Debye temperatures evaluated by this means with temperatures obtained empirically by fitting the law (8.201) to the low-temperature variation of the heat capacity. It can be seen that there is fairly good agreement between the theoretical and empirical Debye temperatures. This suggests that the Debye theory affords a good, though not perfect, representation of the behavior of c_V in solids over the entire temperature range.

Finally, Figure 8.6 shows the actual temperature variation of the molar heat capacities of various solids, as well as that predicted by Debye's theory. The prediction of Einstein's theory is also shown, for the sake of comparison.

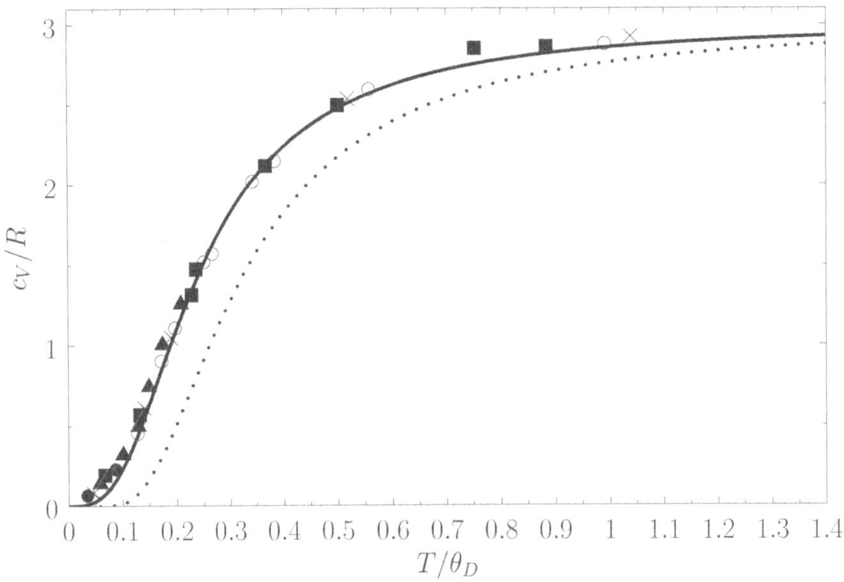

Figure 8.6 The molar heat capacity of various solids. The solid circles, solid triangles, solid squares, open circles, and crosses correspond to Ag, graphite, Al, Al_2O_3, and KCl, respectively. The solid curve shows the prediction of Debye theory. The dotted curve shows the prediction of Einstein theory (assuming that $\theta_E = \theta_D$). Data from [Seitz (1943)].

8.15 Maxwell Velocity Distribution

Consider a molecule of mass m in a gas that is sufficiently dilute for the intermolecular forces to be negligible (i.e., an ideal gas). The energy of the molecule is written

$$\epsilon = \frac{\mathbf{p}^2}{2\,m} + \epsilon^{\rm int}, \tag{8.202}$$

where $\mathbf{p}$ is its momentum vector, and $\epsilon^{\rm int}$ is its internal (i.e., non-translational) energy. The latter energy is due to molecular rotation, vibration, et cetera. Translational degrees of freedom can be treated classically to an excellent approximation, whereas internal degrees of freedom usually require a quantum-mechanical approach. Classically, the probability of finding the molecule in a given internal state with a position vector in the range $\mathbf{r}$ to $\mathbf{r}+d\mathbf{r}$, and a momentum vector in the range $\mathbf{p}$ to $\mathbf{p}+d\mathbf{p}$, is proportional to the number of cells (of "volume" h_0) contained in the corresponding region of phase-space, weighted by the Boltzmann factor. In fact, because classical phase-space is divided up into uniform cells, the number of cells is just proportional to the "volume" of the region under consideration. This "volume" is written $d^3\mathbf{r}\,d^3\mathbf{p}$. Thus, the probability of finding the molecule in a given internal state s is

$$P_s(\mathbf{r},\mathbf{p})\,d^3\mathbf{r}\,d^3\mathbf{p} \propto \exp\left(\frac{-\beta\,p^2}{2\,m}\right)\exp(-\beta\,\epsilon_s^{\rm int})\,d^3\mathbf{r}\,d^3\mathbf{p}, \tag{8.203}$$

where P_s is a probability density defined in the usual manner. The probability $P(\mathbf{r},\mathbf{p})\,d^3\mathbf{r}\,d^3\mathbf{p}$ of finding the molecule in any internal state with position and momentum vectors in the specified range is obtained by summing the previous expression over all possible internal states. The sum over $\exp(-\beta\,\epsilon_s^{\rm int})$ just contributes a constant of proportionality (because the internal states do not depend on $\mathbf{r}$ or $\mathbf{p}$), so

$$P(\mathbf{r},\mathbf{p})\,d^3\mathbf{r}\,d^3\mathbf{p} \propto \exp\left(-\frac{\beta\,p^2}{2\,m}\right)d^3\mathbf{r}\,d^3\mathbf{p}. \tag{8.204}$$

Of course, we can multiply this probability by the total number of molecules, N, in order to obtain the mean number of molecules with position and momentum vectors in the specified range.

Suppose that we now wish to determine $f(\mathbf{r},\mathbf{v})\,d^3\mathbf{r}\,d^3\mathbf{v}$: that is, the mean number of molecules with positions between $\mathbf{r}$ and $\mathbf{r}+d\mathbf{r}$, and velocities in the range $\mathbf{v}$ and $\mathbf{v}+d\mathbf{v}$. Because $\mathbf{v}=\mathbf{p}/m$, it is easily seen that

$$f(\mathbf{r},\mathbf{v})\,d^3\mathbf{r}\,d^3\mathbf{v} = C\exp\left(-\frac{\beta\,m\,v^2}{2}\right)d^3\mathbf{r}\,d^3\mathbf{v}, \tag{8.205}$$

where C is a constant of proportionality. This constant can be determined by the condition

$$\int_{(\mathbf{r})}\int_{(\mathbf{v})} f(\mathbf{r},\mathbf{v})\,d^3\mathbf{r}\,d^3\mathbf{v} = N. \tag{8.206}$$

In other word, the sum over molecules with all possible positions and velocities gives the total number of molecules, N. The integral over the molecular position coordinates just gives the volume, V, of the gas, because the Boltzmann factor is independent of position. The integration over the velocity coordinates can be reduced to the product of three identical integrals (one for v_x, one for v_y, and one for v_z), so we have

$$C\,V\left[\int_{-\infty}^{\infty}\exp\left(-\frac{\beta\,m\,v_z^2}{2}\right)dv_z\right]^3 = N. \tag{8.207}$$

Now,

$$\int_{-\infty}^{\infty}\exp\left(-\frac{\beta\,m\,v_z^2}{2}\right)dv_z = \sqrt{\frac{2}{\beta\,m}}\int_{-\infty}^{\infty}\exp\left(-y^2\right)dy = \sqrt{\frac{2\pi}{\beta\,m}}, \tag{8.208}$$

so $C = (N/V)\,(\beta\,m/2\pi)^{3/2}$. (See Exercise 2.2.) Thus, the properly normalized distribution function for molecular velocities is written

$$f(\mathbf{v})\,d^3\mathbf{r}\,d^3\mathbf{v} = n\left(\frac{m}{2\pi\,k\,T}\right)^{3/2}\,\exp\left(-\frac{m\,v^2}{2\,k\,T}\right)d^3\mathbf{r}\,d^3\mathbf{v}. \tag{8.209}$$

Here, $n = N/V$ is the number density of the molecules. We have omitted the variable $\mathbf{r}$ in the argument of f, because f clearly does not depend on position. In other words, the distribution of molecular velocities is uniform in space. This is hardly surprising, because there is nothing to distinguish one region of space from another in our calculation. The previous distribution is called the *Maxwell velocity distribution*, because it was discovered by James Clark Maxwell in the middle of the nineteenth century. The average number of molecules per unit volume with velocities in the range $\mathbf{v}$ to $\mathbf{v} + d\mathbf{v}$ is obviously $f(\mathbf{v})\,d^3\mathbf{v}$.

Let us consider the distribution of a given component of velocity: the z-component (say). Suppose that $g(v_z)\,dv_z$ is the average number of molecules per unit volume with the z-component of velocity in the range v_z to $v_z + dv_z$, irrespective of the values of their other velocity components. It is fairly obvious that this distribution is obtained from the Maxwell distribution by summing (integrating actually) over all possible values of v_x and v_y, with v_z in the specified range. Thus,

$$g(v_z)\,dv_z = \int_{(v_x)}\int_{(v_y)} f(\mathbf{v})\,d^3\mathbf{v}. \tag{8.210}$$

This gives

$$\begin{aligned} g(v_z)\,dv_z &= n\left(\frac{m}{2\pi\,k\,T}\right)^{3/2}\int_{(v_x)}\int_{(v_y)}\exp\left[-\left(\frac{m}{2\,k\,T}\right)(v_x^2 + v_y^2 + v_z^2)\right]dv_x\,dv_y\,dv_z \\ &= n\left(\frac{m}{2\pi\,k\,T}\right)^{3/2}\,\exp\left(-\frac{m\,v_z^2}{2\,k\,T}\right)\left[\int_{-\infty}^{\infty}\exp\left(-\frac{m\,v_x^2}{2\,k\,T}\right)\right]^2 \\ &= n\left(\frac{m}{2\pi\,k\,T}\right)^{3/2}\,\exp\left(-\frac{m\,v_z^2}{2\,k\,T}\right)\left(\sqrt{\frac{2\pi\,k\,T}{m}}\right)^2, \end{aligned} \tag{8.211}$$

or

$$g(v_z)\,dv_z = n\left(\frac{m}{2\pi\,k\,T}\right)^{1/2}\,\exp\left(-\frac{m\,v_z^2}{2\,k\,T}\right)dv_z. \tag{8.212}$$

Of course, this expression is properly normalized, so that

$$\int_{-\infty}^{\infty} g(v_z)\,dv_z = n. \tag{8.213}$$

It is clear that each component (because there is nothing special about the z-component) of the velocity is distributed with a Gaussian probability distribution (see Section 2.9), centered on a mean value

$$\overline{v_z} = 0, \tag{8.214}$$

with variance

$$\overline{v_z^2} = \frac{k\,T}{m}. \tag{8.215}$$

Equation (8.214) implies that each molecule is just as likely to be moving in the plus z-direction as in the minus z-direction. Equation (8.215) can be rearranged to give

$$\overline{\frac{1}{2}\,m\,v_z^2} = \frac{1}{2}\,k\,T, \tag{8.216}$$

in accordance with the equipartition theorem.

Note that Equation (8.209) can be rewritten

$$\frac{f(\mathbf{v})\,d^3\mathbf{v}}{n} = \left[\frac{g(v_x)\,dv_x}{n}\right]\left[\frac{g(v_y)\,dv_y}{n}\right]\left[\frac{g(v_z)\,dv_z}{n}\right], \tag{8.217}$$

where $g(v_x)$ and $g(v_y)$ are defined in an analogous way to $g(v_z)$. Thus, the probability that the velocity lies in the range $\mathbf{v}$ to $\mathbf{v} + d\mathbf{v}$ is just equal to the product of the probabilities that the velocity components lie in their respective ranges. In other words, the individual velocity components act like statistically independent variables.

Suppose that we now wish to calculate $F(v)\,dv$: that is, the average number of molecules per unit volume with a speed $v = |\mathbf{v}|$ in the range v to $v+dv$. It is obvious that we can obtain this quantity by summing over all molecules with speeds in this range, irrespective of the direction of their velocities. Thus,

$$F(v)\,dv = \int f(\mathbf{v})\,d^3\mathbf{v}, \tag{8.218}$$

where the integral extends over all velocities satisfying

$$v < |\mathbf{v}| < v + dv. \tag{8.219}$$

This inequality is satisfied by a spherical shell of radius v and thickness dv in velocity space. Because $f(\mathbf{v})$ only depends on $|v|$, so $f(\mathbf{v}) \equiv f(v)$, the previous integral is just $f(v)$ multiplied by the volume of the spherical shell in velocity space. So,

$$F(v)\,dv = f(v)\,4\pi\,v^2\,dv, \tag{8.220}$$

which gives

$$F(v)\,dv = 4\pi\,n\left(\frac{m}{2\pi\,k\,T}\right)^{3/2} v^2\,\exp\left(-\frac{m\,v^2}{2\,k\,T}\right)dv. \tag{8.221}$$

This is the famous *Maxwell distribution of molecular speeds*. Of course, it is properly normalized, so that

$$\int_0^\infty F(v)\,dv = n. \tag{8.222}$$

Note that the Maxwell distribution exhibits a maximum at some non-zero value of v. The reason for this is quite simple. As v increases, the Boltzmann factor decreases, but the volume of phase-space available to the molecule (which is proportional to v^2) increases: the net result is a distribution with a non-zero maximum.

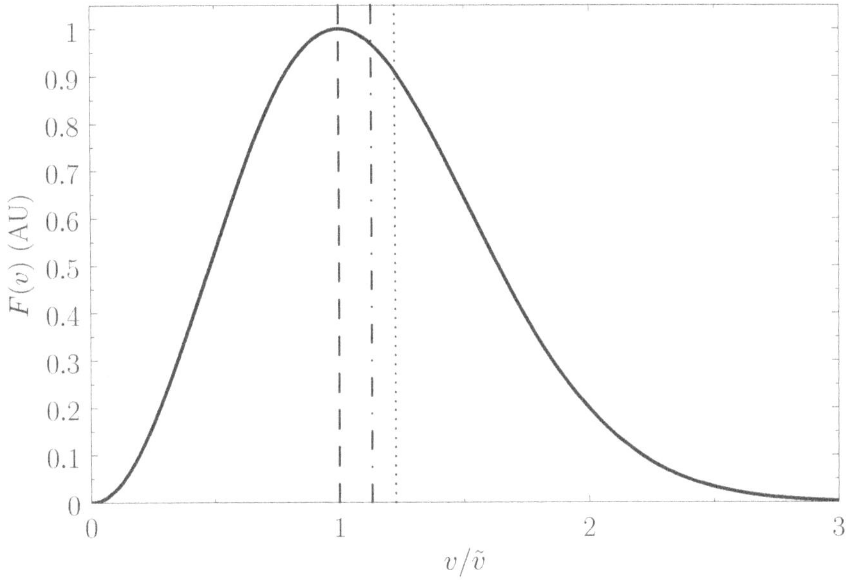

Figure 8.7 The Maxwell velocity distribution as a function of molecular speed, in units of the most probable speed ($\tilde{v}$). The dashed, dash-dotted, and dotted lines indicates the most probable speed, the mean speed, and the root-mean-square speed, respectively.

The mean molecular speed is given by

$$\bar{v} = \frac{1}{n}\int_0^\infty F(v)\,v\,dv. \tag{8.223}$$

Thus, we obtain

$$\bar{v} = 4\pi\left(\frac{m}{2\pi\,k\,T}\right)^{3/2}\int_0^\infty v^3\exp\left(-\frac{m\,v^2}{2\,k\,T}\right)dv, \tag{8.224}$$

or

$$\bar{v} = 4\pi\left(\frac{m}{2\pi\,k\,T}\right)^{3/2}\left(\frac{2\,k\,T}{m}\right)^2\int_0^\infty y^3\exp\left(-y^2\right)dy. \tag{8.225}$$

Now

$$\int_0^\infty y^3\exp\left(-y^2\right)dy = \frac{1}{2} \tag{8.226}$$

(see Exercise 8.2), so

$$\bar{v} = \sqrt{\frac{8}{\pi}\frac{kT}{m}}. \tag{8.227}$$

A similar calculation gives

$$v_{\rm rms} = \left[\overline{v^2}\right]^{1/2} = \sqrt{\frac{3\,kT}{m}}. \tag{8.228}$$

(See Exercise 8.14.) However, this result can also be obtained from the equipartition theorem. Because

$$\overline{\frac{1}{2}\,m\,v^2} = \overline{\frac{1}{2}\,m\,(v_x^2 + v_y^2 + v_z^2)} = 3\left(\frac{1}{2}\,kT\right), \tag{8.229}$$

then Equation (8.228) follows immediately. It is easily demonstrated that the most probable molecular speed (i.e., the maximum of the Maxwell distribution function) is

$$\tilde{v} = \sqrt{\frac{2\,kT}{m}}. \tag{8.230}$$

The speed of sound in an ideal gas is given by

$$c_s = \sqrt{\frac{\gamma\,p}{\rho}}, \tag{8.231}$$

where γ is the ratio of specific heats. This can also be written

$$c_s = \sqrt{\frac{\gamma\,kT}{m}}, \tag{8.232}$$

because $p = n\,k\,T$ and $\rho = n\,m$. It is clear that the various average speeds that we have just calculated are all of order the sound speed (i.e., a few hundred meters per second at room temperature). In ordinary air ($\gamma = 1.4$) the sound speed is about 84% of the most probable molecular speed, and about 74% of the mean molecular speed. Because sound waves ultimately propagate via molecular motion, it makes sense that they travel at slightly less than the most probable and mean molecular speeds.

Figure 8.7 shows the Maxwell velocity distribution as a function of molecular speed in units of the most probable speed. Also shown are the mean speed and the root-mean-square speed.

8.16 Effusion

Consider a dilute gas enclosed in a container. Let us calculate how many molecules per unit time strike a unit area of the container wall.

Consider a wall element of area dA. The z-axis is oriented so as to point along the outward normal of this element. Let $\mathbf{v}$ represent a molecular velocity. The Cartesian components of this velocity are conveniently written $\mathbf{v} = v\,(\sin\theta\,\cos\phi,\ \sin\theta\,\sin\phi,\ \cos\theta)$, where the standard spherical angles θ and ϕ specify the velocity vector's direction. Let us concentrate on those molecules, in the immediate vicinity of the wall element, whose velocities are such that they lie between $\mathbf{v}$ and $\mathbf{v} + d\mathbf{v}$. In other words, the velocities are such that their magnitudes lies between v and $v + dv$, their polar angles lie between θ and $\theta + d\theta$, and their azimuthal angles lie between ϕ and $\phi+d\phi$. Molecules of this type suffer a displacement $\mathbf{v}\,dt$ in the infinitesimal time interval dt. Hence, all molecules that lie within the infinitesimal cylinder of cross-sectional area dA, and length $\mathbf{v}\,dt$ (whose axis subtends an angle θ with the z-axis, and whose two ends are parallel to the wall element), will strike the wall element within the time interval dt, whereas those molecules outside this cylinder will not. (See Figure 8.8.) The volume of the cylinder is $dA\,v\,\cos\theta\,dt$, whereas the number of molecules per unit volume in the prescribed velocity range is $f(\mathbf{v})\,d^3\mathbf{v}$. Hence, the number of molecules of this type that strike the area dA in the time interval dt is $[f(\mathbf{v})\,d^3\mathbf{v}]\,(dA\,v\,\cos\theta\,dt)$. Dividing this expression by the area dA, and the time interval dt, we obtain $\Phi(\mathbf{v})\,d^3\mathbf{v}$, which is defined as the number of molecules, with velocities in the range $\mathbf{v}$ to $\mathbf{v}+d\mathbf{v}$, that strike a unit area of the wall per unit time. Thus,

$$\Phi(\mathbf{v})\,d^3\mathbf{v} = f(\mathbf{v})\,v\,\cos\theta\,d^3\mathbf{v}. \tag{8.233}$$

Let Φ_0 be the total number of molecules that strike a unit area of the wall per unit time. This quantity is simply obtained by summing (8.233) over all possible molecular velocities that would cause molecules to collide with the element of area. This means that we have to sum over all possible speeds, $0 < v < \infty$, all possible azimuthal angles, $0 < \phi < 2\pi$, and all polar angles in the range $0 < \theta < \pi/2$. (Those molecules with polar angles in the range $\pi/2 < \theta < \pi$ have their velocity directed away from the wall, and will, hence, not collide with it.) In other words, we have to sum over all possible velocities, $\mathbf{v}$, subject to the restriction that the velocity component $v_z = v\,\cos\theta$ is positive. Thus, we have

$$\Phi_0 = \int_{v_z>0} f(\mathbf{v})\,v\,\cos\theta\,d^3\mathbf{v}. \tag{8.234}$$

The element of volume in velocity space can be expressed in spherical coordinates:

$$d^3\mathbf{v} = v^2\,dv\,(\sin\theta\,d\theta\,d\phi), \tag{8.235}$$

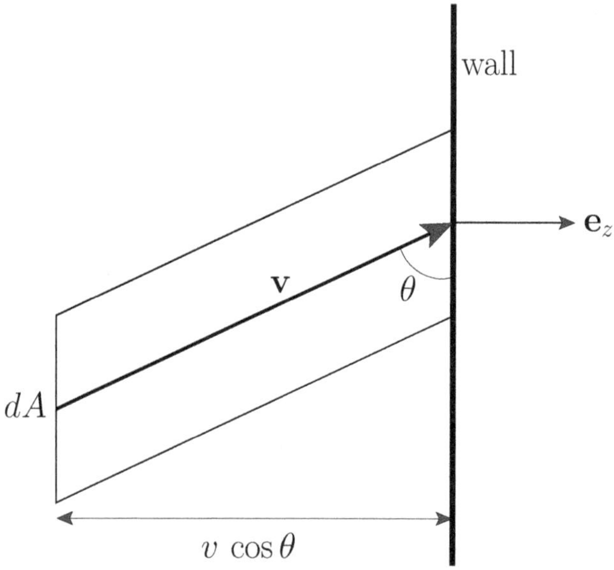

Figure 8.8 All molecules with velocity $\mathbf{v}$ that lie within the cylinder strike the wall in a unit time interval.

where $\sin\theta\, d\theta\, d\phi = d\Omega$ is just an element of solid angle. Moreover, in thermal equilibrium, $f(\mathbf{v}) = f(v)$ is only a function of $|\mathbf{v}|$. Hence, Equation (8.234) becomes

$$\begin{aligned}\Phi_0 &= \int_{v_z>0} f(v)\, v\, \cos\theta\, (v^2\, dv\, \sin\theta\, d\theta\, d\phi)\\ &= \int_0^\infty f(v)\, v^2\, dv \int_0^{\pi/2} \sin\theta\, \cos\theta\, d\theta \int_0^{2\pi} d\phi.\end{aligned} \tag{8.236}$$

The integral over ϕ gives 2π, whereas that over θ reduces to $1/2$. It follows that

$$\Phi_0 = \pi \int_0^\infty f(v)\, v^3\, dv. \tag{8.237}$$

Now, the mean molecular speed is written

$$\bar{v} = \frac{1}{n}\int f(v)\, v\, d^3\mathbf{v} = \frac{1}{n}\int_0^\infty \int_0^\pi \int_0^{2\pi} f(v)\, v\, (v^2\, dv\, \sin\theta\, d\theta\, d\phi), \tag{8.238}$$

which reduces to

$$\bar{v} = \frac{4\pi}{n}\int_0^\infty f(v)\, v^3\, dv. \tag{8.239}$$

Hence, we deduce from Equation (8.237) that

$$\Phi_0 = \frac{1}{4}\,n\,\bar{v}. \tag{8.240}$$

The previous result can be combined with Equation (8.227), and the ideal equation of state, $\bar{p} = n\,k\,T$, to give

$$\Phi_0 = \frac{\bar{p}}{\sqrt{2\pi\,m\,k\,T}}. \tag{8.241}$$

If a sufficiently small hole is made in the wall of the container then the equilibrium of the gas inside the container is disturbed to a negligible extent. [In practice, a sufficiently small hole is one that is small compared to the molecular *mean-free-path* (i.e., the mean distance travelled between collisions) inside the container.] In this case, the number of molecules that emerge through the small hole is the same as the number of molecules that would strike the area occupied by the hole if the latter were closed off. The process by which molecules emerge through a small hole is called *effusion*. The number of molecules that have speeds in the range between v and $v + dv$, and which emerge through a small hole of area A, into the solid angle range $d\Omega$, in the forward direction $\theta = 0$, is given by

$$A\,\Phi(\mathbf{v})\,d^3\mathbf{v} \propto A\,[f(v)\,v\,\cos\theta]\,(v^2\,dv\,d\Omega) \propto f(v)\,v^3\,dv\,d\Omega$$
$$\propto v^3\exp\left(-\frac{m\,v^2}{2\,k\,T}\right)dv\,d\Omega. \tag{8.242}$$

Now, it is difficult to directly verify the Maxwell velocity distribution. However, this distribution can be verified indirectly by measuring the velocity distribution of atoms effusing from a small hole in an oven. As we have just seen, the predicted velocity distribution of the escaping atoms varies like $v^3\,\exp(-m\,v^2/2\,k\,T)$, in contrast to the $v^2\,\exp(-m\,v^2/2\,k\,T)$ variation of the velocity distribution inside the oven. There is good agreement between the experimentally measured velocity distribution of effusing atoms and the theoretical prediction.

As another example, consider a container that is divided into two parts by a partition containing a small hole. The container is filled with gas, but one part of the container is maintained at the temperature T_1, and the other at the temperature T_2. Let us calculate the relationship between the mean gas pressures, $\bar{p}_1$ and $\bar{p}_2$, respectively, in the two parts, when the system is in equilibrium. (That is, when a situation is reached in which neither $\bar{p}_1$ or $\bar{p}_2$, nor the amount of gas in either parts, changes with time.) If the hole is larger than the molecular mean-free-path then the relationship is simply $\bar{p}_1 = \bar{p}_2$—otherwise, the pressure difference would give rise to bulk motion of the gas from one side of the container to the other, until the pressures were equalized. But, if the dimensions of the hole are less than the mean-free-path then we are dealing with effusion through the hole, rather than hydrodynamical flow. In this case, the equilibrium condition requires that the mass of gas on each side remain constant. In other words, the effusion rates in both directions must be equal. Thus, it follows from Equation (8.241) that

$$\frac{\bar{p}_1}{\sqrt{T_1}} = \frac{\bar{p}_2}{\sqrt{T_2}}. \tag{8.243}$$

Thus, the pressures are not equal. Instead, a higher pressure prevails in the part of the container held at a higher temperature.

8.17 Ferromagnetism

Consider a solid substance consisting of N identical atoms arranged in a regular lattice. Each atom has a net electronic angular momentum, $\hbar\,\mathbf{S}$, and an associated magnetic moment, $\boldsymbol{\mu}$. [Here, we use $\mathbf{S}$, rather than $\mathbf{J}$, to denote atomic angular momentum (divided by $\hbar$) to avoid confusion with the exchange energy, J, introduced in Equation (8.246).] The atomic magnetic moment is related to the corresponding total angular momentum via

$$\boldsymbol{\mu} = g\,\mu_B\,\mathbf{S}, \tag{8.244}$$

where μ_B is the Bohr magneton, and the g-factor, g, is a dimensionless number of order unity. (See Section 8.9.) In the presence of an externally applied magnetic field, $\mathbf{B}_0 = B_z\,\mathbf{e}_z$, directed parallel to the z-axis, the Hamiltonian, H_0, representing the interaction of the atoms with this field is written

$$H_0 = -g\,\mu_B \sum_{j=1,N} \mathbf{S}_j\cdot\mathbf{B}_0 = -g\,\mu_B\,B_z \sum_{j=1,N} S_{jz}. \tag{8.245}$$

Each atom is also assumed to interact with neighboring atoms. The interaction in question is not simply the magnetic dipole-dipole interaction due to the magnetic field produced by one atom at the position of another. Such an interaction is, in general, much too small to produce ferromagnetism. Instead, the predominant interaction is known as the *exchange interaction*. This interaction is a quantum-mechanical consequence of the Pauli exclusion principle. (See Section 10.2.) Because electrons cannot occupy the same state, two electrons on neighboring atoms that have parallel spins (here, to simplify the discussion of the origin of the exchange interaction, we are temporarily assuming that the electrons all have zero orbital angular momenta, so that their total angular momenta are solely a consequence of their spin angular momenta) cannot come too close to one another in space (else they would simultaneously occupy identical spin and orbital states, which is forbidden). On the other hand, if the electrons have antiparallel spins then they are in different spin states, so they are allowed to occupy the same orbital state. In other words, there is no exclusion-related restriction on how close they can come together. Because different spatial separations of the electrons give rise to different electrostatic interaction between them, the electrostatic interaction between neighboring atoms (which is much stronger than any magnetic interaction) depends on the relative orientations of their spins. This, then, is the origin of the exchange interaction, which for two atoms labelled j and k can be written in the form

$$H_{jk} = -2\,J\,\mathbf{S}_j \cdot \mathbf{S}_k. \tag{8.246}$$

Here, the so-called *exchange energy*, J, is a parameter (with the dimensions of energy) that depends on the atomic separation, and which measures the strength of the exchange interaction. If $J > 0$ then the interaction energy is lower when spins on neighboring atoms are parallel, rather than antiparallel. This state of affairs favors parallel spin alignment of neighboring atoms. In other words, it tends to produce ferromagnetism. Note that, because the exchange interaction depends on the degree to which the electron orbitals of the two atoms can overlap, so as to occupy approximately the same region of space, the exchange energy, J, falls off very rapidly with increasing atomic separation. Hence, a given atom only interacts appreciably with its n (say) nearest neighbors.

To simplify the interaction problem, we shall replace the previous expression by the simpler functional form

$$H_{jk} = -2\,J\,S_{jz}\,S_{kz}. \tag{8.247}$$

This approximate expression, which is known as the *Ising model*, leaves the essential physical situation intact, while avoiding the complications introduced by vector quantities. The Ising model is equivalent to assuming that all atomic magnetic moments are either directed parallel or anti-parallel to the z-axis.

The Hamiltonian, H', representing the interaction energy between atoms is written in the form

$$H' = \frac{1}{2}\left(2\,J \sum_{j=1,N} \sum_{k=1,n} S_{jz}\,S_{kz}\right), \tag{8.248}$$

where J is the exchange energy for neighboring atoms, and the index k refers to the nearest neighbor shell surrounding the atom j. The factor $1/2$ is introduced because the interaction between the same two atoms is counted twice in performing the sums.

The total Hamiltonian of the atoms is

$$H = H_0 + H'. \tag{8.249}$$

The task in hand is to calculate the mean magnetic moment of the system parallel to the applied magnetic field, $\overline{M_z}$, as a function of its absolute temperature, T, and the applied field, B_z. The presence of the interatomic interactions makes this task difficult, despite the extreme simplicity of expression (8.248). Although the problem has been solved exactly for a linear one-dimensional array of atoms, and for two-dimensional array of atoms when $B_z = 0$, the three-dimensional problem is so complex that no exact solution has ever been found. Thus, in order to make further progress, we must introduce an additional approximation.

Let us focus attention on a particular atom, j, which we shall refer to as the "central atom." The interactions of this atom are described by the Hamiltonian

$$H_j = -g\,\mu_B\,B_z\,S_{jz} - 2\,J\,S_{jz}\sum_{k=1,n} S_{kz}. \tag{8.250}$$

The final term represents the interaction of the central atom with its nearest neighbors. As a simplifying approximation, we shall replace the sum over these neighbors by its mean value. In other words, we shall write

$$2\,J\sum_{k=1,n} S_{kz} \simeq 2\,J\overline{\sum_{k=1,n} S_{kz}} = g\,\mu_B\,B_m, \tag{8.251}$$

where B_m is a parameter with the dimensions of magnetic field-strength, which is called the *Weiss molecular field*, and is to be determined in such a way that it leads to a self-consistent solution of the statistical problem. In terms of this parameter, expression (8.250) becomes

$$H_j = -g\,\mu_B\,(B_z + B_m)\,S_{jz}. \tag{8.252}$$

Thus, the influence of neighboring atoms has been replaced by an effective magnetic field, B_m. The problem presented by expression (8.252) reduces to the elementary one of a single atom, of normalized angular momentum, S, and g-factor, g, placed in an external z-directed magnetic field, $B_z + B_m$. Recall, however, that we already solved this problem in Section 8.9.

According to the analysis of Section 8.9,

$$\overline{S_{jz}} = S\,\mathcal{B}_S(\eta), \tag{8.253}$$

where

$$\eta = \beta\,g\,\mu_B\,(B_z + B_m), \tag{8.254}$$

and $\beta = 1/(k\,T)$. Here, $\mathcal{B}_S(\eta)$ is the Brillouin function, for normalized atomic angular momentum S, that was introduced in Equation (8.113).

Expression (8.253) involves the unknown parameter B_m. To determine the value of this parameter in a self-consistent manner, we note that there is nothing

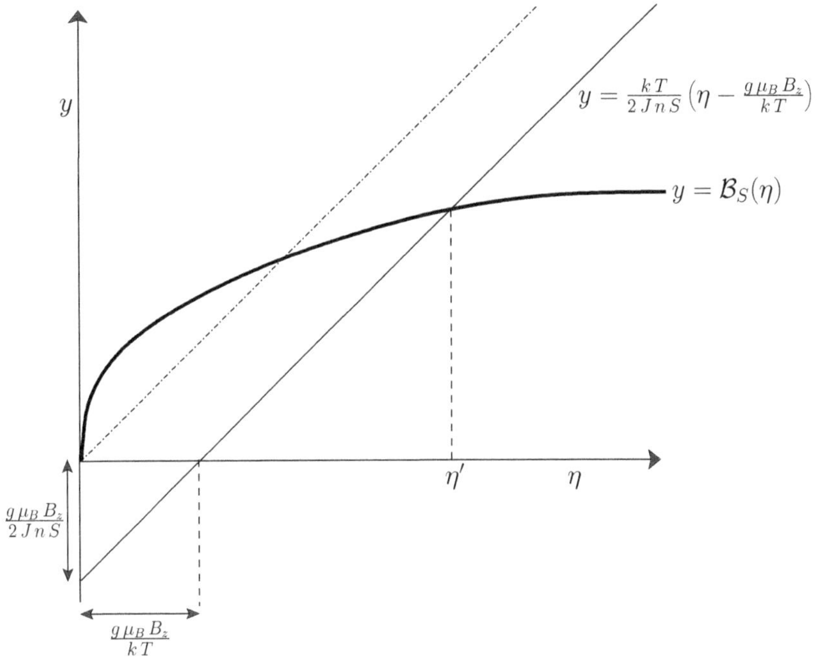

Figure 8.9 Graphical solution of Equation (8.256) that determines the molecular field, B_m, corresponding to the intersection of the curves at $\eta = \eta'$. The dash-dotted straight-line corresponds to the case where the external magnetic field, B_z, is zero.

that distinguishes the central *j*th atom from any of its neighboring atoms. Hence, any one of these atoms might equally well have been considered the central atom. Thus, the corresponding mean value $\overline{S_z}$ is also given by Equation (8.253). It follows from Equations (8.251) and (8.253) that

$$2\,J\,n\,S\,\mathcal{B}_S(\eta) = g\,\mu_B\,B_m. \tag{8.255}$$

The previous two equations can be combined to give

$$\mathcal{B}_S(\eta) = \frac{k\,T}{2\,J\,n\,S}\left(\eta - \frac{g\,\mu_B\,B_z}{k\,T}\right), \tag{8.256}$$

which determines η (and, hence, B_m). Once η has been found, the mean magnetic moment of the system follows from Equation (8.253):

$$\overline{M_z} = g\,\mu_B \sum_j \overline{S_{jz}} = N\,g\,\mu_B\,S\,\mathcal{B}_S(\eta). \tag{8.257}$$

The solution of Equation (8.256) can be found by drawing, on the same graph, both the Brillouin function $y = \mathcal{B}_S(\eta)$ and the straight-line

$$y = \frac{k\,T}{2\,J\,n\,S}\left(\eta - \frac{g\,\mu_B\,B_z}{k\,T}\right), \tag{8.258}$$

and then finding the point of intersection of the two curves. See Figure 8.9.

Consider the case where the external field, B_z, is zero. Equation (8.256) simplifies to give

$$\mathcal{B}_S(\eta) = \left(\frac{k\,T}{2\,J\,n\,S}\right)\eta. \tag{8.259}$$

Given that $\mathcal{B}_S(0) = 0$, it follows that $\eta = 0$ is a solution of the previous equation. This solution is characterized by zero magnetization: that is, $\overline{M_z} = 0$. However, there is also the possibility of another solution with $\eta \neq 0$, and, hence, $\overline{M_z} \neq 0$. (See Figure 8.9.) The presence of such spontaneous magnetization in the absence of an external magnetic field is, of course, the distinguishing feature of ferromagnetic materials. In order to have a solution with $\eta \neq 0$, it is necessary that the curves shown in Figure 8.9 intersect at a point $\eta \neq 0$ when both curves pass through the origin. The condition for the existence of an intersection with $\eta \neq 0$ is that the slope of the curve $y = \mathcal{B}_J(\eta)$ at $\eta = 0$ should exceed that of the straight-line $y = (k\,T/2\,J\,n\,S)\,\eta$. In other words,

$$\left(\frac{d\mathcal{B}_S}{d\eta}\right)_{\eta=0} > \frac{k\,T}{2\,J\,n\,S}. \tag{8.260}$$

However, when $\eta \ll 1$, the Brillouin function takes the simple form specified in Equation (8.119):

$$\mathcal{B}_S(\eta) \simeq \frac{1}{3}\,(S+1)\,\eta. \tag{8.261}$$

Hence, Equation (8.260) becomes

$$\frac{1}{3}\,(S+1) > \frac{k\,T}{2\,J\,n\,S}, \tag{8.262}$$

or

$$T < T_c, \tag{8.263}$$

where

$$k\,T_c = \frac{2\,J\,n\,S\,(S+1)}{3}. \tag{8.264}$$

Thus, we deduce that spontaneous magnetization of a ferromagnetic material is only possible below a certain critical temperature, T_c, known as the *Curie temperature*. The magnetized state, in which all atomic magnetic moments can exploit their mutual exchange energy by being preferentially aligned parallel to one another, has a lower magnetic energy than the unmagnetized state in which $\eta = 0$. Thus, at temperatures below the Curie temperature, the magnetized state is stable, whereas the unmagnetized state is unstable.

As the temperature, T, is decreased below the Curie temperature, T_c, the slope of the dash-dotted straight-line in Figure 8.9 decreases so that it intersects the curve $y = \mathcal{B}_S(\eta)$ at increasingly large values of η. As $T \to 0$, the intersection occurs at $\eta \to \infty$. Because, $\mathcal{B}_S(\infty) = 1$ [see Equation (8.117)], it follows from Equation (8.257) that $\overline{M_z} \to N\,g\,\mu_B\,S$ as $T \to 0$. This corresponds to a state in which all atomic magnetic moments are aligned completely parallel to one another. For temperatures, T, less than the Curie temperature, T_c, we can use Equations (8.257) and (8.259) to compute $\overline{M_z}(T)$ in the absence of an external magnetic field. We then obtain a magnetization curve of the general form shown in Figure 8.10.

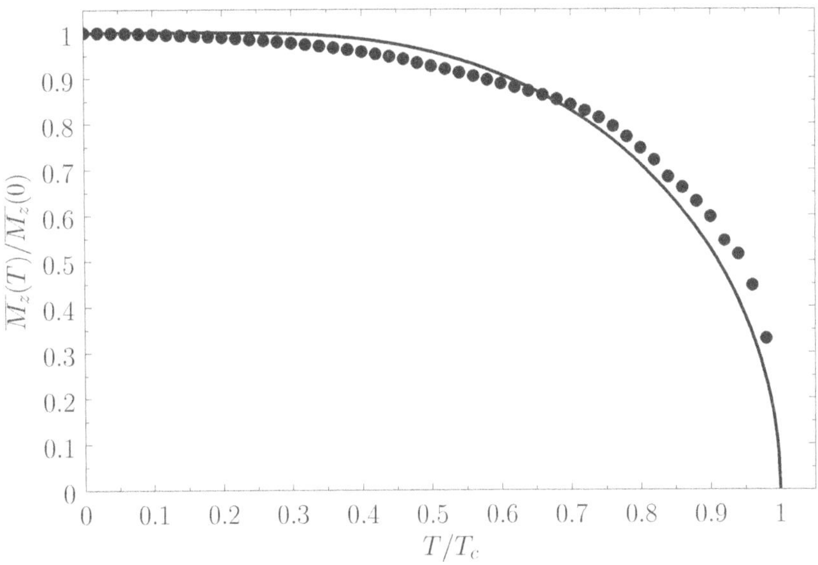

Figure 8.10 Spontaneous magnetization of a ferromagnet as a function of the temperature in the presence of zero external magnetic field. The solid curve is the prediction of molecular field theory for $J = 1/2$ and $g = 2$. The points are experimental data for nickel. Data from [Kaul and Thompson (1969)].

Let us, finally, investigate the magnetic susceptibility of a ferromagnetic substance, in the presence of a small external magnetic field, at temperatures above the Curie temperature. In this case, the crossing point of the two curves in Figure 8.9 lies at small η. Hence, we can use the approximation (8.261) to write Equation (8.256) in the form

$$\frac{1}{3}(S+1)\,\eta = \frac{k\,T}{2\,J\,n\,S}\left(\eta - \frac{g\,\mu_B\,B_z}{k\,T}\right). \tag{8.265}$$

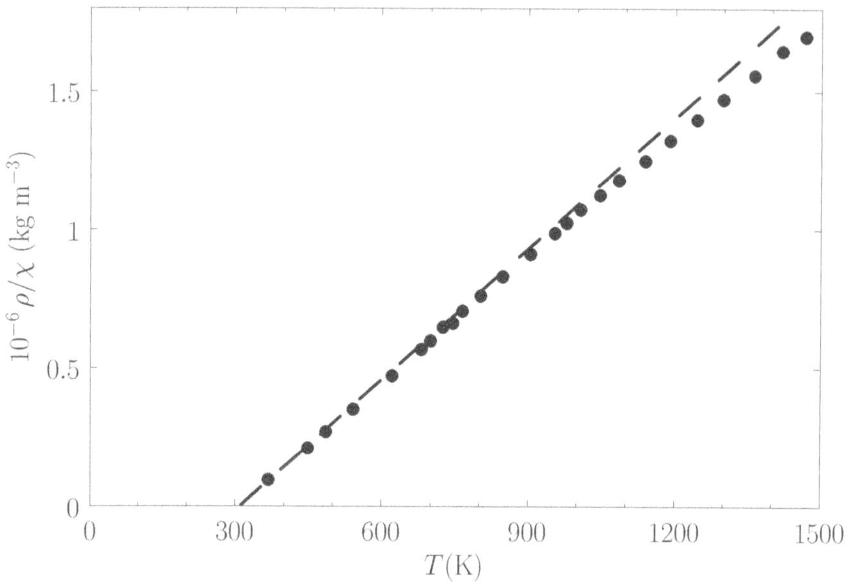

Figure 8.11 Plot of ρ/χ (where ρ is the mass density) versus temperature for gadolinium above its Curie temperature. The curve is (apart from some slight departures at high temperatures) a straight-line, in accordance with the Curie-Weiss law. The intercept of the line with the temperature axis gives $T_c = 310$K. The metal becomes ferromagnetic below 289K. Data from [Arajs and Colvin (1961)].

Solving the previous equation for η gives

$$\eta = \frac{g\,\mu_B\,B_z}{k\,(T - T_c)}, \tag{8.266}$$

where use has been made of Equation (8.264). Now, making use of the approximation (8.261), Equation (8.257) reduces to

$$\overline{M_z} = \frac{1}{3}\,N\,g\,\mu_B\,S\,(S+1)\,\eta. \tag{8.267}$$

The previous two equations can be combined to give

$$\chi \simeq \frac{\mu_0\,\overline{M_z}}{V\,B_z} = \frac{N\,\mu_0\,\mu_B^{\,2}\,g^2\,S\,(S+1)}{3\,V\,k\,(T - T_c)}, \tag{8.268}$$

where χ is the dimensionless magnetic susceptibility of the substance, and V is its volume. This result is known as the *Curie-Weiss law*. It differs from Curie's law, $\chi \propto 1/T$ (see Section 8.3), because of the presence of the parameter T_c in the denominator. According to the Curie-Weiss law, χ becomes infinite when $T \to T_c$: that is, at the Curie temperature, when the substance becomes ferromagnetic.

Experimentally, the Curie-Weiss law is well obeyed at temperatures significantly above the Curie temperature. See Figure 8.11. It is, however, not true that the temperature, T_c, that occurs in this law is exactly the same as the Curie temperature at which the substance becomes ferromagnetic.

The so-called *Weiss molecular-field model* (which was first put forward by Pierre Weiss in 1907) that we have just described is remarkably successful at explaining the major features of ferromagnetism. Nevertheless, there are some serious discrepancies between the predictions of this simple model and experimental data. In particular, the magnetic contribution to the specific heat is observed to have a very sharp discontinuity at the Curie temperature (in the absence of an external field), whereas the molecular field model predicts a much less abrupt change. Needless to say, more refined models have been devised that considerably improve the agreement with experimental data.

Exercises

8.1 Demonstrate that the geometric series

$$S_n = \sum_{k=0,n-1} a\,r^k$$

can be summed to give

$$S_n = a\left(\frac{1-r^n}{1-r}\right).$$

Here, $r \neq 1$. Hence, deduce that

$$S_\infty = \frac{a}{1-r},$$

assuming that $0 < r < 1$.

8.2 Let

$$I(n) = \int_0^\infty x^n\,{\rm e}^{-\alpha\,x^2}\,dx.$$

Demonstrate that

$$I(n) = -\frac{\partial I(n-2)}{\partial\alpha}.$$

Furthermore, show that

$$I(0) = \frac{\sqrt{\pi}}{2\,\alpha^{1/2}}$$

(see Exercise 2.2), and

$$I(1) = \frac{1}{2\,\alpha}.$$

Hence, deduce that

$$I(2) = \frac{\sqrt{\pi}}{4\,\alpha^{3/2}},$$

$$I(3) = \frac{1}{2\,\alpha^{2}},$$

$$I(4) = \frac{3\,\sqrt{\pi}}{8\,\alpha^{5/2}},$$

$$I(5) = \frac{1}{\alpha^{3}}.$$

8.3 A sample of mineral oil is placed in an external magnetic field B. Each proton has spin $1/2$, and a magnetic moment μ. It can, therefore, have two possible energies, $\epsilon = \mp\mu\,B$, corresponding to the two possible orientations of its spin. An applied radio-frequency field can induce transitions between these two energy levels if its frequency ν satisfies the Bohr condition $h\,\nu = 2\,\mu\,B$. The power absorbed from this radiation field is then proportional to the difference in the number of nuclei in these two energy levels. Assume that the protons in the mineral oil are in thermal equilibrium at a temperature T that is sufficiently high that $\mu\,B \ll k\,T$. How does the absorbed power depend on the temperature, T, of the sample? From [Reif (1965)].

8.4 Consider an assembly of N_0 weakly-interacting magnetic atoms per unit volume, held at temperature T. According to classical physics, each atomic magnetic moment, $\boldsymbol{\mu}$, can be orientated so as to make an arbitrary angle θ with respect to the z-direction (say). In the absence of an external magnetic field, the probability that the angle lies between θ and $\theta + d\theta$ is simply proportional to the solid angle, $2\pi\,\sin\theta\,d\theta$, enclosed in this range. In the presence of a magnetic field of strength B_z, directed parallel to the z-axis, this probability is further proportional to the Boltzmann factor, $\exp(-\beta\,E)$, where $\beta = 1/(k\,T)$, and $E = -\boldsymbol{\mu}\cdot\mathbf{B}$ is the magnetic energy of the atom.

(a) Show that the classical mean magnetization is

$$\overline{M_z} = N_0\,\mu\,\mathcal{L}(x),$$

where $x = \mu\,B_z/(k\,T)$, and

$$\mathcal{L}(x) = \coth x - \frac{1}{x}$$

is known as the *Langevin function*.

(b) Demonstrate that the corresponding quantum mechanical expression for a collection of atoms with overall angular momentum $J\hbar$ is

$$\overline{M_z} = N_0\,\mu\,\mathcal{B}_J(x),$$

where $x = \mu\,B_z/(k\,T)$, $\mu = g\,J\,\mu_B$, and

$$\mathcal{B}_J(x) = \left[\left(\frac{2\,J+1}{2\,J}\right)\coth\left(\frac{2\,J+1}{2\,J}\,x\right) - \frac{1}{2\,J}\,\coth\left(\frac{x}{2\,J}\right)\right]$$

is the Brillouin function.

(c) Show, finally, that the previous two expressions are identical in the classical limit $J \gg 1$. (This is the classical limit because the spacing between adjacent magnetic energy levels, $E_m = \mu\,B_z\,m/J$, where m is an integer lying between $-J$ and $+J$, is $\Delta E_m = \mu\,B_z/J$, which tends to zero as $J \to \infty$.)

8.5 Consider a spin-1/2 (i.e., $J = 1/2$ and $g = 2$) paramagnetic substance containing N non-interacting atoms.

(a) Show that the overall magnetic partition function, Z, is such that

$$\ln Z = N\,\ln\left[2\,\cosh\left(\frac{\mu_B\,B_z}{k\,T}\right)\right],$$

where μ_B is the Bohr magneton, B_z the magnetic field-strength, and T the absolute temperature.

(b) Demonstrate that the mean magnetic energy of the system is

$$\overline{E} = -N\,\epsilon\,\tanh\left(\frac{\epsilon}{k\,T}\right),$$

where $\epsilon = \mu_B\,B_z$. Show that $\overline{E} \to -N\,\epsilon$ as $T \to 0$, and $\overline{E} \to 0$ as $T \to \infty$. Plot $\overline{E}/(N\,\epsilon)$ versus $k\,T/\epsilon$.

(c) Demonstrate that the magnetic contribution to the specific heat of the substance is

$$C = N\,k\left(\frac{\epsilon}{k\,T}\right)^2 \mathrm{sech}^2\left(\frac{\epsilon}{k\,T}\right),$$

and that

$$C \simeq N\,k\left(\frac{2\,\epsilon}{k\,T}\right)^2 \exp\left(-\frac{2\,\epsilon}{k\,T}\right)$$

when $k\,T \ll \epsilon$, whereas

$$C \simeq N\,k\left(\frac{\epsilon}{k\,T}\right)^2$$

when $k\,T \gg \epsilon$. Plot $C/(N\,k)$ versus $k\,T/\epsilon$. The sharp peak that is evident when $k\,T \sim \epsilon$ is known as the *Schottky anomaly*.

(d) Show that the magnetic contribution to the entropy of the substance is
$$S = Nk\left\{\ln\left[2\cosh\left(\frac{\epsilon}{kT}\right)\right] - \left(\frac{\epsilon}{kT}\right)\tanh\left(\frac{\epsilon}{kT}\right)\right\},$$
and demonstrate that $S \to 0$ as $T \to 0$ and $S \to Nk\ln 2$ as $T \to \infty$. Plot $S/(Nk)$ versus kT/ϵ.

8.6 The nuclei of atoms in a certain crystalline solid have spin one. According to quantum theory, each nucleus can therefore be in any one of three quantum states labeled by the quantum number m, where $m = 1, 0$, or -1. This quantum number measures the projection of the nuclear spin along a crystal axis of the solid. Because the electric charge distribution in the nucleus is not spherically symmetric, but ellipsoidal, the energy of a nucleus depends on its spin orientation with respect to the internal electric field existing at its location. Thus a nucleus has the same energy $E = \epsilon$ in the state $m = 1$ and the state $m = -1$, compared with energy $E = 0$ in the state $m = 0$.

(a) Find an expression, as a function of absolute temperature, T, of the nuclear contribution to the molar internal energy of the solid.
(b) Find an expression, as a function of T, of the nuclear contribution to the molar entropy of the solid.
(c) By directly counting the total number of accessible states, calculate the nuclear contribution to the molar entropy of the solid at very low temperatures. Calculate it also at high temperatures. Show that the expression in part (b) reduces properly to these values as $T \to 0$ and $T \to \infty$.
(d) Make a qualitative graph showing the temperature dependence of the nuclear contribution to the molar heat capacity of the solid. Calculate the temperature dependence explicitly. What is the temperature dependence for large values of T?

From [Reif (1965)]

8.7 A dilute solution of a macromolecule (large molecules of biological interest) at temperature T is placed in an ultracentrifuge rotating with angular velocity ω. The centripetal acceleration $\omega^2 r$ acting on a particle of mass m may then be replaced by an equivalent centrifugal force $m\omega^2 r$ in the rotating frame of reference.

(a) Find how the relative density, $\rho(r)$, of molecules varies with their radial distance, r, from the axis of rotation.
(b) Show qualitatively how the molecular weight of the macromolecules can be determined if the density ratio ρ_1/ρ_2 at the radii r_1 and r_2 is measured by optical means.

From [Reif (1965)].

8.8 Consider a homogeneous mixture of inert monatomic ideal gases at absolute temperature T in a container of volume V. Let there be ν_1 moles of gas 1, ν_2 moles of gas 2, ..., and ν_k moles of gas k.

(a) By considering the classical partition function of this system, derive its equation of state. In other words, find an expression for its total mean pressure, $\overline{p}$.

(b) How is this total pressure, $\overline{p}$, of the gas related to the so-called *partial pressure*, $\overline{p}_i$, that the ith gas would produce if it alone occupied the entire volume at this temperature?

8.9 Monatomic molecules adsorbed on a surface are free to move on this surface, and can be treated as a classical ideal two-dimensional gas. At absolute temperature T, what is the heat capacity per mole of molecules thus adsorbed on a surface of fixed size? From [Reif (1965)]

8.10 Consider a system in thermal equilibrium with a heat bath held at absolute temperature T. The probability of observing the system in some state r of energy E_r is is given by the canonical probability distribution:

$$P_r = \frac{\exp(-\beta\, E_r)}{Z},$$

where $\beta = 1/(k\, T)$, and

$$Z = \sum_r \exp(-\beta\, E_r)$$

is the partition function.

(a) Demonstrate that the entropy can be written

$$S = -k \sum_r P_r \ln P_r.$$

(b) Demonstrate that the mean Helmholtz free energy is related to the partition function according to

$$Z = \exp\left(-\beta\, \overline{F}\right).$$

8.11 Show that the logarithm of the classical partition function of an ideal gas consisting of N identical molecules of mass m, held in a container of volume V, and in thermal equilibrium with a heat bath held at absolute temperature T, is

$$\ln Z = N\left[\ln\left(\frac{V}{N}\right) - \frac{3}{2}\,\ln\beta + \sigma\right],$$

where

$$\sigma = \frac{3}{2}\ln\left(\frac{2\pi m}{h_0^2}\right) + 1.$$

Here, $\beta = kT$, and h_0 parameterizes how finely classical phase-space is partitioned. Demonstrate that:

(a)

$$\overline{E} = \frac{3}{2}\nu R T.$$

(b)

$$\overline{H} = \frac{5}{2}\nu R T.$$

(c)

$$\overline{F} = -\nu R T\left[\ln\left(\frac{V}{N}\right) - \frac{3}{2}\ln\beta + \sigma\right].$$

(d)

$$\overline{G} = -\nu R T\left[\ln\left(\frac{V}{N}\right) - \frac{3}{2}\ln\beta + \sigma - 1\right].$$

(e)

$$\frac{\Delta^* E}{\overline{E}} = \left(\frac{2}{3N}\right)^{1/2}.$$

8.12 Use the Debye approximation to calculate the contribution of lattice vibrations to the thermodynamic functions of a solid.

(a) To be more specific, show that

$$\ln Z = -\frac{9}{8}N\frac{\theta_D}{T} - 3N\ln\left(1 - e^{-\theta_D/T}\right) + N D\left(\frac{\theta_D}{T}\right),$$

$$\overline{E} = \frac{9}{8}N k\,\theta_D + 3NkT\,D\left(\frac{\theta_D}{T}\right),$$

$$S = Nk\left[-3\ln\left(1 - e^{-\theta_D/T}\right) + 4D\left(\frac{\theta_D}{T}\right)\right].$$

Here, N is the number of atoms in the solid, T the absolute temperature, Z the partition function, $\overline{E}$ the mean energy, S the entropy, $D(y) \equiv (3/y^3)\int_0^y x^3\,dx/(e^x - 1)$, $\theta_D = \hbar\,\omega_D/k$, and ω_D is the Debye frequency.

(b) Show that in the limit $T \ll \theta_D$,

$$\frac{\ln Z}{N} \simeq -\frac{9}{8}\left(\frac{\theta_D}{T}\right) + \frac{\pi^4}{5}\left(\frac{T}{\theta_D}\right)^3,$$

$$\frac{\overline{E}}{NkT} \simeq \frac{9}{8}\left(\frac{\theta_D}{T}\right) + \frac{3\pi^4}{5}\left(\frac{T}{\theta_D}\right)^3,$$

$$\frac{S}{Nk} \simeq \frac{4\pi^4}{5}\left(\frac{T}{\theta_D}\right)^3.$$

[Hint: $\int_0^y x^3\, dx/(e^x - 1) = \pi^4/15$.]

(c) Show that in the limit $T \gg \theta_D$,

$$\frac{\ln Z}{N} \simeq -\frac{9}{8}\left(\frac{\theta_D}{T}\right) - 3\ln\left(\frac{\theta_D}{T}\right) + 1,$$

$$\frac{\overline{E}}{NkT} \simeq \frac{9}{8}\left(\frac{\theta_D}{T}\right) + 3,$$

$$\frac{S}{Nk} \simeq -3\ln\left(\frac{\theta_D}{T}\right) + 4.$$

(d) Further, show that

$$\bar{p} = \frac{\gamma\overline{E}}{V},$$

where $\bar{p}$ is the mean pressure, V the volume, and

$$\gamma = -\frac{d\ln\theta_D}{d\ln V} = \frac{1}{3}.$$

8.13 For the quantized lattice waves (phonons) in the Debye theory of specific heats, the frequency, ω, of a propagating wave is related to its wavevector, $\mathbf{k}$, by the dispersion relation $\omega = c_s k$, where c_s is the velocity of sound. On the other hand, in a ferromagnetic solid at low temperatures, quantized waves of magnetization (spin waves) have their frequencies, ω, related to their wavevectors, $\mathbf{k}$, according to the dispersion relation $\omega = A k^2$, where A is a constant. Show that, at low temperatures, the contribution of spin waves to the heat capacity of the ferromagnet varies as $T^{3/2}$.

8.14 Verify directly that

$$\overline{v^2} = \frac{3kT}{m}$$

for a Maxwellian velocity distribution. Here, m is the molecular mass, and T the absolute temperature.

8.15 Show that the mean speed of molecules effusing through a small hole in a gas-filled container is $3\pi/8 = 1.18$ times larger than the mean speed of the molecules within the container.

8.16 A vessel is closed off by a porous partition through which gases can pass by effusion and then be pumped off to some collecting chamber. The vessel is filled with dilute gas containing two types of molecule which differ because they contain different atomic isotopes, and thus have the different masses, m_1 and m_2. The concentrations of these molecules are c_1 and c_2, respectively, and are maintained constant within the vessel by constantly replenishing the supply of gas in it.

(a) Let c_1' and c_2' be the concentrations of the two types of molecule in the collecting chamber. What is the ratio c_1'/c_2'?

(b) By using the gas UF_6, one can attempt to separate U^{235} from U^{237}, the first of these isotopes being the one that undergoes nuclear fission reactions. The molecules in the vessel are then $U^{235}F_6^{19}$ and $U^{238}F_6^{19}$. The concentrations of these molecules in the vessel corresponds to the natural abundances of the two isotopes: $c_{238} = 99.3$ percent, and $c_{235} = 0.7$ percent. What is the ratio, c_{235}'/c_{238}', of the two isotopic concentrations in the gas collected after effusion, compared to the original concentration ratio, c_{235}/c_{238}?

From [Reif (1965)]

8.17 Show that the mean force per unit area exerted on the walls of a container enclosing a Maxwellian gas is

$$\bar{p} = \int_{v_z>0} 2\,m\,v_z\,\Phi(\mathbf{v})\,d^3\mathbf{v} = \int_{v_z>0} 2\,m\,v_z^2\,f(v)\,d^3\mathbf{v},$$

where m is the molecular mass, and the outward normal to the wall element is directed in the z-direction. Hence, deduce that

$$\bar{p} = \frac{1}{3}\,m\,n\,\overline{v^2} = n\,k\,T,$$

where n is the molecular concentration, and T the absolute gas temperature.

8.18 Consider a spin-1/2 ferromagnetic material consisting of N identical atoms with $S = 1/2$ and $g = 2$. Let each atom have n nearest neighbors.

(a) Show that

$$\mathcal{B}_{1/2}(\eta) = \tanh\left(\frac{\eta}{2}\right),$$

where $\mathcal{B}_S(\eta)$ is a Brillouin function.

(b) Use the molecular field approximation to demonstrate that

$$m = \tanh\left[\frac{\beta}{\beta_c}\,(m + \beta_c\,\mu_B\,B_z)\right],$$

where $m = \overline{M}_z/(N\,\mu_B)$, $\beta_c = 1/(k\,T_c)$, and $k\,T_c = J\,n/2$.

(c) Show that for T slightly less than T_c, and in the absence of an external magnetic field,

$$m \simeq \sqrt{3}\left(1 - \frac{T}{T_c}\right)^{1/2},$$

(d) Demonstrate that exactly at the critical temperature,

$$m \simeq \left(\frac{3\,\mu_B\,B_z}{k\,T_c}\right)^{1/3}.$$

(e) Finally, show that for T slightly larger than T_c,

$$m \simeq \frac{\mu_B\,B_z}{k\,(T - T_c)}.$$

[Hint: At small arguments $\tanh(z) \simeq z - z^3/3$.]

Chapter 9

Chemical Equilibria

9.1 Introduction

Up to now, we have concentrated on interacting systems that can exchange energy with one another, or can do mechanical work on one another, or some combination of the two. In this chapter, we shall generalize our discussion to treat systems that can also exchange particles with one another. In addition, we shall consider systems consisting of multiple chemical species that can undergo chemical reactions with one another.

9.2 Grand Canonical Probability Distribution

Let us start by generalizing the discussion of Section 8.2. Consider a system A of fixed volume V that is in contact with a large reservoir A' with which it can exchange not only energy, but also particles. As before, we assume weak interaction between systems A and A', so that their energies, E and E', respectively, are additive. In other words,

$$E + E' = E^{(0)}, \tag{9.1}$$

where $E^{(0)}$ is the fixed total energy of the combined system. If N is the number of particles in A, and N' is the number of particles in A', then

$$N + N' = N^{(0)}, \tag{9.2}$$

where $N^{(0)}$ is the fixed number of particles in the combined system. (Obviously, we are treating the particles as indestructible objects.)

Let us calculate the probability, P_r, of finding system A in one particular microstate, r, of energy E_r and number of particles N_r. Let $\Omega'(E', N')$ be the number of states accessible to the reservoir A' when it has an energy in the range near E' and contains N' particles. If A is in a particular state r then the number of states accessible to the combined system is $\Omega'(E', N')$. Hence, according to the principle of equal a priori probabilities,

$$P_r(E_r, N_r) = C'\,\Omega'(E', N') = C'\,\Omega'(E^{(0)} - E_r, N^{(0)} - N_r), \tag{9.3}$$

where C' is a constant of proportionality that is independent of E_r and N_r, and use has been made of Equations (9.1) and (9.2).

We can now take into account the fact that system A is far smaller than system A'. It follows that $E_r \ll E^{(0)}$ and $N_r \ll N^{(0)}$, so the slowly-varying logarithm of P_r can be Taylor expanded about $E' = E^{(0)}$ and $N' = N^{(0)}$. Thus,

$$\ln P_r = \ln C' + \ln\,\Omega'(E^{(0)}, N^{(0)}) - \left(\frac{\partial \ln \Omega'}{\partial E'}\right)_0 E_r - \left(\frac{\partial \ln \Omega'}{\partial N'}\right)_0 N_r, \tag{9.4}$$

where the derivatives are evaluated for $E' = E^{(0)}$ and $N' = N^{(0)}$. In fact, let

$$\beta \equiv \left(\frac{\partial \ln \Omega'}{\partial E'}\right)_{E'=E^{(0)}, N'=N^{(0)}}, \tag{9.5}$$

$$\alpha \equiv \left(\frac{\partial \ln \Omega'}{\partial N'}\right)_{E'=E^{(0)}, N'=N^{(0)}}. \tag{9.6}$$

It follows that

$$P_r = C\,\exp(-\beta\,E_r - \alpha\,N_r), \tag{9.7}$$

where C is a constant that is independent of E_r and N_r. This probability distribution is known as the *grand canonical distribution*. The parameter β is the usual temperature parameter. (See Section 5.3.) Hence, $T = 1/(k\beta)$ is the absolute temperature of the reservoir, where k is the Boltzmann constant. The quantity $\mu = -\alpha\,k\,T$, which has units of energy, is known as the *chemical potential* of the reservoir. Using the standard definition $S' = k\,\ln\Omega'$ (see Section 5.6), it is clear that

$$\mu = -T\left(\frac{\partial S'}{\partial N'}\right)_{E'}. \tag{9.8}$$

It should be obvious, by analogy with the discussion in Section 8.4, that if one considers a system where only the mean energy $\overline{E}$ and the mean number of particles $\overline{N}$ of some system A are known then the distribution of systems in the ensemble is also described by the grand canonical distribution. However, the parameters β and α no longer characterize a reservoir. Instead, they are determined by the conditions that the system A has the specified mean energy $\overline{E}$, and the mean number of particle $\overline{N}$: that is,

$$\overline{E} = \frac{\sum_r E_r\,\mathrm{e}^{-\beta\,E_r - \alpha\,N_r}}{\sum_r \mathrm{e}^{-\beta\,E_r - \alpha\,N_r}}, \tag{9.9}$$

$$\overline{N} = \frac{\sum_r N_r\,\mathrm{e}^{-\beta\,E_r - \alpha\,N_r}}{\sum_r \mathrm{e}^{-\beta\,E_r - \alpha\,N_r}}. \tag{9.10}$$

Here, the sum is over all possible states of system A, irrespective of the energy or the number of particles.

9.3 Systems of Several Components

Consider an isolated homogeneous system, of energy E and volume V, that consists of c different types of molecule. Let N_i be the number of molecules of type i. The entropy of the system can be written

$$S = S(E, V, N_1, N_2, \cdots, N_c). \tag{9.11}$$

Suppose that the numbers of molecules of each species change as the result of a chemical reaction. Assuming that the process is infinitesimal and quasi-static, the entropy change is given by

$$dS = \left(\frac{\partial S}{\partial E}\right)_{V,N} dE + \left(\frac{\partial S}{\partial V}\right)_{E,N} dV + \sum_{i=1,c} \left(\frac{\partial S}{\partial N_i}\right)_{E,V,N} dN_i. \tag{9.12}$$

Here, the subscript N denotes the fact that all the molecule numbers are kept constant when taking the partial derivatives. In the case of a derivative such as $\partial S/\partial N_i$, the subscript denotes that all molecule numbers except for N_i are kept constant when taking the derivative.

Now, in the simple case in which all of the molecule numbers are kept fixed, we have the usual result

$$dS = \frac{dE + p\,dV}{T}. \tag{9.13}$$

[See Equation (6.5).] A comparison of the coefficients of dE and dV between the previous two equations (with $dN_i = 0$ for all i) reveals that

$$\left(\frac{\partial S}{\partial E}\right)_{V,N} = \frac{1}{T}, \tag{9.14}$$

$$\left(\frac{\partial S}{\partial V}\right)_{E,N} = \frac{p}{T}. \tag{9.15}$$

Let us introduce the quantity

$$\mu_i \equiv -T \left(\frac{\partial S}{\partial N_i}\right)_{E,V,N}. \tag{9.16}$$

This quantity is called the *chemical potential per molecule* of the ith chemical species, and has the units of energy. Note that this definition is consistent with Equation (9.8). Thus, Equation (9.12) becomes

$$dS = \frac{1}{T}\,dE + \frac{p}{T}\,dV - \sum_{i=1,c} \frac{\mu_i}{T}\,dN_i, \tag{9.17}$$

which can be rearranged to give

$$dE = T\,dS - p\,dV + \sum_{i=1,c} \mu_i\,dN_i. \tag{9.18}$$

The preceding expression is the generalization of the fundamental thermodynamic relation $dE = T\,dS - p\,dV$ for the case in which the numbers of the various types of constituent molecules are allowed to vary.

The chemical potential, μ_i, can be written in many forms that are equivalent to Equation (9.16). For example, suppose that all of the independent variables except for N_j are kept constant in Equation (9.18). It follows that $dS = dV = 0$ and $dN_i = 0$ for $i \neq j$. Hence, we obtain

$$\mu_j = \left(\frac{\partial E}{\partial N_j}\right)_{S,V,N}. \tag{9.19}$$

Alternatively, we can write Equation (9.18) in the form (see Section 6.11)

$$d(E - T\,S) \equiv dF = -S\,dT - p\,dV + \sum_{i=1,c} \mu_i\,dN_i. \tag{9.20}$$

If all independent variables other than N_j are kept constant then we find that

$$\mu_j = \left(\frac{\partial F}{\partial N_j}\right)_{T,V,N}. \tag{9.21}$$

We can also write Equation (9.18) in the form (see Section 6.12)

$$d(E - T\,S + p\,V) \equiv dG = -S\,dT + V\,dp + \sum_{i=1,c} \mu_i\,dN_i, \tag{9.22}$$

from which we deduce that

$$\mu_j = \left(\frac{\partial G}{\partial N_j}\right)_{T,p,N}. \tag{9.23}$$

Suppose that there is only a single chemical species present. Let us denote this species as species j. It follows that

$$G = G(T, p, N_j). \tag{9.24}$$

However, the Gibbs free energy, G, is an extensive quantity. Thus, if all of the independent extensive quantities are multiplied by some scale factor α then G must be multiplied by the same factor. In other words,

$$G(T, p, \alpha\,N_j) = \alpha\,G(T, p, N_j). \tag{9.25}$$

This implies that G is proportional to N_j. Hence, we can write

$$G(T, p, N_j) = N_j\,g_j(T, p), \tag{9.26}$$

where $g_j(T, p)$ is the Gibbs free energy per molecule of molecular species j (in the absence of any other molecular species). Hence, we deduce that

$$\mu_j(T, p) = \left(\frac{\partial G}{\partial N_j}\right)_{T,p} = g_j(T, p). \tag{9.27}$$

In other words, the chemical potential per molecule (in the absence of any other molecular species) is equal to the Gibbs free energy per molecule (in the absence of any other molecular species).

Let us try to generalize Equation (9.26) to allow for multiple chemical species. Consider the total energy

$$E = E(S, V, N_1, N_2, \cdots, N_c). \tag{9.28}$$

If we increase all extensive variables by the same scale factor then the previous equation must remain valid. In other words, if $S \to \alpha\, S$, $V \to \alpha\, V$, and $N_i \to \alpha\, N_i$ then $E \to \alpha\, E$. That is,

$$E(\alpha\, S, \alpha\, V, \alpha\, N_1, \cdots, \alpha\, N_m) = \alpha\, E(S, V, N_1, \cdots, N_c). \tag{9.29}$$

In particular, suppose that $\alpha = 1 + \gamma$, where $|\gamma| \ll 1$. The left-hand side of the previous equation becomes $E(S + \gamma\, S, V + \gamma\, V, N_1 + \gamma\, N_1, \cdots)$, which can be expanded to give

$$E + \left(\frac{\partial E}{\partial S}\right)_{V,N} \gamma\, S + \left(\frac{\partial E}{\partial V}\right)_{S,N} \gamma\, V + \sum_{i=1,c} \left(\frac{\partial E}{\partial N_i}\right)_{S,V,N} \gamma\, N_i = (1 + \gamma)\, E. \tag{9.30}$$

Hence, we deduce that

$$E = \left(\frac{\partial E}{\partial S}\right)_{V,N} S + \left(\frac{\partial E}{\partial V}\right)_{S,N} V + \sum_{i=1,c} \left(\frac{\partial E}{\partial N_i}\right)_{S,V,N} N_i. \tag{9.31}$$

However, the derivatives are also given by the respective coefficients of dS, dV, and dN_i in Equation (9.18). It follows that

$$E = T\, S - p\, V + \sum_{i=1,c} \mu_i\, N_i, \tag{9.32}$$

which can be rearranged to give

$$G \equiv E - T\, S + p\, V = \sum_{i=1,c} \mu_i\, N_i. \tag{9.33}$$

Note that when there are multiple molecular species, $\mu_i = \mu_i(T, p, N_j/N_i)$ for $j \neq i$. In other words, the chemical potential per molecule of species i is affected by the presence of molecules of other chemical species. (See Section 9.6.)

It follows from Equation (9.32) that

$$dE = T\, dS + S\, dT - p\, dV - V\, dp + \sum_{i=1,c} \mu_i\, dN_i + \sum_{i=1,c} N_i\, d\mu_i. \tag{9.34}$$

Comparison with Equation (9.18) reveals the general result

$$S\, dT - V\, dp + \sum_{i=1,c} N_i\, d\mu_i = 0, \tag{9.35}$$

which is known as the *Gibbs-Duhem relation*.

9.4 Equilibrium Between Phases

Consider an isolated system consisting of c chemical species and π distinct phases. Let E_j, S_j, T_j, p_j, for $j = 1,\pi$, denote the energy, entropy, temperature, and pressure, respectively, of the jth phase. Let N_{ij} and μ_{ij}, for $i = 1, c$ and $j = 1,\pi$, represent the number of molecules of type i in the jth phase, and the chemical potential of molecules of type i in the jth phase, respectively. The system is assumed to be isolated from its surroundings, so its total energy is fixed:

$$\sum_{j=1,\pi} E_j = E^{(0)}. \tag{9.36}$$

Likewise, the total volume of the system is assumed to be fixed:

$$\sum_{j=1,\pi} V_j = V^{(0)}. \tag{9.37}$$

Finally, assuming that there are no chemical reactions, the total numbers of molecules of each chemical species contained in the system are fixed:

$$\sum_{j=1,\pi} N_{ij} = N_i^{(0)}, \tag{9.38}$$

for $i = 1, c$.

Consider an infinitesimal change. According to Equation (9.17),

$$dS_j = \frac{dE_j}{T_j} + \frac{p_j}{T_j}\,dV_j - \sum_{i=1,c} \frac{\mu_{ij}}{T_j}\,dN_{ij}, \tag{9.39}$$

for $j = 1,\pi$. Thus, the total entropy change is

$$dS = \sum_{j=1,\pi} dS_j = \sum_{j=1,\pi} \frac{dE_j}{T_j} + \sum_{j=1,\pi} \frac{p_j}{T_j}\,dV_j - \sum_{j=1,\pi}\sum_{i=1,c} \frac{\mu_{ij}}{T_j}\,dN_{ij}. \tag{9.40}$$

In equilibrium, the system is in a maximum entropy state (see Section 7.2), which implies that $dS = 0$. According to Equations (9.36)–(9.38),

$$\sum_{j=1,\pi} dE_j = 0, \tag{9.41}$$

$$\sum_{j=1,\pi} dV_j = 0, \tag{9.42}$$

$$\sum_{j=1,\pi} dN_{ij} = 0, \tag{9.43}$$

for $i = 1, c$. Suppose that

$$T_1 = T_2 = \cdots = T_\pi, \tag{9.44}$$

$$p_1 = p_2 = \cdots = p_\pi, \tag{9.45}$$

$$\mu_{i1} = \mu_{i2} = \cdots = \mu_{i\pi}, \tag{9.46}$$

for $i = 1, c$. In this case, Equation (9.40) reduces to

$$dS = \frac{1}{T_\pi} \sum_{j=1,\pi} dE_j + \frac{p_\pi}{T_\pi} \sum_{j=1,\pi} dV_j - \sum_{i=1,c} \frac{\mu_{i\pi}}{T_\pi} \sum_{j=1\pi} dN_{ij}. \tag{9.47}$$

Making use of Equations (9.41)–(9.43), we deduce that $dS = 0$. In other words, the conditions (9.44)–(9.46) guarantee that we are in a maximum entropy state. Hence, these conditions also guarantee that the various different phases that make up the system are in equilibrium with one another.

The equilibrium conditions (9.44)–(9.46) are not particularly surprising. In order to be in *thermal equilibrium* with one another, the various different phases all have to have the same temperature. This ensures that heat does not flow from one phase to another. In order to be in *mechanical equilibrium* with one another, the various different phases all have to have the same pressure. This ensures that the volume of one particular phase does not grow at the expense of the volumes of the other phases. Finally, in order to be in *diffusive equilibrium* with one another, the chemical potentials of a given chemical species have to be the same in all phases. This ensures that the number of molecules of this species in a given phase does not grow at the expense of the numbers of molecules of the same species in the other phases.

Suppose that the system is not in thermal equilibrium. In this case, we would expect heat to flow from one phase to another until the temperatures of all the phases are the same. The direction of heat flow is always such as to increase the overall entropy of the system. Hence, it is clear from Equation (9.40) that heat flows from phases with higher temperatures to phases with lower temperatures.

Suppose that the system is not in mechanical equilibrium. In this case, we would expect the volumes of the various phases to adjust until the pressures of all the phases are the same. The volume changes are always such as to increase the overall entropy of the system. Hence, it is clear from Equation (9.40) that phases with higher pressures expand at the expense of phases with lower pressures.

Suppose that the system is not in diffusive equilibrium. In this case, we would expect molecules to diffuse from one phase to another until the chemical potentials of molecules of a particular type are the same in all the phases. The direction of molecular diffusion is always such as to increase the overall entropy of the system. Hence, it is clear from Equation (9.40) that molecules diffuse from phases with higher chemical potentials to phases with lower chemical potentials.

9.5 Gibbs Phase Rule

Let us continue the discussion of the previous section. The chemical makeup of the jth phase can be specified in terms of the *mole fractions* of the various different chemical species:

$$x_{ij} = \frac{N_{ij}}{\sum_{i=1,c} N_{ij}}. \tag{9.48}$$

Thus, x_{ij} is the fraction of molecules of type i in phase j. Note that

$$\sum_{i=1,c} x_{ij} = 1. \tag{9.49}$$

In other words, there are only $c-1$ independent mole fractions in each phase. Thus, in order to specify the internal state of phase j, we need to specify its temperature, T_j, its pressure, p_j, and the $c-1$ mole fractions x_{ij}, for $i = 1, c-1$. Hence, the internal state of each phase is determined by $c+1$ parameters. It follows that the internal state of the whole system is determined by $\pi\,(c+1)$ parameters. However, the system is subject to the equilibrium constraints (9.44)–(9.46). The condition that the temperatures of all the phases are equal is equivalent to $\pi-1$ interrelations between the T_j. Likewise, the condition that the pressures of all the phases are equal is equivalent to $\pi-1$ interrelations between the p_j. Finally, the condition that each chemical species has the same chemical potential in every phase is equivalent to $(\pi-1)\,c$ interrelations between the μ_{ij}. Hence, there are effectively $(\pi-1)\,(c+2)$ constraints on the system. The number of degrees of freedom of the system, f, is defined as the number of parameters that can be independently varied when the system is in equilibrium. Obviously, f is the difference between the number of parameters needed to specify the internal state of the system and the number of constraints on the system. Thus, $f = \pi\,(c+1) - (\pi-1)\,(c+2)$, which leads to

$$f = 2 + c - \pi. \tag{9.50}$$

This result is known as the *Gibbs phase rule*. It only holds in the absence of chemical reactions.

Let us consider a number of examples:

(1) A homogeneous fluid with a single chemical constituent and one phase. Here, $c = 1$, $\pi = 1$, and, hence, $f = 2$. Thus, we can vary T and p arbitrarily. An example of this type of system is an ideal gas.

(2) Water in equilibrium with its saturated vapour. In this case, $c = 1$, $\pi = 2$, and, hence, $f = 1$. Thus, once we choose the temperature, the pressure is determined, and vice versa. (See Section 7.9.)

(3) Ice, liquid water, and water vapour in equilibrium. In this case, $c = 1$, $\pi = 3$, and, hence, $f = 0$. Thus, the three phases of water can only coexist in equilibrium at a particular temperature and pressure. (See Section 7.8.)

9.6 Dilute Solutions

A solution is a mixture of different chemical species in which one species–the *solvent*—has a much larger mole fraction than the other species—the *solutes*. A solution is regarded as *dilute* if the solute molecules are so much less abundant than the solvent molecules that every solute molecule is always surrounded by solvent molecules, and never interacts directly with other solute molecules.

Consider a two-component dilute solution in which the solvent molecules are of type 0, and the solute molecules are of type 1. Let the solution contain N_0 solvent molecules, and N_1 solute molecules. We wish to determine the Gibbs free energy of the solution (because this will allow us to calculate the chemical potentials of the solvent and the solute molecules).

Suppose that we start with a pure solvent of type 0 molecules. In this case, according to Equation (9.27), the Gibbs free energy per molecule is equal to the chemical potential, $\mu_0(T, p)$, of pure type 0 molecules. Here, T and p are the temperature and pressure of the solution, respectively. Hence, the Gibbs free energy of the whole system is simply

$$G = N_0\,\mu_0(T, p). \tag{9.51}$$

Let us add a single type 1 molecule to the system, holding the temperature and the pressure fixed. Given that $G = E - T\,S + p\,V$, and T and p are constant, the increase in the Gibbs free energy of the system is

$$dG = dE - T\,dS + p\,dV. \tag{9.52}$$

Now, dE and $p\,dV$ are independent of N_0. Essentially, they depend on how the type 1 molecule interacts with its nearest neighbors, regardless of how many other type 0 molecules are present in the system. However, the situation is more complicated for the $T\,dS$ term. Part of dS is independent of N_0, but part comes from our freedom in choosing where to put the type 1 molecule. The number of choices is obviously proportional to N_0, so introducing the type 1 molecule increases the number of microstates by a factor proportional to N_0, and the increase in the entropy of the system, dS, consequently includes a term of the form $k\,\ln N_0$. (Here, we are assuming that the constant of proportionality is negligible compared to N_0.) The change in the Gibbs free energy of the system can thus be written

$$dG = f_1(T, p) - \ln N_0\,k\,T, \tag{9.53}$$

where $f_1(T, p)$ is independent of N_0.

Next, suppose that we add two type 1 molecules to the pure solvent. We can apply the preceding argument twice to arrive at

$$dG = 2\,f_1(T, p) - 2\,\ln N_0\,k\,T. \tag{9.54}$$

Unfortunately, the previous expression is incorrect because it fails to take into account the fact that the two type 1 molecules are identical. Interchanging the two molecules does not result in a new microstate, so we need to divide the increase in the number of microstates by 2. With this modification, we get

$$dG = 2\, f_1(T, p) - 2\, \ln N_0\, k\, T + \ln 2\, k\, T. \tag{9.55}$$

The generalization to many type 1 molecules is now straightforward. In the expression for dG we get $N_1\, f(T, p)$ and $N_1\, (-\ln N_0\, k\, T)$. To account for the fact that the type 1 molecules are identical, we get a term of the form $\ln(N_1!)\, k\, T \simeq N_1\, (\ln N_1 - 1)\, k\, T$. Adding these terms to the Gibbs free energy of the pure solvent, (9.51), we arrive at

$$G = N_0\, \mu_0(T, p) + N_1\, f_1(T, p) - (N_1\, \ln N_0)\, k\, T + (N_1\, \ln N_1)\, k\, T - N_1\, k\, T. \tag{9.56}$$

This expression is valid in the limit $N_1 \ll N_0$. The generalization to the case in which there are multiple solute species is straightforward:

$$\begin{aligned} G = {} & N_0\, \mu_0(T, p) \\ & + \sum_{i=1,c} [N_i\, f_i(T, p) - (N_i\, \ln N_0)\, k\, T + (N_i\, \ln N_i)\, k\, T - N_i\, k\, T]. \end{aligned} \tag{9.57}$$

Here, c is the number of solute species, and N_i is the number of molecules of solute species i present in the solution.

The chemical potentials per molecule of the solvent and solute species follows from the previous equation and Equation (9.23):

$$\mu_0 = \left(\frac{\partial G}{\partial N_0}\right)_{T,p,N} = \mu_0(T, p) - \left(\frac{\sum_{i=1,c} N_i}{N_0}\right) k\, T, \tag{9.58}$$

$$\mu_i = \left(\frac{\partial G}{\partial N_i}\right)_{T,p,N} = f_i(T, p) + \ln\left(\frac{N_i}{N_0}\right) k\, T, \tag{9.59}$$

for $i = 1, c$. It can be seen that adding more solute decreases the chemical potential per molecule of the solvent, but increases the chemical potential per molecule of the relevant solutes.

9.7 Molality

The *molality* of a solution, which is usually denoted m, is defined as the number of moles of solute per mass of solvent, and has units of moles per kilogram. The molality is proportional to N_i/N_0, where N_i is the number of solute molecules, and N_0 the number of solvent molecules. The constant of proportionality can be absorbed into the function $f_i(T, p)$ that appears in Equation (9.59) to give a new function called $\mu_i^\circ(T, p)$. The chemical potential of the solute can then be written

$$\mu_i = \mu_i^\circ(T, p) + \ln m_i \, k\, T, \tag{9.60}$$

where m_i is the molality of the solute (in moles per kilogram), and μ_i° is the chemical potential of the solute under the standard condition $m_i = 1$. Values of μ° can be found in standard chemical reference books [Rumble *et al.* (2017)]. Given that $N_i/N_0 \ll \ln(N_i/N_0)$ in a dilute solution, it is often a good approximation to write Equation (9.58) in the form

$$\mu_0 = \mu_0(T, p) \equiv \mu_0^\circ(T, p), \tag{9.61}$$

where $\mu_0^\circ(T, p)$ is the chemical potential of the solvent under the standard condition that it is pure.

9.8 Osmosis

Consider a solution that is separated from a pure solvent by a membrane that allows solvent molecules to pass through it, but not solute molecules. An example of such a semipermeable membrane is the membrane that surrounds any plant or animal cell, which is permeable to water and other very small molecules, but not to larger molecules or charged ions. Other semipermeable membranes are used in industry; for instance, in the desalination of water.

According to Equation (9.58), at a given temperature and pressure, the chemical potential of the solvent molecules in the solution is less than that in the pure solvent. Now, molecules tend to diffuse in the direction of lower chemical potential. Hence, the solvent molecules will spontaneously diffuse from the pure solvent into the solution. This process is known as *osmosis*.

In order to attain an equilibrium in which there is no osmosis, it is necessary to apply an additional pressure to the solution. The requisite pressure is such that the chemical potentials of the solvent molecules are the same on both sides of the membrane. Let p_0 be the pressure in the pure solvent, and p_1 the pressure in the solution. Making use of Equation (9.58), the condition for equilibrium is

$$\mu_0(T, p_0) = \mu_0(T, p_1) - \frac{N_1}{N_0} \, k\, T \tag{9.62}$$

Here, N_0 and N_1 are the number of solvent molecules and the total number of solute molecules (summing over all solute species), respectively, in the solution. Assuming that the pressure difference is relatively small, we can write

$$\mu_0(T, p_1) \simeq \mu_0(T, p_0) + (p_1 - p_0)\, \frac{\partial \mu_0}{\partial p}. \tag{9.63}$$

Combining the previous two equations, we obtain

$$(p_1 - p_0)\,\frac{\partial\mu_0}{\partial p} = \frac{N_1}{N_0}\,k\,T. \tag{9.64}$$

Recall that the chemical potential of a pure substance is just the Gibbs free energy per particle, G/N [see Equations (9.26) and (9.27)]. However, according to Equation (6.116),

$$\left(\frac{\partial G}{\partial p}\right)_T = V. \tag{9.65}$$

It follows that

$$\frac{\partial\mu_0}{\partial p} = \frac{V}{N}. \tag{9.66}$$

In other words, $\partial\mu_0/\partial p$ is equal to the volume per molecule of the pure solvent. However, if the solution is dilute then its volume per solvent molecule is essentially the same as that of the pure solvent. Thus, we can take V and N in the previous equation to be the volume of the solution and the number of solvent molecules in the solution (i.e., N_0), respectively. Equation (9.64) then gives

$$p_1 - p_0 = \frac{N_1\,k\,T}{V} = \frac{\nu_1\,R\,T}{V}, \tag{9.67}$$

where $\nu_1 = N_1/N_A$ is the number of moles of solute molecules in the solution. The pressure difference specified in the previous equation is known as *osmotic pressure*, and is the excess pressure required in the solution to prevent osmosis. Equation (9.67) is known as the *van't Hoff formula*, and states that the osmotic pressure is the same as the pressure of an ideal gas with the same concentration as the solute.

Incidentally, if we apply a pressure to the solution that exceeds the osmotic pressure then we get *reverse osmosis*; a process in which the solvent molecules diffuse from the solution into the pure solvent. Reverse osmosis is used commercially to purify water.

9.9 Boiling and Freezing Points

Let us employ the analysis of Section 9.6 to investigate how the presence of impurities can shift the boiling and freezing points of a substance.

Consider, first, the case of a dilute solution at its boiling point. The solution is in equilibrium with its gas phase. Let us assume, for the sake of simplicity, that the solute does not evaporate. (This is an excellent approximation for the case of salt in water.) It follows that the gas contains no solute. Hence, we only need to consider the equilibrium condition for the solvent:

$$\mu_{\rm liq}(T,p) = \mu_{\rm gas}(T,p). \tag{9.68}$$

Making use of Equation (9.58), we get

$$\mu_0(T,p) - \frac{N_1}{N_0}\,k\,T = \mu_{\rm gas}(T,p), \tag{9.69}$$

where $\mu_0(T,p)$ is the chemical potential of the pure solvent. Here, N_0 and N_1 are the numbers of solvent and solute molecules in the solution.

The procedure is to expand each μ function about a nearby point in which the pure solvent would be in equilibrium. Because μ depends on both temperature and pressure, we can hold either quantity fixed while allowing the other to vary. Let us first vary the pressure. Let p_0 be the vapor pressure of the pure solvent at temperature T. It follows that

$$\mu_0(T,p_0) = \mu_{\rm gas}(T,p_0). \tag{9.70}$$

Expanding Equation (9.69), we get

$$\mu_0(T,p_0) + (p-p_0)\,\frac{\partial\mu_0}{\partial p} - \frac{N_1}{N_0}\,k\,T = \mu_{\rm gas}(T,p_0) + (p-p_0)\,\frac{\partial\mu_{\rm gas}}{\partial p}. \tag{9.71}$$

However, the firsts terms on the left- and right-hand sides of the previous equation cancel. Furthermore, each $\partial\mu/\partial p$ is equal to the volume per molecule in that phase. Hence, we obtain

$$(p-p_0)\left(\frac{V}{N}\right)_{\rm liq} - \frac{N_1}{N_0}\,k\,T = (p-p_0)\left(\frac{V}{N}\right)_{\rm gas}. \tag{9.72}$$

Now, the volume per molecule in the gas phase is simply $k\,T/p_0$, whereas that in the liquid phase is negligible by comparison. Hence, we deduce that

$$\frac{p}{p_0} = 1 - \frac{N_1}{N_0}. \tag{9.73}$$

We conclude that the vapor pressure is reduced by a fraction equal to the ratio of the numbers of solute and solvent molecules. The preceding relation is known as *Raoult's law*. The reduction in the vapor pressure is due to the dilutation of solvent molecules by non-volatile solute molecules.

Let us now hold the pressure fixed, and vary the temperature. Let T_0 be the boiling point of the pure solvent at pressure p. It follows that

$$\mu_0(T_0,p) = \mu_{\rm gas}(T_0,p). \tag{9.74}$$

Expanding Equation (9.69), we get

$$\mu_0(T_0,p) + (T-T_0)\,\frac{\partial\mu_0}{\partial T} - \frac{N_1}{N_0}\,k\,T = \mu_{\rm gas}(T_0,p) + (T-T_0)\,\frac{\partial\mu_{\rm gas}}{\partial T}. \tag{9.75}$$

Again, the first terms on the left- and right-hand sides cancel. Furthermore, each $\partial\mu/\partial T$ is equal to minus the entropy per particle for that phase [because $\partial G/\partial T = -S$; see Equation (6.115)]. Hence

$$-(T - T_0)\left(\frac{S}{N}\right)_{\rm liq} - \frac{N_1}{N_0}\,k\,T = -(T - T_0)\left(\frac{S}{N}\right)_{\rm gas}. \tag{9.76}$$

Let us set N (and N_0) on either side of the previous equation equal to Avagadro's number, N_A. It follows that each S now refers to N_A molecules of solvent. Now, the difference in entropy between the gas and the liquid is L/T_0, where L is the latent heat (per mole) of vaporization. Thus,

$$T - T_0 = \frac{N_1\,k\,T\,T_0}{L} \simeq \frac{\nu_1\,R\,T_0^{\,2}}{L}, \tag{9.77}$$

where we have approximated T and T_0 on the right-hand side. It follows that the presence of impurities raises the boiling point of a substance.

In order to investigate the effect of impurities on the melting point of a substance, we simply take Equation (9.76), and replace the gas phase by the solid phase. In other words,

$$-(T - T_0)\left(\frac{S}{N}\right)_{\rm liq} - \frac{N_1}{N_0}\,k\,T = -(T - T_0)\left(\frac{S}{N}\right)_{\rm solid}. \tag{9.78}$$

If we again set $N = N_0 = N_A$ on both sides of the equation then the difference between the entropy of the liquid and the solid phase is L/T_0, where T_0 is the melting point of the pure substance, and L is the latent heat (per mole) of fusion. Hence, we obtain

$$T - T_0 \simeq -\frac{\nu_1\,R\,T_0^{\,2}}{L}. \tag{9.79}$$

It follows that the presence of impurities lowers the melting point of a substance.

9.10 General Conditions for Chemical Equilibrium

Consider a homogeneous system (i.e., a system consisting of a single phase) that contains c different types of molecule. Let us designate the chemical symbols of these molecules as B_1, B_2, $\cdots$, B_c. Consider a chemical reaction between the molecules by which they are transformed into one another. This reaction must be consistent with the conservation of the total number of atoms of each kind. A properly balanced chemical equation ensures that this is the case.

For instance, suppose that the system consists of H_2, O_2, and H_2O molecules in the gas phase. The molecules can transform into each other through the chemical reaction

$$2\,{\rm H_2} + {\rm O_2} \rightleftarrows 2\,{\rm H_2O}.$$

This chemical equation is properly balanced because the total numbers of hydrogen and oxygen atoms are the same on both sides of the equation.

Let b_i denote the coefficient of B_i in the chemical equation. It follows that b_i is a small integer designating the number of B_i molecules involved in the chemical reaction. For the sake of convenience, we shall adopt the convention that b_i is positive for any product molecule formed as the result of the reaction, whereas b_i is negative for a reagent molecule that is used up by the reaction. (The direction in which a given chemical reaction is assumed to proceed is arbitrary.) For example, the reaction

$$2\,H_2 + O_2 \rightarrow 2\,H_2O.$$

would be written in the standard form as

$$-2\,H_2 - O_2 + 2\,H_2O = 0.$$

It follows that $b_{H_2} = -2$, $b_{O_2} = -1$, and $b_{H_2O} = +2$. By analogy, a general chemical reaction takes the form

$$\sum_{i=1,c} b_i\,B_i = 0. \tag{9.80}$$

Let N_i denote the number of B_i molecules in the system. The various N_i can change as a result of the chemical reaction. However, the change must be such as to conserve the number of atoms of a given type. In other words, the change must be such that Equation (9.80) is satisfied. It follows that the changes in the N_i must be proportional to the corresponding numbers of molecules appearing in the balance equation. In other words,

$$dN_i = \lambda\,b_i, \tag{9.81}$$

for $i = 1, c$. Here, $dN_i > 0$ for molecules that are formed as a result of the reaction, and $dN_i < 0$ for molecules that are used up by the reaction.

Consider an equilibrium state in which the molecules confined within an isolated enclosure of volume V can react chemically with one another in accordance with the reaction (9.80). Let E denote the total energy of the system. The equilibrium condition is that $S = S(E, V, N_1, \cdots, N_c)$ attains a maximum value. In other words, $dS = 0$. According to Equation (9.17), under the assumed constraints that V and E are constant, this condition becomes

$$\sum_{i=1,c} \mu_i\,dN_i = 0. \tag{9.82}$$

Combining the previous two equations, we arrive at the general condition for chemical equilibrium:

$$\sum_{i=1,c} b_i\,\mu_i = 0. \tag{9.83}$$

9.11 Dissociation of Water

Consider the chemical reaction

$$-\mathrm{H_2O} + \mathrm{H}^+ + \mathrm{OH}^- = 0,$$

by which water molecules dissociate to give H^+ and OH^- ions. Assuming that the solution is dilute (which is a very good approximation under normal circumstances), the chemical potentials are given by Equations (9.60) and (9.61):

$$\mu_{\mathrm{H_2O}} = \mu^{\circ}_{\mathrm{H_2O}}, \tag{9.84}$$

$$\mu_{\mathrm{H}^+} = \mu^{\circ}_{\mathrm{H}^+} + \ln m_{\mathrm{H}^+}\, k\,T, \tag{9.85}$$

$$\mu_{\mathrm{OH}^-} = \mu^{\circ}_{\mathrm{OH}^-} + \ln m_{\mathrm{OH}^-}\, k\,T. \tag{9.86}$$

Here, m_{H^+} and m_{OH^-} are the molalities of the H^+ and OH^- ions, respectively, in moles per kilogram of water. Moreover, $\mu^{\circ}_{\mathrm{H_2O}}$, $\mu^{\circ}_{\mathrm{H}^+}$, and $\mu^{\circ}_{\mathrm{OH}^-}$ are the chemical potentials of H_2O, H^+, and OH^-, respectively, in their standard states. These states are pure liquid for H_2O, and unit molality solutions for H^+ and OH^-.

The general condition for chemical equilibrium, (9.83), gives

$$-\mu_{\mathrm{H_2O}} + \mu_{\mathrm{H}^+} + \mu_{\mathrm{OH}^-} = 0. \tag{9.87}$$

Making use of Equations (9.84)–(9.86), we obtain

$$-\mu^{\circ}_{\mathrm{H_2O}} + \mu^{\circ}_{\mathrm{H}^+} + \ln m_{\mathrm{H}^+}\, k\,T + \mu^{\circ}_{\mathrm{OH}^-} + \ln m_{\mathrm{OH}^-}\, k\,T = 0. \tag{9.88}$$

Rearranging, and multiplying through by N_A, we get

$$R\,T\,\ln(m^+_{\mathrm{H}}\, m^-_{\mathrm{OH}}) = -N_A\left(\mu^{\circ}_{\mathrm{H}^+} + \mu^{\circ}_{\mathrm{OH}^-} - \mu^{\circ}_{\mathrm{H_2O}}\right) \equiv -\Delta G^{\circ}, \tag{9.89}$$

where ΔG° is the change in the Gibbs free energy [see Equation (9.33)] in a reaction in which a mole of water molecules dissociate. This quantity can be looked up in a standard chemical reference book [Rumble *et al.* (2017)]. The previous equation implies that

$$m^+_{\mathrm{H}}\, m^-_{\mathrm{OH}} = K, \tag{9.90}$$

where the equilibrium constant, K, takes the form

$$K = \mathrm{e}^{-\Delta G^{\circ}/R\,T}. \tag{9.91}$$

The value of ΔG° for the dissociation of one mole of water at room temperature and atmospheric pressure is 79.9 kJ [Rumble *et al.* (2017)]. Hence, the equilibrium constant of the reaction at this temperature and pressure is

$$K = \exp\left[\frac{-\left(79.9\times 10^3\right)}{(8.314)\,(298)}\right] = 1.0\times 10^{-14}. \tag{9.92}$$

If all of the H^+ and OH^- ions come from the dissociation of water molecules then they must be equally abundant. It follows that

$$m_{H^+} = m_{OH^-} = 1.0\times 10^{-7}. \qquad (9.93)$$

Now, the pH is defined as minus the base-10 logarithm of the molality of H^+ ions:

$$\text{pH} = -\log_{10} m_{H^+}. \qquad (9.94)$$

It follows that the pH of pure water is 7. If other substances are dissolved in water then they can modify the pH. If the pH is less than 7 (indicating an excess of H^+ ions) then the solution is termed *acidic*. On the other hand, if the pH is greater than 7 (indicating an excess of OH^+ ions) then the solution is termed *basic*.

9.12 Ideal Gas Mixture

Consider a gas that can be considered to be ideal, and is made up of c chemical species. Suppose that the gas is confined in a container of volume V that is held at absolute temperature T.

Let the possible states of the kth molecule in the gas be labeled by s_k, and let $\epsilon_k(s_k)$ denote the energy of the molecule in state s_k. Because there is negligible interaction between different molecules in an ideal gas, the total energy of the gas in one of its possible microstates can be written in the form

$$E = \epsilon_1(s_1) + \epsilon_2(s_2) + \epsilon_3(s_3) + \cdots, \qquad (9.95)$$

where the number of terms in the sum is equal to the total number of molecules. Hence, the partition function becomes

$$Z = \frac{\sum_{s_1,s_2,s_3,\cdots} \exp\left(-\beta\,[\epsilon_1(s_1) + \epsilon_2(s_2) + \epsilon_3(s_3) + \cdots]\right)}{N_1!\,N_2!N_3!\cdots N_c!}, \qquad (9.96)$$

where we have divided by the $N_1!\,N_2!\,N_3!\cdots N_c!$ possible permutations of like molecules among themselves to avoid the Gibbs paradox. (See Section 8.8.) Now,

$$\sum_{s_1,s_2,\cdots} e^{-\beta\,[\epsilon_1(s_1)+\epsilon_2(s_2)+\cdots]} = \left(\sum_{s_1} e^{-\beta\,\epsilon(s_1)}\right)\left(\sum_{s_2} e^{-\beta\,\epsilon(s_2)}\right)\cdots. \qquad (9.97)$$

In the product on the right-hand side of the previous equation there are N_i equal factors for all molecules of type i, each of these factors taking the form

$$\zeta_i = \sum_{s_i} e^{-\beta\,\epsilon_i(s_i)}, \qquad (9.98)$$

where the sum is over all of the possible states, s_i, of molecules of type i, whose energies are denoted the $\epsilon_i(s_i)$. Hence,

$$\sum_{s_1,s_2\cdots} e^{-\beta\,[\epsilon_1(s_1)+\epsilon_2(s_2)+\cdots]} = \zeta_1^{N_1}\,\zeta_2^{N_2}\cdots\zeta_C^{N_c}. \qquad (9.99)$$

It follows that the partition function (9.96) factorizes to give

$$Z = Z_1\, Z_2 \cdots Z_c, \tag{9.100}$$

where

$$Z_i = \frac{\zeta_i^{N_i}}{N_i!} \tag{9.101}$$

is the partition function of a gas of N_i molecules of type i occupying the given volume V in the absence of the other gases.

Equation (9.100) implies that

$$\ln Z = \sum_{i=1,c} \ln Z_i. \tag{9.102}$$

A variety of important results follow from this equation. For example, because the mean energy of a system is given by $\overline{E} = -(\partial \ln Z/\partial\beta)$ (see Section 8.5), we deduce that

$$\overline{E} = \sum_{i=1,c} \overline{E}_i, \tag{9.103}$$

where $\overline{E}_i$ is the mean energy of the ith gas occupying the volume V in the absence of the other gases. Furthermore, because the mean pressure of a system is given by $\overline{p} = \beta^{-1}\,(\partial \ln Z/\partial V)$, we find that

$$\overline{p} = \sum_{i=1,c} \overline{p}_i, \tag{9.104}$$

where $\overline{p}_i$ is the mean pressure that would be exerted by the ith gas if it occupied the volume V in the absence of the other gases. The quantity $\overline{p}_i$ is known as the *partial pressure* of the ith gas.

Now, the equation of state of the ith gas occupying the volume V by itself is simply

$$\overline{p}_i = n_i\, k\, T, \tag{9.105}$$

where $n_i = N_i/V$. It follows that the equation of state of the gas mixture is

$$\overline{p} = n\, k\, T, \tag{9.106}$$

where

$$n = \sum_{i=1,c} n_i. \tag{9.107}$$

Finally, because the entropy of system is given by $S = k\,(\ln Z + \beta\,\overline{E})$, we deduce, with the aid of Equation (9.103), that

$$S = \sum_{i=1,c} S_i, \tag{9.108}$$

where S_i is the entropy of the ith gas occupying the volume V in the absence of the other gases.

9.13 Chemical Potentials of Ideal Gas Mixture

Continuing the discussion of the previous section, the Helmholtz free energy and entropy of the system are given by $F = \overline{E} - T\,S$ and $S = k\,(\ln Z + \beta\,\overline{E})$, respectively, so it follows that $F = -k\,T\,\ln Z$. Hence,

$$F = \sum_{i=1,c} F_i, \tag{9.109}$$

where, according to Equations (9.101) and (9.102),

$$F_i = -k\,T\,\ln Z_i = -k\,T\,(N_i\,\ln\zeta_i - \ln N_i!) \tag{9.110}$$

is the Helmholtz free energy of the ith gas occupying the volume V in the absence of the other gases. Here, $\zeta_i = \zeta_i(T, V)$ is the partition function of a single molecule of species i, and is, therefore, independent of N_i. We can write

$$F(T, V) = -k\,T \sum_{i=1,c} N_i\,(\ln\zeta_i - \ln N_i - 1), \tag{9.111}$$

where use has been made of Stirling's approximation, $\ln N! = N\,\ln N - N$.

According to Equation (9.21), the chemical potential of gas species i takes the form

$$\mu_i(T, V) = \left(\frac{\partial F}{\partial N_i}\right)_{T,V,N} = -k\,T\,\ln\left(\frac{\zeta_i}{N_i}\right), \tag{9.112}$$

where use has been made of the previous equation. Furthermore, employing the analysis of Section 8.7, the partition function of a single molecule of species i can be written

$$\zeta_i(T, V) = \frac{1}{h^3}\int d^3\mathbf{r}\int d^3\mathbf{p}\,\mathrm{e}^{-\beta\,p^2/2m_i}\sum_{s_i}^{(\mathrm{int})}\mathrm{e}^{-\beta\,\epsilon_i(s_i)}, \tag{9.113}$$

where we have treated the translational degrees of freedom of the molecule classically, and the remaining sum is over the internal (i.e., vibration, rotation, electronic, and nuclear) states of the molecule. The parameter h measures the size of the cells into which classical phase space is divided. As we shall see in Section 10.10, if h is set equal to Planck's constant then the classical calculation gives the same results as a corresponding quantum mechanical calculation. The parameter m_i is the mass of a molecule of species i. Performing the integrals, which are straightforward, we get

$$\zeta_i(T, V) = V\left(\frac{2\pi\,m_i\,k\,T}{h^2}\right)^{3/2}\zeta_i^{(\mathrm{int})}(T), \tag{9.114}$$

where

$$\zeta_i^{(\mathrm{int})}(T) = \sum_{s_i}^{(\mathrm{int})} \mathrm{e}^{-\beta\,\epsilon_i(s_i)}. \tag{9.115}$$

Combining Equations (9.112) and (9.114), our final expression for the chemical potential of the i species becomes

$$\mu_i(T,V) = -k\,T\,\ln\left(\frac{\zeta_i^{(\mathrm{int})}}{n_i}\right) - \frac{3}{2}\,k\,T\,\ln\left(\frac{2\pi\,m_i\,k\,T}{h^2}\right), \tag{9.116}$$

where $n_i = N_i/V$ is the number density of molecules of species i.

9.14 Law of Mass Action

Combining the general condition for chemical equilibrium (9.83) with Equation (9.112), we arrive at

$$\sum_{i=1,c} b_i\,\mu_i = \varDelta F_0 + k\,T\sum_{i=1,c} b_i\,\ln N_i = 0, \tag{9.117}$$

where

$$\varDelta F_0 = -k\,T\sum_{i=1,c} b_i\,\ln\zeta_i \tag{9.118}$$

is known as the *standard free energy change of the reaction.* Note that $\varDelta F_0$ is a function of T and V, but is independent of the N_i. The equilibrium condition becomes

$$\sum_{i=1,c} \ln N_i^{b_i} = \ln(N_1^{b_1}\,N_2^{b_2}\cdots N_c^{b_c}) = -\frac{\varDelta F_0}{k\,T}, \tag{9.119}$$

or

$$N_1^{b_1}\,N_2^{b_2}\cdots N_c^{b_c} = K_N(T,V), \tag{9.120}$$

where

$$K_N(T,V) = \mathrm{e}^{-\varDelta F_0/k\,T} = \zeta_1^{b_1}\,\zeta_2^{b_2}\cdots\zeta_c^{b_c}. \tag{9.121}$$

The quantity K_N is independent of the numbers of molecules present in the system, and is known as the *equilibrium constant.* In general, K_N is a function of T and V through the dependence of the molecular partition functions, ζ_i, on these quantities. Equation (9.120) is known as the *law of mass action.*

For example, consider the chemical reaction

$$-2\,H_2 - O_2 + 2\,H_2O = 0.$$

The law of mass action yields

$$N_{H_2}^{-2}\,N_{O_2}^{-1}\,N_{H_2O}^{\,2} = K_N, \tag{9.122}$$

or

$$\frac{N_{H_2O}^{\,2}}{N_{H_2}^{\,2}\,N_{O_2}} = K_N(T,V). \tag{9.123}$$

Now, according to Equation (9.114),

$$\zeta_i(T,V) = V\,\zeta_i'(T), \tag{9.124}$$

where

$$\zeta_i'(T) = \left(\frac{2\pi\,m_i\,k\,T}{h^2}\right)^{3/2}\,\zeta_i^{(\rm int)}(T). \tag{9.125}$$

Making use of Equations (9.121) and (9.124), the law of mass action, (9.120), can be re-expressed in the form

$$n_1^{\,b_1}\,n_2^{\,b_2}\cdots n_c^{\,b_c} = K_n(T), \tag{9.126}$$

where $n_i = N_i/V$, and

$$K_n(T) = \zeta_1'^{\,b_1}\,\zeta_2'^{\,b_2}\cdots\zeta_3'^{\,b_3}. \tag{9.127}$$

Note that

$$K_N(T,V) = V^{\,b}\,K_n(T), \tag{9.128}$$

where $b = \sum_{i=1,c} b_i$.

9.15 Temperature Dependence of Equilibrium Constant

Equations (9.125) and (9.127) imply that

$$\frac{d\ln K_n}{dT} = \frac{\sum_{i=1,c} b_i\left[\epsilon_i^{(\rm trans)} + \overline{\epsilon}_i^{(\rm int)}\right]}{k\,T^2}, \tag{9.129}$$

where

$$\epsilon_i^{(\rm trans)} = \frac{3}{2}\,k\,T, \tag{9.130}$$

is the translational energy of the ith molecule, and

$$\overline{\epsilon}_i^{(\rm int)} = \frac{\sum_{s_i}^{(\rm int)}\epsilon_i(s_i)\,{\rm e}^{-\beta\,\epsilon_i(s_i)}}{\sum_{s_i}^{(\rm int)}{\rm e}^{-\beta\,\epsilon_i(s_i)}} \tag{9.131}$$

is the mean internal energy of the ith molecule. It follows that

$$\frac{d\ln K_n}{dT} = \frac{\Delta E}{R\,T^2}, \tag{9.132}$$

where

$$\Delta E = \sum_{i=1,c} b_i\, N_A \left[\epsilon_i^{(\rm trans)} + \overline{\epsilon}_i^{(\rm int)}\right] \tag{9.133}$$

is the mean increase in internal energy of the system when $|b_i|$ moles of each reactant are converted into b_i moles of each reaction product. Given that the reaction is carried out at constant volume, the first law of thermodynamics implies that ΔE is the heat absorbed by the system from its surroundings per reaction. It follows that if $\Delta E > 0$ then the reaction is *endothermic* (i.e., the reaction causes a spontaneous flow of heat into the system). On the other hand, if $\Delta E < 0$ then the reaction is *exothermic* (i.e., the reaction causes a spontaneous flow of heat into the surroundings). Thus, according to Equation (9.132), the equilibrium constant of an endothermic/exothermic reaction increases/decreases with temperature, which is in accordance with Le Châtelier's principle.

9.16 Saha Equation

Consider a system consisting of a single species of atom in a gaseous phase that is held at a constant temperature T. Consider the chemical reaction

$$-{\rm A} + {\rm e}^- + {\rm A}^+ = 0,$$

where A represents an atom, $\rm e^-$ represents an electron, and $\rm A^+$ represents the atom's first ionized state. Clearly, in this particular reaction, the atom is stripped of a single electron and becomes an ion. In other words, the atom is ionized. We wish to determine the relative equilibrium number densities of atoms, ions, and electrons as a functions of T.

According to Equations (9.126) and (9.127), the law of mass action yields

$$\frac{n_{\rm A^+}\, n_{\rm e^-}}{n_{\rm A}} = \frac{\zeta'_{\rm A^+}\, \zeta'_{\rm e^-}}{\zeta'_{\rm A}}, \tag{9.134}$$

where n_i is the number density of species i, and ζ'_i is defined in Equations (9.115) and (9.125). For the case of the electron, which has no internal electronic states, we obtain

$$\zeta'_{\rm e^-} = 2\left(\frac{2\pi\, m\, k\, T}{h^2}\right)^{3/2}, \tag{9.135}$$

where m is the electron mass. Here, the factor of 2 arises because the electron, which is a spin 1/2 particle, possesses two possible spin states for each translational state. For the case of the atom,

$$\zeta'_{\rm A} = g_{\rm A}\left(\frac{2\pi\, M\, k\, T}{h^2}\right)^{3/2} {\rm e}^{-\epsilon_{\rm A}/k\,T}, \tag{9.136}$$

where M is the atomic mass, $\epsilon_{\rm A}$ is the energy of the atom's ground electronic state, and $g_{\rm A}$ the degeneracy of this state. Here, we are neglecting the excited electronic states of the atom (because the relevant Boltzmann factors are much smaller than the ground-state Boltzmann factor). Finally, for the case of the ion,

$$\zeta'_{\rm A^+} = g_{\rm A^+} \left[\frac{2\pi\,(M-m)\,k\,T}{h^2}\right]^{3/2} {\rm e}^{-\epsilon_{\rm A^+}/k\,T}, \tag{9.137}$$

where $\epsilon_{\rm A^+}$ is the energy of the ion's ground electronic state, and $g_{\rm A^+}$ the degeneracy of this state. We are again neglecting the excited electronic states. Given that $m \ll M$, the previous four equations can be combined to give

$$\frac{n_{\rm A^+}\,n_{\rm e^-}}{n_{\rm A}} = \frac{2\,g_{\rm A^+}}{g_{\rm A}} \left(\frac{2\pi\,m\,k\,T}{h^2}\right)^{3/2} {\rm e}^{-\epsilon/k\,T}, \tag{9.138}$$

where $\epsilon = \epsilon_{\rm A^+} - \epsilon_{\rm A}$ is the energy that must be provided to the atom in order to ionize it; this quantity is called the *ionization energy*. The previous expression is known as *Saha's equation*.

Consider the ionization of hydrogen:

$$-{\rm H} + {\rm e}^- + {\rm H}^+ = 0.$$

The multiplicity of ${\rm H}^+$ (i.e., proton) states is $g_{\rm H^+} = 2$, because the hydrogen ion is a spin 1/2 particle that possesses two spin states per translational state. The multiplicity of ground-state hydrogen atom states is $g_{\rm H} = 4$, because there are four possible ways of arranging the nuclear and the electron spins (spin up/spin up, spin up/spin down, spin down/spin up, and spin down/spin down). Hence, the Saha equation yields

$$\frac{n_{\rm H^+}\,n_{\rm e^-}}{n_{\rm H}} = \left(\frac{2\pi\,m\,k\,T}{h^2}\right)^{3/2} {\rm e}^{-\epsilon/k\,T}, \tag{9.139}$$

where $\epsilon = 13.6$ eV [Rumble *et al.* (2017)].

Exercises

9.1 Show that Equations (9.58) and (9.59) are consistent with Equation (9.33), which states that

$$G = N_0\,\mu_0 + \sum_{i=1,c} N_i\,\mu_i.$$

9.2 Show that if there are r independent chemical reactions then the Gibbs phase rule generalizes to

$$f = 2 + c - r - \pi.$$

9.3 Determine the number of degrees of freedom at equilibrium of a chemically reactive system containing solid sulphur, S, and the three gases O_2, SO_2, and SO_3. Two chemical reactions are present in the system:

$$S + O_2 \rightleftarrows SO_2,$$

$$2\,S + 3\,O_2 \rightleftarrows 2\,SO_3.$$

9.4 Show that the chemical potential of an ideal gas at fixed temperature T varies with pressure as

$$\mu = \mu^\circ + \ln\left(\frac{p}{p^\circ}\right) RT,$$

where μ° is the value of μ at the reference point (p°, T).

9.5 (a) Show that, at constant temperature and pressure, the Gibbs-Duhem equation leads to the relation

$$\sum_{i=1,c} x_i\, d\mu_i = 0,$$

where x_i is the mole fraction of the ith component.

(b) Show that for a two-component system

$$\left(\frac{d\mu_1}{d\ln x_1}\right)_{T,p} = \left(\frac{d\mu_2}{d\ln x_2}\right)_{T,p}.$$

(c) Show that for a two-component liquid phase whose vapor pressure can be treated as an ideal gas mixture,

$$\left(\frac{dp_1}{d\ln x_1}\right)_{T,p} = \left(\frac{dp_2}{d\ln x_2}\right)_{T,p},$$

where p_1 and p_2 are the partial pressures of the components of the vapor, and x_1 and x_2 are the mole fractions of the liquid. The previous relation is known as the *Duhem-Margules equation.*

(From [Carter (2001)])

9.6 At fixed temperature, $T = 1200^\circ$ K, the gases

$$CO_2 + H_2 \rightleftarrows CO + H_2O$$

are in chemical equilibrium in a vessel of volume V. If the volume of the vessel is increased, its temperature being maintained constant, does the relative concentration of CO_2 increase, decrease, or stay the same? (From [Reif (1965)])

9.7 Consider the following reaction between ideal gases:

$$\sum_{i=1,c} b_i B_i = 0.$$

Let T be the temperature, and let p_i be the partial pressure of the ith species. Show that the law of mass action can be expressed in the form

$$p_1^{b_1} p_2^{b_2} \cdots p_c^{b_c} = K_p(T),$$

where K_p only depends on T. (From [Reif (1965)])

9.8 Show that if the chemical reaction of the preceding problem is carried out at constant total pressure then the heat of reaction per mole (i.e., the heat that must be supplied to transform $|b_i|$ moles of each of the reactants into b_i moles of each of the reaction products) is given by the enthalpy change

$$\Delta H = \sum_{i=1,c} b_i h_i,$$

where h_i is the enthalpy per mole of the ith gas at the given temperature and pressure. (From [Reif (1965)])

9.9 Show that

$$\frac{d \ln K_p}{dT} = \frac{\Delta H}{R T^2}.$$

This result is known as the *van't Hoff equation*. (From [Reif (1965)])

Chapter 10

Quantum Statistics

10.1 Introduction

Previously, we investigated the statistical thermodynamics of ideal gases using a rather ad hoc combination of classical and quantum mechanics. (See, for example, Sections 8.7 and 8.8.) In fact, we employed classical mechanics to deal with the translational degrees of freedom of the constituent particles, and quantum mechanics to deal with the non-translational degrees of freedom. Let us now discuss ideal gases from a purely quantum-mechanical standpoint. It turns out that this approach is necessary in order to treat either very low-temperature or very high-density gases. Furthermore, it also allows us to investigate completely non-classical "gases," such as photons, or the conduction electrons in a metal.

10.2 Symmetry Requirements in Quantum Mechanics

Consider a gas consisting of N identical, non-interacting, structureless particles, enclosed within a container of volume V. Let Q_i denote collectively all the coordinates of the ith particle. In other words, the three Cartesian coordinates that determine its spatial position, as well as the spin coordinate that determines its internal state. Let s_i be an index labelling the possible quantum states of the ith particle. It follows that each possible value of s_i corresponds to a specification of the three momentum components of the particle, as well as the direction of its spin orientation. According to quantum mechanics, the overall state of the system when the ith particle is in state s_i, et cetera, is completely determined by the complex wavefunction

$$\Psi_{s_1,\cdots,s_N}(Q_1, Q_2, \cdots, Q_N). \tag{10.1}$$

In particular, the probability of an observation of the system finding the ith particle with coordinates in the range Q_i to $Q_i + dQ_i$, et cetera, is simply

$$|\Psi_{s_1,\cdots,s_N}(Q_1, Q_2, \cdots, Q_N)|^2 \, dQ_1 \, dQ_2 \cdots dQ_N. \tag{10.2}$$

One of the fundamental postulates of quantum mechanics is the essential indistinguishability of particles of the same species. What this means, in practice, is that we cannot label particles of the same species. In other words, a proton is just a proton—we cannot meaningfully talk of proton number 1, or proton number 2, et cetera. Note that no such constraint arises in classical mechanics. Thus, in classical mechanics, particles of the same species are regarded as being distinguishable, and can, therefore, be labelled. Of course, the quantum-mechanical approach is the correct one.

Suppose that we interchange the ith and jth particles: that is,

$$Q_i \leftrightarrow Q_j, \tag{10.3}$$

$$s_i \leftrightarrow s_j. \tag{10.4}$$

If the particles are truly indistinguishable then nothing has changed. We have a particle in quantum state s_i and a particle in quantum state s_j both before and after the particles are swapped. Thus, the probability of observing the system in a given state also cannot have changed: that is,

$$|\Psi(\cdots Q_i \cdots Q_j \cdots)|^2 = |\Psi(\cdots Q_j \cdots Q_i \cdots)|^2. \tag{10.5}$$

Here, we have omitted the subscripts $s_1, \cdots, s_N$ for the sake of clarity. Equation (10.5) implies that

$$\Psi(\cdots Q_i \cdots Q_j \cdots) = A\,\Psi(\cdots Q_j \cdots Q_i \cdots), \tag{10.6}$$

where A is a complex constant of modulus unity: that is, $|A|^2 = 1$.

Suppose that we interchange the ith and jth particles a second time. Swapping the ith and jth particles twice leaves the system completely unchanged. In other words, it is equivalent to doing nothing at all to the system. Thus, the wavefunctions before and after this process must be identical. It follows from Equation (10.6) that

$$A^2 = 1. \tag{10.7}$$

Of course, the only solutions to the previous equation are $A = \pm 1$.

We conclude, from the previous discussion, that the wavefunction, Ψ, is either completely symmetric under the interchange of identical particles, or it is completely anti-symmetric. In other words, either

$$\Psi(\cdots Q_i \cdots Q_j \cdots) = +\Psi(\cdots Q_j \cdots Q_i \cdots), \tag{10.8}$$

or

$$\Psi(\cdots Q_i \cdots Q_j \cdots) = -\Psi(\cdots Q_j \cdots Q_i \cdots). \tag{10.9}$$

In 1940, the Nobel prize-winning physicist Wolfgang Pauli demonstrated, via arguments involving relativistic invariance, that the wavefunction associated with a collection of identical integer-spin (i.e., spin $0, 1, 2$, etc.) particles satisfies Equation (10.8), whereas the wavefunction associated with a collection of identical half-integer-spin (i.e., spin $1/2, 3/2, 5/2$, etc.) particles satisfies Equation (10.9). The former type of particles are known as *bosons* [after the Indian physicist S.N. Bose, who first put forward Equation (10.8) on empirical grounds]. The latter type of particles are called *fermions* (after the Italian physicist Enrico Fermi, who first studied the properties of fermion gases). Common examples of bosons are photons and He^4 atoms. Common examples of fermions are protons, neutrons, and electrons.

Consider a gas made up of identical bosons. Equation (10.8) implies that the interchange of any two particles does not lead to a new state of the system. Bosons must, therefore, be considered as genuinely indistinguishable when enumerating the different possible states of the gas. Note that Equation (10.8) imposes no restriction on how many particles can occupy a given single-particle quantum state, s.

Consider a gas made up of identical fermions. Equation (10.9) implies that the interchange of any two particles does not lead to a new physical state of the system (because $|\Psi|^2$ is invariant). Hence, fermions must also be considered genuinely indistinguishable when enumerating the different possible states of the gas. Consider the special case where particles i and j lie in the same quantum state. In this case, the act of swapping the two particles is equivalent to leaving the system unchanged, so

$$\Psi(\cdots Q_i \cdots Q_j \cdots) = \Psi(\cdots Q_j \cdots Q_i \cdots). \tag{10.10}$$

However, Equation (10.9) is also applicable, because the two particles are fermions. The only way in which Equations (10.9) and (10.10) can be reconciled is if

$$\Psi = 0 \tag{10.11}$$

wherever particles i and j lie in the same quantum state. This is another way of saying that it is impossible for any two particles in a gas of identical fermions to lie in the same single-particle quantum state. This proposition is known as the *Pauli exclusion principle*, because it was first proposed by Wolfgang Pauli in 1924 on empirical grounds.

Consider, for the sake of comparison, a gas made up of identical classical particles. In this case, the particles must be considered distinguishable when enumerating the different possible states of the gas. Furthermore, there are no constraints on how many particles can occupy a given quantum state.

According to the previous discussion, there are three different sets of rules that can be used to enumerate the states of a gas made up of identical particles. For a boson gas, the particles must be treated as being indistinguishable, and there is no limit to how many particles can occupy a given quantum state. This set of rules is called *Bose-Einstein statistics*, after S.N. Bose and A. Einstein, who first developed them. For a fermion gas, the particles must be treated as being indistinguishable, and there can never be more than one particle in any given quantum state. This set of rules is called *Fermi-Dirac statistics*, after E. Fermi and P.A.M. Dirac, who first developed them. Finally, for a classical gas, the particles must be treated as being distinguishable, and there is no limit to how many particles can occupy a given quantum state. This set of rules is called *Maxwell-Boltzmann statistics*, after J.C. Maxwell and L. Boltzmann, who first developed them.

10.3 Illustrative Example

Consider a very simple gas made up of two identical particles. Suppose that each particle can be in one of three possible quantum states, $s = 1, 2, 3$. Let us enumerate the possible states of the whole gas according to Maxwell-Boltzmann, Bose-Einstein, and Fermi-Dirac statistics, respectively.

For the case of Maxwell-Boltzmann statistics, the two particles are considered to be distinguishable. Let us denote them A and B. Furthermore, any number of particles can occupy the same quantum state. The possible different states of the gas are shown in Table 10.1. There are clearly nine distinct states.

For the case of Bose-Einstein statistics, the two particles are considered to be indistinguishable. Let us denote them both as A. Furthermore, any number of particles can occupy the same quantum state. The possible different states of the gas are shown in Table 10.2. There are clearly six distinct states.

Finally, for the case of Fermi-Dirac statistics, the two particles are considered to be indistinguishable. Let us again denote them both as A. Furthermore, no more than one particle can occupy a given quantum state. The possible different states of the gas are shown in Table 10.3. There are clearly only three distinct states.

It follows, from the previous example, that Fermi-Dirac (FD) statistics are more restrictive (i.e., there are less possible states of the system) than Bose-Einstein (BE) statistics, which are, in turn, more restrictive than Maxwell-

Table 10.1 Two particles distributed amongst three states according to Maxwell-Boltzmann statistics.

1	2	3
AB	$\cdots$	$\cdots$
$\cdots$	AB	$\cdots$
$\cdots$	$\cdots$	AB
A	B	$\cdots$
B	A	$\cdots$
A	$\cdots$	B
B	$\cdots$	A
$\cdots$	A	B
$\cdots$	B	A

Table 10.2 Two particles distributed amongst three states according to Bose-Einstein statistics.

1	2	3
AA	$\cdots$	$\cdots$
$\cdots$	AA	$\cdots$
$\cdots$	$\cdots$	AA
A	A	$\cdots$
A	$\cdots$	A
$\cdots$	A	A

Table 10.3 Two particles distributed amongst three states according to Fermi-Dirac statistics.

1	2	3
A	A	$\cdots$
A	$\cdots$	A
$\cdots$	A	A

Boltzmann (MB) statistics. Let

$$\xi = \frac{\text{probability that the two particles are found in the same state}}{\text{probability that the two particles are found in different states}}. \tag{10.12}$$

For the case under investigation,

$$\xi_{\rm MB} = 1/2, \tag{10.13}$$

$$\xi_{\rm BE} = 1, \tag{10.14}$$

$$\xi_{\rm FD} = 0. \tag{10.15}$$

We conclude that in Bose-Einstein statistics there is a greater relative tendency for particles to cluster in the same state than in classical (i.e., Maxwell-Boltzmann) statistics. On the other hand, in Fermi-Dirac statistics there is a lesser tendency for particles to cluster in the same state than in classical statistics.

10.4 Formulation of Statistical Problem

Consider a gas consisting of N identical non-interacting particles occupying volume V, and in thermal equilibrium at absolute temperature T. Let us label the possible quantum states of a single particle by r (or s). Let the energy of a particle in state r be denoted ϵ_r. Let the number of particles in state r be written n_r. Finally, let us label the possible quantum states of the whole gas by R.

The particles are assumed to be non-interacting, so the total energy of the gas in state R, where there are n_r particles in quantum state r, et cetera, is simply

$$E_R = \sum_r n_r\,\epsilon_r, \tag{10.16}$$

where the sum extends over all possible quantum states, r. Furthermore, because the total number of particles in the gas is known to be N, we must have

$$N = \sum_r n_r. \tag{10.17}$$

In order to calculate the thermodynamic properties of the gas (i.e., its internal energy or its entropy), it is necessary to calculate its partition function,

$$Z = \sum_R \mathrm{e}^{-\beta\,E_R} = \sum_R \mathrm{e}^{-\beta\,(n_1\,\epsilon_1+n_2\,\epsilon_2+\cdots)}. \tag{10.18}$$

Here, the sum is over all possible states, R, of the whole gas. That is, over all the various possible values of the numbers $n_1, n_2, \cdots$.

Now, $\exp[-\beta\,(n_1\,\epsilon_1 + n_2\,\epsilon_2 + \cdots)]$ is the relative probability of finding the gas in a particular state in which there are n_1 particles in state 1, n_2 particles in state 2, et cetera. Thus, the mean number of particles in quantum state s can be written

$$\bar{n}_s = \frac{\sum_R n_s\,\exp[-\beta\,(n_1\,\epsilon_1 + n_2\,\epsilon_2 + \cdots)]}{\sum_R \exp[-\beta\,(n_1\,\epsilon_1 + n_2\,\epsilon_2 + \cdots)]}. \tag{10.19}$$

A comparison of Equations (10.18) and (10.19) yields the result

$$\bar{n}_s = -\frac{1}{\beta}\frac{\partial \ln Z}{\partial \epsilon_s}. \tag{10.20}$$

Here, $\beta \equiv 1/(k\,T)$.

10.5 Fermi-Dirac Statistics

Let us, first of all, consider Fermi-Dirac statistics. According to Equation (10.19), the average number of particles in quantum state s can be written

$$\bar{n}_s = \frac{\sum_{n_s} n_s \, \mathrm{e}^{-\beta\, n_s\, \epsilon_s} \sum_{n_1,n_2,\cdots}^{(s)} \mathrm{e}^{-\beta\,(n_1\,\epsilon_1+n_2\,\epsilon_2+\cdots)}}{\sum_{n_s} \mathrm{e}^{-\beta\, n_s\, \epsilon_s} \sum_{n_1,n_2,\cdots}^{(s)} \mathrm{e}^{-\beta\,(n_1\,\epsilon_1+n_2\,\epsilon_2+\cdots)}}. \tag{10.21}$$

Here, we have rearranged the order of summation, using the multiplicative properties of the exponential function. Note that the first sums in the numerator and denominator only involve n_s, whereas the last sums omit the particular state s from consideration (this is indicated by the superscript s on the summation symbol). Of course, the sums in the previous expression range over all values of the numbers $n_1, n_2, \cdots$ such that $n_r = 0$ and 1 for each r, subject to the overall constraint that

$$\sum_r n_r = N. \tag{10.22}$$

Let us introduce the function

$$Z_s(N) = \sum_{n_1,n_2,\cdots}^{(s)} \mathrm{e}^{-\beta\,(n_1\,\epsilon_1+n_2\,\epsilon_2+\cdots)}, \tag{10.23}$$

which is defined as the partition function for N particles distributed over all quantum states, excluding state s, according to Fermi-Dirac statistics. By explicitly performing the sum over $n_s = 0$ and 1, the expression (10.21) reduces to

$$\bar{n}_s = \frac{0 + \mathrm{e}^{-\beta\,\epsilon_s}\, Z_s(N-1)}{Z_s(N) + \mathrm{e}^{-\beta\,\epsilon_s}\, Z_s(N-1)}, \tag{10.24}$$

which yields

$$\bar{n}_s = \frac{1}{[Z_s(N)/Z_s(N-1)]\, \mathrm{e}^{\beta\,\epsilon_s} + 1}. \tag{10.25}$$

In order to make further progress, we must somehow relate $Z_s(N-1)$ to $Z_s(N)$. Suppose that $\Delta N \ll N$. It follows that $\ln Z_s(N - \Delta N)$ can be Taylor expanded to give

$$\ln Z_s(N - \Delta N) \simeq \ln Z_s(N) - \frac{\partial \ln Z_s}{\partial N}\,\Delta N = \ln Z_s(N) - \alpha_s\,\Delta N, \tag{10.26}$$

where

$$\alpha_s \equiv \frac{\partial \ln Z_s}{\partial N}. \tag{10.27}$$

As always, we Taylor expand the slowly-varying function $\ln Z_s(N)$, rather than the rapidly-varying function $Z_s(N)$, because the radius of convergence of the latter

Taylor series is too small for the series to be of any practical use. Equation (10.26) can be rearranged to give

$$Z_s(N - \Delta N) = Z_s(N)\, e^{-\alpha_s \Delta N}. \tag{10.28}$$

Now, because $Z_s(N)$ is a sum over very many different quantum states, we would not expect the logarithm of this function to be sensitive to which particular state, s, is excluded from consideration. Let us, therefore, introduce the approximation that α_s is independent of s, so that we can write

$$\alpha_s \simeq \alpha \tag{10.29}$$

for all s. It follows that the derivative (10.27) can be expressed approximately in terms of the derivative of the full partition function $Z(N)$ (in which the N particles are distributed over all quantum states). In fact,

$$\alpha \simeq \frac{\partial \ln Z}{\partial N}. \tag{10.30}$$

Making use of Equation (10.28), with $\Delta N = 1$, plus the approximation (10.29), the expression (10.25) reduces to

$$\bar{n}_s = \frac{1}{e^{\alpha+\beta\,\epsilon_s} + 1}. \tag{10.31}$$

This is called the *Fermi-Dirac distribution*. The parameter α is determined by the constraint that $\sum_r \bar{n}_r = N$: that is,

$$\sum_r \frac{1}{e^{\alpha+\beta\,\epsilon_r} + 1} = N. \tag{10.32}$$

Note that $\bar{n}_s \to 0$ if ϵ_s becomes sufficiently large. On the other hand, because the denominator in Equation (10.31) can never become less than unity, no matter how small ϵ_s becomes, it follows that $\bar{n}_s \leq 1$. Thus,

$$0 \leq \bar{n}_s \leq 1, \tag{10.33}$$

in accordance with the Pauli exclusion principle.

Equations (10.20) and (10.30) can be integrated to give

$$\ln Z = \alpha\, N + \sum_r \ln\left(1 + e^{-\alpha-\beta\,\epsilon_r}\right), \tag{10.34}$$

where use has been made of Equation (10.31).

10.6 Photon Statistics

Up to now, we have assumed that the number of particles, N, contained in a given system is a fixed number. This is a reasonable assumption if the particles possess non-zero mass, because we are not generally considering relativistic systems in this book (i.e., we are assuming that the particle energies are much less than their rest-mass energies). However, this assumption breaks down for the case of photons, which are zero-mass bosons. In fact, photons enclosed in a container of volume V, maintained at temperature T, can readily be absorbed or emitted by the walls. Thus, for the special case of a gas of photons, there is no requirement that limits the total number of particles.

It follows, from the previous discussion, that photons obey a simplified form of Bose-Einstein statistics in which there is an unspecified total number of particles. This type of statistics is called *photon statistics*.

Consider the expression (10.21). For the case of photons, the numbers $n_1, n_2, \cdots$ assume all values $n_r = 0, 1, 2, \cdots$ for each r, without any further restriction. It follows that the sums $\sum^{(s)}$ in the numerator and denominator are identical and, therefore, cancel. Hence, Equation (10.21) reduces to

$$\bar{n}_s = \frac{\sum_{n_s} n_s \, \mathrm{e}^{-\beta\, n_s\, \epsilon_s}}{\sum_{n_s} \mathrm{e}^{-\beta\, n_s\, \epsilon_s}}. \tag{10.35}$$

However, the previous expression can be rewritten

$$\bar{n}_s = -\frac{1}{\beta}\frac{\partial}{\partial \epsilon_s}\left(\ln \sum_{n_s} \mathrm{e}^{-\beta\, n_s\, \epsilon_s}\right). \tag{10.36}$$

Now, the sum on the right-hand side of the previous equation is an infinite geometric series, which can easily be evaluated. In fact,

$$\sum_{n_s=0,\infty} \mathrm{e}^{-\beta\, n_s\, \epsilon_s} = 1 + \mathrm{e}^{-\beta\, \epsilon_s} + \mathrm{e}^{-2\beta\, \epsilon_s} + \cdots = \frac{1}{1 - \mathrm{e}^{-\beta\, \epsilon_s}}. \tag{10.37}$$

(See Exercise 8.1.) Thus, Equation (10.36) gives

$$\bar{n}_s = \frac{1}{\beta}\frac{\partial}{\partial \epsilon_s} \ln\left(1 - \mathrm{e}^{-\beta\, \epsilon_s}\right) = \frac{\mathrm{e}^{-\beta\, \epsilon_s}}{1 - \mathrm{e}^{-\beta\, \epsilon_s}}, \tag{10.38}$$

or

$$\bar{n}_s = \frac{1}{\mathrm{e}^{\beta\, \epsilon_s} - 1}. \tag{10.39}$$

This is known as the *Planck distribution*, after the German physicist Max Planck who first proposed it in 1900 on purely empirical grounds.

Equation (10.20) can be integrated to give

$$\ln Z = -\sum_r \ln\left(1 - \mathrm{e}^{-\beta\, \epsilon_r}\right), \tag{10.40}$$

where use has been made of Equation (10.39).

10.7 Bose-Einstein Statistics

Let us now consider Bose-Einstein statistics. The particles in the system are assumed to be massive, so the total number of particles, N, is a fixed number.

Consider the expression (10.21). For the case of massive bosons, the numbers $n_1, n_2, \cdots$ assume all values $n_r = 0, 1, 2, \cdots$ for each r, subject to the constraint that $\sum_r n_r = N$. Performing explicitly the sum over n_s, this expression reduces to

$$\bar{n}_s = \frac{0 + \mathrm{e}^{-\beta\,\epsilon_s}\,Z_s(N-1) + 2\,\mathrm{e}^{-2\beta\,\epsilon_s}\,Z_s(N-2) + \cdots}{Z_s(N) + \mathrm{e}^{-\beta\,\epsilon_s}\,Z_s(N-1) + \mathrm{e}^{-2\beta\,\epsilon_s}\,Z_s(N-2) + \cdots}, \tag{10.41}$$

where $Z_s(N)$ is the partition function for N particles distributed over all quantum states, excluding state s, according to Bose-Einstein statistics [cf., Equation (10.23)]. Using Equation (10.28), and the approximation (10.29), the previous equation reduces to

$$\bar{n}_s = \frac{\sum_s n_s\,\mathrm{e}^{-n_s\,(\alpha+\beta\,\epsilon_s)}}{\sum_s \mathrm{e}^{-n_s\,(\alpha+\beta\,\epsilon_s)}}. \tag{10.42}$$

Note that this expression is identical to (10.35), except that $\beta\,\epsilon_s$ is replaced by $\alpha + \beta\,\epsilon_s$. Hence, an analogous calculation to that outlined in the previous section yields

$$\bar{n}_s = \frac{1}{\mathrm{e}^{\alpha+\beta\,\epsilon_s} - 1}. \tag{10.43}$$

This is called the *Bose-Einstein distribution*. Note that $\bar{n}_s$ can become very large in this distribution. The parameter α is again determined by the constraint on the total number of particles: that is,

$$\sum_r \frac{1}{\mathrm{e}^{\alpha+\beta\,\epsilon_r} - 1} = N. \tag{10.44}$$

Equations (10.20) and (10.30) can be integrated to give

$$\ln Z = \alpha\,N - \sum_r \ln\left(1 - \mathrm{e}^{-\alpha-\beta\,\epsilon_r}\right), \tag{10.45}$$

where use has been made of Equation (10.43).

Note that photon statistics correspond to the special case of Bose-Einstein statistics in which the parameter α takes the value zero, and the constraint (10.44) does not apply.

10.8 Maxwell-Boltzmann Statistics

For the purpose of comparison, it is instructive to consider the purely classical case of Maxwell-Boltzmann statistics. The partition function is written

$$Z = \sum_R \mathrm{e}^{-\beta\,(n_1\,\epsilon_1 + n_2\,\epsilon_2 + \cdots)}, \tag{10.46}$$

where the sum is over all distinct states R of the gas, and the particles are treated as distinguishable. For given values of $n_1, n_2, \cdots$ there are

$$\frac{N!}{n_1!\,n_2!\,\cdots} \tag{10.47}$$

possible ways in which N distinguishable particles can be put into individual quantum states such that there are n_1 particles in state 1, n_2 particles in state 2, et cetera. Each of these possible arrangements corresponds to a distinct state for the whole gas. Hence, Equation (10.46) can be written

$$Z = \sum_{n_1, n_2, \cdots} \frac{N!}{n_1!\,n_2!\,\cdots}\,\mathrm{e}^{-\beta\,(n_1\,\epsilon_1 + n_2\,\epsilon_2 + \cdots)}, \tag{10.48}$$

where the sum is over all values of $n_r = 0, 1, 2, \cdots$ for each r, subject to the constraint that

$$\sum_r n_r = N. \tag{10.49}$$

Now, Equation (10.48) can be written

$$Z = \sum_{n_1, n_2, \cdots} \frac{N!}{n_1!\,n_2!\,\cdots}\,(\mathrm{e}^{-\beta\,\epsilon_1})^{n_1}\,(\mathrm{e}^{-\beta\,\epsilon_2})^{n_2}\,\cdots, \tag{10.50}$$

which, by virtue of Equation (10.49), is just the result of expanding a polynomial. In fact,

$$Z = (\mathrm{e}^{-\beta\,\epsilon_1} + \mathrm{e}^{-\beta\,\epsilon_2} + \cdots)^N, \tag{10.51}$$

or

$$\ln Z = N\,\ln\left(\sum_r \mathrm{e}^{-\beta\,\epsilon_r}\right). \tag{10.52}$$

Note that the argument of the logarithm is simply the single-particle partition function

Equations (10.20) and (10.52) can be combined to give

$$\bar{n}_s = N\,\frac{\mathrm{e}^{-\beta\,\epsilon_s}}{\sum_r \mathrm{e}^{-\beta\,\epsilon_r}}. \tag{10.53}$$

This is known as the *Maxwell-Boltzmann distribution*. It is, of course, just the result obtained by applying the canonical distribution to a single particle. (See Chapter 8.) The previous expression can also be written in the form

$$\bar{n}_s = e^{-\alpha-\beta\,\epsilon_r}, \tag{10.54}$$

where

$$\sum_r e^{-\alpha-\beta\,\epsilon_r} = N. \tag{10.55}$$

The Bose-Einstein, Maxwell-Boltzmann, and Fermi-Dirac distributions are illustrated in Figure 10.1.

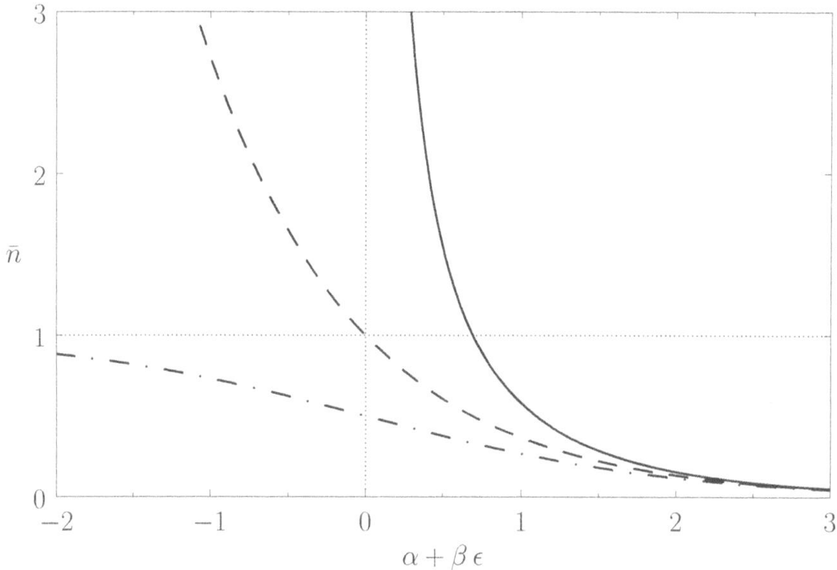

Figure 10.1 A comparison of the Bose-Einstein (solid curve), Maxwell-Boltzmann (dashed curve), and Fermi-Dirac (dash-dotted curve) distributions.

10.9 Quantum Statistics in Classical Limit

The preceding analysis regarding the quantum statistics of ideal gases is summarized in the following statements. The mean number of particles occupying quantum state s is given by

$$\bar{n}_s = \frac{1}{e^{\alpha+\beta\,\epsilon_s} \pm 1}, \tag{10.56}$$

where the upper sign corresponds to Fermi-Dirac statistics and the lower sign corresponds to Bose-Einstein statistics. The parameter α is determined via

$$\sum_r \bar{n}_r = \sum_r \frac{1}{\mathrm{e}^{\alpha+\beta\,\epsilon_r} \pm 1} = N. \tag{10.57}$$

Finally, the partition function of the gas is given by

$$\ln Z = \alpha\,N \pm \sum_r \ln\left(1 \pm \mathrm{e}^{-\alpha-\beta\,\epsilon_r}\right). \tag{10.58}$$

Let us investigate the magnitude of α in some important limiting cases. Consider, first of all, the case of a gas at a given temperature when its concentration is made sufficiently low: that is, when N is made sufficiently small. The relation (10.57) can only be satisfied if each term in the sum over states is made sufficiently small; that is, if $\bar{n}_r \ll 1$ or $\alpha + \beta\,\epsilon_r \gg 1$ for all states r. (See Figure 10.1.)

Consider, next, the case of a gas made up of a fixed number of particles when its temperature is made sufficiently large. That is, when β is made sufficiently small. In the sum in Equation (10.57), the terms of appreciable magnitude are those for which $\alpha + \beta\,\epsilon_r < 1$. (See Figure 10.1.) Thus, it follows that as $\beta \to 0$ an increasing number of terms with large values of ϵ_r contribute substantially to this sum. In order to prevent the sum from exceeding N, the parameter α must become large enough that each term is made sufficiently small: that is, it is again necessary that $\bar{n}_r \ll 1$ or $\alpha + \beta\,\epsilon_r \gg 1$ for all states r.

The previous discussion suggests that if the concentration of an ideal gas is made sufficiently low, or the temperature is made sufficiently high, then α must become so large that

$$\exp(\alpha + \beta\,\epsilon_r) \gg 1 \tag{10.59}$$

for all r. Equivalently, this means that the number of particles occupying each quantum state must become so small that

$$\bar{n}_r \ll 1 \tag{10.60}$$

for all r. It is conventional to refer to the limit of sufficiently low concentration, or sufficiently high temperature, in which Equations (10.59) and Equations (10.60) are satisfied, as the *classical limit*.

According to Equations (10.56) and (10.59), both the Fermi-Dirac and Bose-Einstein distributions reduce to

$$\bar{n}_s = \exp(-\alpha - \beta\,\epsilon_s) \tag{10.61}$$

in the classical limit, whereas the constraint (10.57) yields

$$\sum_r \exp(-\alpha - \beta\,\epsilon_r) = N. \tag{10.62}$$

The previous expressions can be combined to give

$$\bar{n}_s = N\,\frac{e^{-\beta\,\epsilon_s}}{\sum_r e^{-\beta\,\epsilon_r}}. \tag{10.63}$$

It follows that in the classical limit of sufficiently low density, or sufficiently high temperature, the quantum distribution functions, whether Fermi-Dirac or Bose-Einstein, reduce to the Maxwell-Boltzmann distribution. It is easily demonstrated that the physical criterion for the validity of the classical approximation is that the mean separation between particles should be much greater than their mean de Broglie wavelengths. (See Exercise 10.21.)

Let us now consider the behavior of the partition function (10.58) in the classical limit. We can expand the logarithm to give

$$\ln Z = \alpha\,N \pm \sum_r \left(\pm e^{-\alpha-\beta\,\epsilon_r}\right) = \alpha\,N + N. \tag{10.64}$$

However, according to Equation (10.62),

$$\alpha = -\ln N + \ln\left(\sum_r e^{-\beta\,\epsilon_r}\right). \tag{10.65}$$

It follows that

$$\ln Z = -N\,\ln N + N + N\,\ln\left(\sum_r e^{-\beta\,\epsilon_r}\right). \tag{10.66}$$

Note that this expression does not match the partition function $Z_{\rm MB}$ computed in Equation (10.52) from Maxwell-Boltzmann statistics, which gives

$$\ln Z_{\rm MB} = N\,\ln\left(\sum_r e^{-\beta\,\epsilon_r}\right). \tag{10.67}$$

In fact,

$$\ln Z \simeq \ln Z_{\rm MB} - \ln N!, \tag{10.68}$$

or

$$Z = \frac{Z_{\rm MB}}{N!}, \tag{10.69}$$

where use has been made of Stirling's approximation ($N! \simeq N\,\ln N - N$), because N is large. Here, the factor $N!$ simply corresponds to the number of different permutations of the N particles: permutations that are physically meaningless when the particles are identical. Recall, that we had to introduce precisely this factor, in an ad hoc fashion, in Section 8.8, in order to avoid the non-physical consequences of the Gibbs paradox. Clearly, there is no Gibbs paradox when an ideal gas is treated properly via quantum mechanics.

A gas in the classical limit, where the typical de Broglie wavelength of the constituent particles is much smaller than the typical inter-particle spacing, is said to be *non-degenerate*. In the opposite limit, where the concentration and temperature are such that the typical de Broglie wavelength becomes comparable with the typical inter-particle spacing, and the actual Fermi-Dirac or Bose-Einstein distributions must be employed, the gas is said to be *degenerate*.

10.10 Quantum-Mechanical Treatment of Ideal Gas

Let us calculate the partition function of an ideal gas from quantum mechanics, making use of Maxwell-Boltzmann statistics. Obviously, such a partition function is only applicable when the gas is non-degenerate. According to Equations (10.67) and (10.69), we can write the partition function in the form

$$Z = \frac{\zeta^N}{N!}, \tag{10.70}$$

where N is the number of constituent molecules,

$$\zeta = \sum_r \mathrm{e}^{-\beta\,\epsilon_r} \tag{10.71}$$

is the partition function of an individual molecule, and the factor $N!$ is necessary to take into account the fact that the molecules are indistinguishable according to quantum theory.

Suppose that the gas is enclosed in a parallelepiped with sides of lengths L_x, L_y, and L_z. It follows that the de Broglie wavenumbers of the constituent molecules are quantized such that

$$k_x = n_x\,\frac{\pi}{L_x}, \tag{10.72}$$

$$k_y = n_y\,\frac{\pi}{L_y}, \tag{10.73}$$

$$k_z = n_z\,\frac{\pi}{L_z}, \tag{10.74}$$

where n_x, n_y, and n_z are independent integers. (See Section C.10.) Thus, the allowed energy levels of a single molecule are

$$\epsilon_r = \frac{\hbar^2\,(k_x^2 + k_y^2 + k_z^2)}{2\,m}, \tag{10.75}$$

where m is the molecular mass. Hence, we deduce that

$$\zeta = \sum_{k_x,k_y,k_z} \exp\left[-\frac{\beta\,\hbar^2}{2\,m}\,(k_x^2 + k_y^2 + k_z^2)\right], \tag{10.76}$$

where the sum is over all possible values of k_x, k_y, k_z. The previous expression can also be written

$$\zeta = \left(\sum_{k_x} \mathrm{e}^{-(\beta\hbar^2/2m)\,k_x^2}\right)\left(\sum_{k_y} \mathrm{e}^{-(\beta\hbar^2/2m)\,k_y^2}\right)\left(\sum_{k_z} \mathrm{e}^{-(\beta\hbar^2/2m)\,k_z^2}\right). \tag{10.77}$$

Now, in the previous equation, the successive terms in the sum over k_x (say) correspond to a very small increment, $\Delta k_x = 2\pi/L_x$, in k_x (in fact, the increment can be made arbitrarily small by increasing L_x), and, therefore, differ very little from each other. Hence, it is an excellent approximation to replace the sums in Equation (10.77) by integrals. Given that $1 = \Delta n_x = (L_x/2\pi)\,dk_x$, we can write

$$\sum_{k_x} \mathrm{e}^{-(\beta\hbar^2/2m)\,k_x^2} \simeq \int_{-\infty}^{\infty} \mathrm{e}^{-(\beta\hbar^2/2m)\,k_x^2}\left(\frac{L_x}{2\pi}\,dk_x\right) = \frac{L_x}{2\pi}\left(\frac{2\pi\,m}{\beta\,\hbar^2}\right)^{1/2}$$
$$= \frac{L_x}{2\pi\,\hbar}\left(\frac{2\pi\,m}{\beta}\right)^{1/2}. \tag{10.78}$$

(See Exercise 2.2.) Hence, Equation (10.77) becomes

$$\zeta = \frac{V}{(2\pi\,\hbar)^3}\left(\frac{2\pi\,m}{\beta}\right)^{3/2} = V\left(\frac{2\pi\,m}{\beta\,h^2}\right)^{3/2}, \tag{10.79}$$

where $V = L_x\,L_y\,L_z$ is the volume of the gas.

It follows from Equation (10.70) that

$$\ln Z = N\,(\ln\zeta - \ln N + 1), \tag{10.80}$$

where use has been made of Stirling's approximation. Thus, we obtain

$$\ln Z = N\left[\ln\left(\frac{V}{N}\right) - \frac{3}{2}\,\ln\beta + \frac{3}{2}\,\ln\left(\frac{2\pi\,m}{h^2}\right) + 1\right], \tag{10.81}$$

and

$$\overline{E} = -\frac{\partial\ln Z}{\partial\beta} = \frac{3}{2}\,\frac{N}{\beta} = \frac{3}{2}\,N\,k\,T, \tag{10.82}$$

and, finally,

$$S = k\left(\ln Z + \beta\,\overline{E}\right) = N\,k\left[\ln\left(\frac{V}{N}\right) + \frac{3}{2}\,\ln T + \sigma_0\right], \tag{10.83}$$

where

$$\sigma_0 = \frac{3}{2}\,\ln\left(\frac{2\pi\,m\,k}{h^2}\right) + \frac{5}{2}. \tag{10.84}$$

These results are exactly the same as those obtained in Sections 8.7 and 8.8, using classical mechanics, except that the arbitrary parameter h_0 is replaced by Planck's constant, $h = 6.61 \times 10^{-34}$ J s.

10.11 Derivation of van der Waals Equation of State

We shall now extend the analysis of the previous section to derive an approximate equation of state for a non-ideal gas. Let us focus attention on a single molecule. It seems reasonable to suppose that this molecule moves in an effective potential, $U_e(\mathbf{r})$, due to all the other molecules in the gas (which are assumed to remain unaffected by the presence of the molecule under consideration). Under these circumstances, the partition function of the gas reduces to that of N independent particles, each with a kinetic energy $\mathbf{p}^2/(2\,m)$, and a potential energy U_e.

Proceeding classically (except that the molecules are treated as indistinguishable, and the arbitrary parameter h_0 is replaced by Planck's constant), we obtain

$$Z = \frac{1}{N!}\left[\int\!\!\int \mathrm{e}^{-\beta\,(p^2/2\,m+U_e)}\,\frac{d^3\mathbf{r}\,d^3\mathbf{p}}{h^3}\right]^N. \tag{10.85}$$

(See Sections 8.7 and 10.10.) Thus,

$$Z = \frac{1}{N!}\left[\int \exp\left(\frac{-\beta\,p_x^2}{2\,m}\right)\frac{dp_x}{h}\int \exp\left(\frac{-\beta\,p_y^2}{2\,m}\right)\frac{dp_y}{h}\right.$$
$$\left.\times\int \exp\left(\frac{-\beta\,p_z^2}{2\,m}\right)\frac{dp_z}{h}\int \mathrm{e}^{-\beta\,U_e}\,d^3\mathbf{r}\right]^N, \tag{10.86}$$

which yields

$$Z = \frac{1}{N!}\left[\left(\frac{2\pi\,m}{\beta\,h^2}\right)^{3/2}\int \mathrm{e}^{-\beta\,U_e}\,d^3\mathbf{r}\right]^N. \tag{10.87}$$

(See Exercise 2.2.)

The remaining integral extends over the volume, V, of the container. To make further progress, we note that there are regions in which $U_e \to \infty$ because of the strong repulsion between molecules when they approach one another too closely. Thus, the integrand vanishes in these regions, which are assumed to have a total volume $V_x < V$. In the remaining volume, $V - V_x$, where U_e is assumed to vary relatively slowly with inter-molecular separation, we shall replace U_e by some effective constant average value, $\overline{U}_e$. Thus, the previous equation becomes

$$Z = \frac{1}{N!}\left[\left(\frac{2\pi\,m}{h^2\,\beta}\right)^{3/2}(V - V_x)\,\mathrm{e}^{-\beta\,\overline{U}_e}\right]^N. \tag{10.88}$$

It remains to estimate the values of the parameters $\overline{U}_e$ and V_x. The total mean potential energy of the molecules is $N\,\overline{U}_e$. But, because there are $(1/2)\,N\,(N-1) \simeq (1/2)\,N^2$ pairs of molecules in the gas, it follows that the total mean potential energy is also $(1/2)\,N^2\,\bar{u}$, where u is the potential energy of interaction between a given pair of molecules. Equating the different expressions for the total mean potential energy of the gas, we obtain $N\,\overline{U}_e = (1/2)\,N^2\,\overline{u}$, or

$$\overline{U}_e = \frac{1}{2}\,N\,\overline{u}. \tag{10.89}$$

To estimate the mean potential energy, $\bar{u}$, between a pair of molecules, let us make the simplistic assumption that the interaction potential takes the form

$$u(R) = \begin{cases} \infty & R < R_0 \\ -u_0\,(R_0/R)^s & R > R_0 \end{cases}, \tag{10.90}$$

where R is separation between the molecular centers, $u_0 > 0$, and $s > 0$. Thus, the molecules act as weakly-attracting hard spheres of radius $(1/2)\,R_0$. The choice of exponent $s = 6$ is usually the most appropriate. (See Exercise 10.8.) Concentrating attention on a particular molecule, and making the crudest possible approximation, we shall say that another molecule is equally likely to be anywhere in the container that is a distance R greater than R_0 from the given molecule. Thus, the probability of the distance lying between R and $R + dR$ is $(4\pi\,R^2\,dR)/V$, and

$$\bar{u} = \frac{1}{V}\int_0^\infty u(R)\,4\pi\,R^2\,dR = -\frac{4\pi\,u_0}{V}\int_{R_0}^\infty \left(\frac{R_0}{R}\right)^s R^2\,dR. \tag{10.91}$$

Let us assume that $s > 3$: in other words, the potential $u(R)$ falls off sufficiently rapidly for the previous integral to converge properly. Thus, Equation (10.89) becomes

$$\overline{U}_e = \frac{1}{2}\,N\,\overline{u} = -a'\,\frac{N}{V}, \tag{10.92}$$

where

$$a' = \frac{2\pi}{3}\,R_0^3\left(\frac{3}{s-3}\right)u_0. \tag{10.93}$$

By Equation (10.90), the distance of closest approach between molecules is R_0. Thus, in each encounter between a pair of molecules, there is a volume excluded to one molecule, by virtue of the presence of the other molecule, that is equal to the volume of a sphere of radius R_0. Because there are $(1/2)\,N\,(N-1) \simeq (1/2)\,N^2$ pairs of molecules, the total excluded volume is $(1/2)\,N^2\,(4/3)\,\pi\,R_0^3$. But, for the sake of consistency, this volume must be equal to $N\,V_x$, because V_x was understood to be the excluded volume per molecule. Thus, it follows that

$$V_x = b'\,N, \tag{10.94}$$

where

$$b' = \frac{2\pi}{3}\,R_0^3 = 4\left[\frac{4\pi}{3}\left(\frac{R_0}{2}\right)^3\right] \tag{10.95}$$

is four times the volume of a hard-sphere molecule.

This completes our crude evaluation of the partition function. The equation of state of the gas is given by

$$\bar{p} = \frac{1}{\beta}\frac{\partial \ln Z}{\partial V} = \frac{1}{\beta}\frac{\partial}{\partial V}\left[N\,\ln(V - V_x) - N\beta\,\overline{U}_e\right]. \tag{10.96}$$

Making use of Equations (10.92) and (10.94), we obtain

$$\bar{p} = \frac{k\,T\,N}{V - b'\,N} - a'\,\frac{N^2}{V^2}, \tag{10.97}$$

or

$$\left(\bar{p} + a'\,\frac{N^2}{V^2}\right)\left(\frac{V}{N} - b'\right) = k\,T. \tag{10.98}$$

Now,

$$\frac{N}{V} = \frac{\nu\,N_A}{V} = \frac{N_A}{v}, \tag{10.99}$$

where $v = V/\nu$ is the molar volume. Thus, our approximate equation of state becomes

$$\left(\bar{p} + \frac{a}{v^2}\right)(v - b) = R\,T, \tag{10.100}$$

where

$$a = N_A^2\,a', \tag{10.101}$$

$$b = N_A\,b'. \tag{10.102}$$

Of course, Equation (10.100) is identical to the well-known van der Waals equation of state. (See Section 6.15.) Our analysis allows us to relate the constants, a and b, appearing in this equation of state, to the inter-molecular potential.

10.12 Planck Radiation Law

Let us now consider the application of statistical thermodynamics to electromagnetic radiation. According to Maxwell's theory, an electromagnetic wave is a coupled self-sustaining oscillation of electric and magnetic fields that propagates though a vacuum at the speed of light, $c = 3\times 10^8\ \mathrm{m\,s^{-1}}$. The electric component of the wave can be written

$$\mathbf{E} = \mathbf{E}_0\exp[\,\mathrm{i}\,(\mathbf{k}\cdot\mathbf{r} - \omega\,t)], \tag{10.103}$$

where $\mathbf{E}_0$ is a constant, $\mathbf{k}$ is the wavevector that determines the wavelength and direction of propagation of the wave, and ω is the angular frequency. The dispersion relation

$$\omega = k\,c \tag{10.104}$$

ensures that the wave propagates at the speed of light. Note that this dispersion relation is very similar to that of sound waves in solids. [See Equation (8.177).] Electromagnetic waves always propagate in the direction perpendicular to the coupled electric and magnetic fields (i.e., electromagnetic waves are transverse waves). This means that $\mathbf{k}\cdot\mathbf{E}_0 = 0$. Thus, once $\mathbf{k}$ is specified, there are only two possible independent directions for the electric field. These correspond to the two independent polarizations of electromagnetic waves.

Consider an enclosure whose walls are maintained at fixed temperature T. What is the nature of the steady-state electromagnetic radiation inside the enclosure? Suppose that the enclosure is a parallelepiped with sides of lengths L_x, L_y, and L_z. Alternatively (and more conveniently), suppose that the radiation field inside the enclosure is periodic in the x-, y-, and z-directions, with periodicity lengths L_x, L_y, and L_z, respectively. As long as the smallest of these lengths, L, say, is much greater than the longest wavelength of interest in the problem, $\lambda = 2\pi/k$, then these assumptions should not significantly affect the nature of the radiation inside the enclosure. We find, just as in our earlier discussion of sound waves (see Section 8.14), that the periodicity constraints ensure that there are only a discrete set of allowed wavevectors (i.e., a discrete set of allowed modes of oscillation of the electromagnetic field inside the enclosure). Let $\rho(\mathbf{k})\,d^3\mathbf{k}$ be the number of allowed modes per unit volume with wavevectors in the range $\mathbf{k}$ to $\mathbf{k}+d\mathbf{k}$. We know, by analogy with Equation (8.183), that

$$\rho(\mathbf{k})\,d^3\mathbf{k} = \frac{d^3\mathbf{k}}{(2\pi)^3}. \tag{10.105}$$

The number of modes per unit volume for which the magnitude of the wavevector lies in the range k to $k + dk$ is just the density of modes, $\rho(\mathbf{k})$, multiplied by the "volume" in $\mathbf{k}$-space of the spherical shell lying between radii k and $k + dk$. Thus,

$$\rho_k(k)\,dk = \frac{4\pi\,k^2\,dk}{(2\pi)^3} = \frac{k^2}{2\pi^2}\,dk. \tag{10.106}$$

Finally, the number of modes per unit volume whose frequencies lie between ω and $\omega + d\omega$ is, by Equation (10.104),

$$\sigma(\omega)\,d\omega = 2\,\rho_k(k)\,\frac{dk}{d\omega}\,d\omega = 2\,\frac{\omega^2}{2\pi^2\,c^3}\,d\omega. \tag{10.107}$$

Here, the additional factor 2 is to take account of the two independent polarizations of the electromagnetic field for a given wavevector, $\mathbf{k}$.

Let us consider the situation classically. By analogy with sound waves, we can treat each allowable mode of oscillation of the electromagnetic field as an independent harmonic oscillator. According to the equipartition theorem (see Section 8.10), each mode possesses a mean energy $k\,T$ in thermal equilibrium at temperature T. In fact, $(1/2)\,k\,T$ resides with the oscillating electric field, and another $(1/2)\,k\,T$ with the oscillating magnetic field. Thus, the classical energy density of electromagnetic radiation (i.e., the energy per unit volume associated with modes whose frequencies lie in the range ω to $\omega + d\omega$) is

$$\overline{u}(\omega)\,d\omega = k\,T\,\sigma(\omega)\,d\omega = \frac{k\,T}{\pi^2\,c^3}\,\omega^2\,d\omega. \tag{10.108}$$

This result is known as the *Rayleigh-Jeans radiation law*, after Lord Rayleigh and James Jeans who first proposed it in the late nineteenth century.

According to Debye theory (see Section 8.14), the energy density of sound waves in a solid is analogous to the Rayleigh-Jeans law, with one very important difference. In Debye theory, there is a cut-off frequency (the Debye frequency) above which no modes exist. This cut-off comes about because of the discrete nature of solids (i.e., because solids are made up of atoms instead of being continuous). It is, of course, impossible to have sound waves whose wavelengths are much less than the inter-atomic spacing. On the other hand, electromagnetic waves propagate through a vacuum, which possesses no discrete structure. It follows that there is no cut-off frequency for electromagnetic waves, and so the Rayleigh-Jeans law holds for all frequencies. This immediately poses a severe problem. The total classical energy density of electromagnetic radiation is given by

$$U = \int_0^\infty \overline{u}(\omega)\,d\omega = \frac{k\,T}{\pi^2\,c^3}\int_0^\infty \omega^2\,d\omega. \tag{10.109}$$

This is an integral that obviously does not converge. Thus, according to classical physics, the total energy density of electromagnetic radiation inside an enclosed cavity is infinite. This is clearly an absurd result, and was recognized as such in the latter half of the nineteenth century. In fact, this prediction is known as the *ultra-violet catastrophe*, because the Rayleigh-Jeans law usually starts to diverge badly from experimental observations (by over-estimating the amount of radiation) in the ultra-violet region of the spectrum.

So, how do we obtain a sensible answer? As usual, quantum mechanics comes to our rescue. According to quantum mechanics, each allowable mode of oscillation of the electromagnetic field corresponds to a photon state with energy and momentum

$$\epsilon = \hbar\,\omega, \tag{10.110}$$

$$\mathbf{p} = \hbar\,\mathbf{k}, \tag{10.111}$$

respectively. Incidentally, it follows from Equation (10.104) that

$$\epsilon = p\,c, \tag{10.112}$$

which implies that photons are massless particles that move at the speed of light. According to the Planck distribution, (10.39), the mean number of photons occupying a photon state of frequency ω is

$$n_\omega(\omega) = \frac{1}{e^{\beta\hbar\omega} - 1}. \tag{10.113}$$

Hence, the mean energy of such a state is given by

$$\bar{\epsilon}(\omega) = \hbar\,\omega\; n_\omega(\omega) = \frac{\hbar\,\omega}{e^{\beta\hbar\omega} - 1}. \tag{10.114}$$

Note that low-frequency states (i.e., $\hbar\,\omega \ll k\,T$) behave classically: that is,

$$\bar{\epsilon} \simeq k\,T. \tag{10.115}$$

On the other hand, high-frequency states (i.e., $\hbar\,\omega \gg k\,T$) are completely "frozen out": that is,

$$\bar{\epsilon} \ll k\,T. \tag{10.116}$$

The reason for this is simply that it is very difficult for a thermal fluctuation to create a photon with an energy greatly in excess of $k\,T$, because $k\,T$ is the characteristic energy associated with such fluctuations.

According to the previous discussion, the true energy density of electromagnetic radiation inside an enclosed cavity is written

$$\bar{u}(\omega)\,d\omega = \bar{\epsilon}(\omega)\,\sigma(\omega)\,d\omega, \tag{10.117}$$

giving

$$\overline{u}(\omega)\,d\omega = \frac{\hbar}{\pi^2\,c^3}\,\frac{\omega^3\,d\omega}{\exp(\beta\,\hbar\,\omega) - 1}. \tag{10.118}$$

This is famous result is known as the *Planck radiation law*. The Planck law approximates to the classical Rayleigh-Jeans law for $\hbar\,\omega \ll k\,T$, peaks at about $\hbar\,\omega \simeq 3\,k\,T$ (see Exercise 10.15), and falls off exponentially for $\hbar\,\omega \gg k\,T$. (See Figure 10.2.) The exponential fall off at high frequencies ensures that the total energy density remains finite.

10.13 Black-Body Radiation

Suppose that we were to make a small hole in the wall of our enclosure, and observe the emitted radiation. A small hole is the best approximation in physics to a *black-body*, which is defined as an object that absorbs, and, therefore, emits, radiation perfectly at all wavelengths. What is the power radiated by the hole? Well, the power density inside the enclosure can be written

$$\overline{u}(\omega)\,d\omega = \hbar\,\omega\;n(\omega)\,d\omega, \tag{10.119}$$

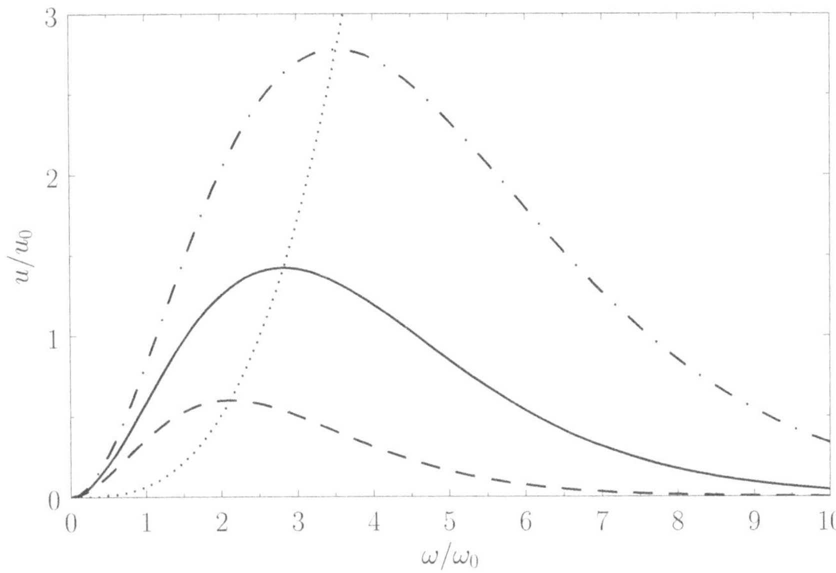

Figure 10.2 The Planck radiation law. Here, $\omega_0 = k\,T_0/k$ and $u_0 = \hbar\,\omega_0^3/(\pi^2\,c^3)$, where T_0 is an arbitrary scale temperature. The dashed, solid, and dash-dotted curves show $\bar{u}/u_0$ for $T/T_0 = 0.75$, 1.0, and 1.25, respectively. The dotted curve shows the locus of the peak emission frequency.

where $n(\omega)$ is the mean number of photons per unit volume whose frequencies lie in the range ω to $\omega + d\omega$. The radiation field inside the enclosure is isotropic (we are assuming that the hole is sufficiently small that it does not distort the field). It follows that the mean number of photons per unit volume whose frequencies lie in the specified range, and whose directions of propagation make an angle in the range θ to $\theta + d\theta$ with the normal to the hole, is

$$n(\omega, \theta)\,d\omega\,d\theta = \frac{1}{2}\,n(\omega)\,d\omega\,\sin\theta\,d\theta, \tag{10.120}$$

where $\sin\theta$ is proportional to the solid angle in the specified range of directions, and

$$\int_0^{\pi} n(\omega, \theta)\,d\omega\,d\theta = n(\omega)\,d\omega. \tag{10.121}$$

Photons travel at the velocity of light, so the power per unit area escaping from the hole in the frequency range ω to $\omega + d\omega$ is

$$P(\omega)\,d\omega = \int_0^{\pi/2} c\,\cos\theta\,\hbar\,\omega\,n(\omega, \theta)\,d\omega\,d\theta, \tag{10.122}$$

where $c \cos\theta$ is the component of the photon velocity in the direction of the hole. This gives

$$P(\omega)\,d\omega = c\,\overline{u}(\omega)\,d\omega\,\frac{1}{2}\int_0^{\pi/2}\cos\theta\,\sin\theta\,d\theta = \frac{c}{4}\,\overline{u}(\omega)\,d\omega, \tag{10.123}$$

so

$$P(\omega)\,d\omega = \frac{\hbar}{4\pi^2\,c^2}\frac{\omega^3\,d\omega}{\exp(\beta\,\hbar\,\omega)-1} \tag{10.124}$$

is the power per unit area radiated by a black-body in the frequency range ω to $\omega + d\omega$.

A black-body is very much an idealization. The power spectra of real radiating bodies can deviate quite substantially from black-body spectra. Nevertheless, we can make some useful predictions using this model. The black-body power spectrum peaks when $\hbar\,\omega \simeq 3\,k\,T$. (See Exercise 10.15.) This means that the peak radiation frequency scales linearly with the temperature of the body. In other words, hot bodies tend to radiate at higher frequencies than cold bodies. This result (in particular, the linear scaling) is known as *Wien's displacement law*. It allows us to estimate the surface temperatures of stars from their colors (surprisingly enough, stars are fairly good black-bodies). Table 10.4 shows some stellar temperatures determined by this method (in fact, the whole emission spectrum is fitted to a black-body spectrum). It can be seen that the apparent colors (which correspond quite well to the colors of the peak radiation) scan the whole visible spectrum, from red to blue, as the stellar surface temperatures gradually rise.

Table 10.4 Physical properties of some well-known stars.

Name	Constellation	Surface Temp. (K)	Color
Antares	Scorpio	3300	Very Red
Aldebaran	Taurus	3800	Reddish Yellow
Sun		5770	Yellow
Procyon	Canis Minor	6570	Yellowish White
Sirius	Canis Major	9250	White
Rigel	Orion	11,200	Bluish White

Probably the most famous black-body spectrum is cosmological in origin. Just after the "big bang," the universe was essentially a "fireball," with the energy associated with radiation completely dominating that associated with matter. The early universe was also fairly well described by equilibrium statistical thermodynamics, which means that the radiation had a black-body spectrum. As the universe expanded, the radiation was gradually Doppler shifted to ever larger wavelengths (in other words, the radiation did work against the expansion of the universe, and, thereby, lost energy—see Exercise 10.13), but its spectrum remained invariant. Nowadays, this primordial radiation is detectable as a faint microwave background that pervades the whole universe. The microwave background was discovered accidentally by Penzias and Wilson in 1961. Until recently, it was difficult to measure the full spectrum with any degree of precision, because of strong microwave absorption and scattering by the Earth's atmosphere. However, all of this changed when the COBE satellite was launched in 1989. It took precisely nine minutes to measure the perfect black-body spectrum reproduced in Figure 10.3. This data can be fitted to a black-body curve of characteristic temperature 2.735 K. In a very real sense, this can be regarded as the "temperature of the universe."

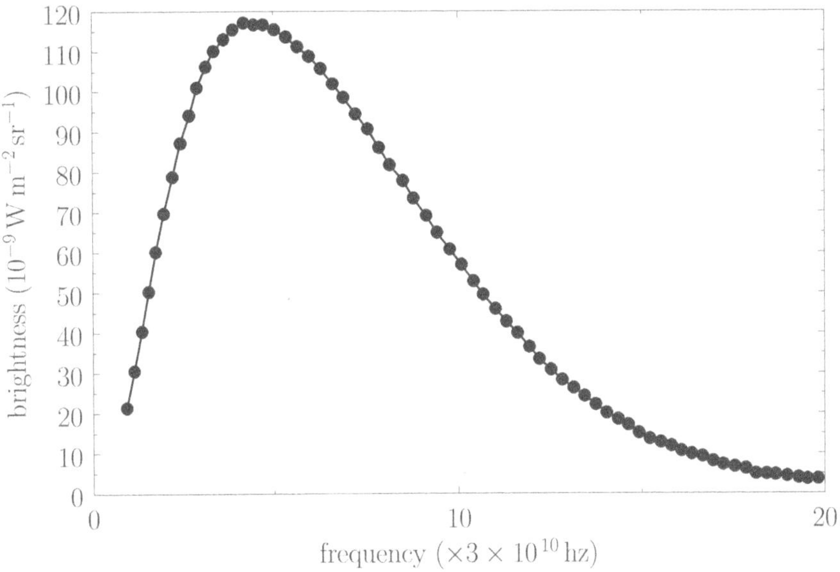

Figure 10.3 Cosmic background radiation spectrum measured by the Far Infrared Absolute Spectrometer (FIRAS) aboard the Cosmic Background Explorer satellite (COBE). The fit is to a black-body spectrum of characteristic temperature 2.735 ± 0.06K. Data from [Mather *et al.* (1990)].

10.14 Stefan-Boltzmann Law

The total power radiated per unit area by a black-body at all frequencies is given by

$$P_{\rm tot}(T) = \int_0^\infty P(\omega)\,d\omega = \frac{\hbar}{4\pi^2\,c^2}\int_0^\infty \frac{\omega^3\,d\omega}{\exp(\hbar\,\omega/k\,T) - 1}, \tag{10.125}$$

or

$$P_{\rm tot}(T) = \frac{k^4\,T^4}{4\pi^2\,c^2\,\hbar^3}\int_0^\infty \frac{\eta^3\,d\eta}{\exp\eta - 1}, \tag{10.126}$$

where $\eta = \hbar\,\omega/k\,T$. The previous integral can easily be looked up in standard mathematical tables. In fact,

$$\int_0^\infty \frac{\eta^3\,d\eta}{\exp\eta - 1} = \frac{\pi^4}{15}. \tag{10.127}$$

(See Exercise 10.2.) Thus, the total power radiated per unit area by a black-body is

$$P_{\rm tot}(T) = \frac{\pi^2}{60}\frac{k^4}{c^2\,\hbar^3}\,T^4 = \sigma\,T^4. \tag{10.128}$$

This T^4 dependence of the radiated power is called the *Stefan-Boltzmann law*, after Josef Stefan, who first obtained it experimentally, and Ludwig Boltzmann, who first derived it theoretically. The parameter

$$\sigma = \frac{\pi^2}{60}\frac{k^4}{c^2\,\hbar^3} = 5.67\times 10^{-8}\ {\rm W\,m^{-2}\,K^{-4}}, \tag{10.129}$$

is called the *Stefan-Boltzmann constant*.

We can use the Stefan-Boltzmann law to estimate the temperature of the Earth from first principles. The Sun is a ball of glowing gas of radius $R_\odot \simeq 7\times 10^5$ km and surface temperature $T_\odot \simeq 5770\,$K. Its luminosity is

$$L_\odot = 4\pi\,R_\odot^2\,\sigma\,T_\odot^4, \tag{10.130}$$

according to the Stefan-Boltzmann law. The Earth is a globe of radius $R_\oplus \simeq 6000$ km located an average distance $r_\oplus \simeq 1.5\times 10^8$ km from the Sun. The Earth intercepts an amount of energy

$$P_\oplus = L_\odot\,\frac{\pi\,R_\oplus^2/r_\oplus^2}{4\pi} \tag{10.131}$$

per second from the Sun's radiative output: that is, the power output of the Sun reduced by the ratio of the solid angle subtended by the Earth at the Sun to the total solid angle 4π. The Earth absorbs this energy, and then re-radiates it at longer wavelengths. The luminosity of the Earth is

$$L_\oplus = 4\pi\,R_\oplus^2\,\sigma\,T_\oplus^4, \tag{10.132}$$

according to the Stefan-Boltzmann law, where $T_\oplus$ is the average temperature of the Earth's surface. Here, we are ignoring any surface temperature variations between polar and equatorial regions, or between day and night. In steady-state, the luminosity of the Earth must balance the radiative power input from the Sun, so equating $L_\oplus$ and $P_\oplus$ we arrive at

$$T_\oplus = \left(\frac{R_\odot}{2\,r_\oplus}\right)^{1/2} T_\odot. \tag{10.133}$$

Remarkably, the ratio of the Earth's surface temperature to that of the Sun depends only on the Earth-Sun distance and the solar radius. The previous expression yields $T_\oplus \simeq 279\,\text{K}$ or 6°C (or 43°F). This is slightly on the cold side, by a few degrees, because of the greenhouse action of the Earth's atmosphere, which was neglected in our calculation. Nevertheless, it is quite encouraging that such a crude calculation comes so close to the correct answer.

10.15 Conduction Electrons in Metal

The conduction electrons in a metal are non-localized (i.e., they are not tied to any particular atoms). In conventional metals, each atom contributes a fixed number of such electrons (corresponding to its valency). To a first approximation, it is possible to neglect the mutual interaction of the conduction electrons, because this interaction is largely shielded out by the stationary ions. The conduction electrons can, therefore, be treated as an ideal gas. However, the concentration of such electrons in a metal far exceeds the concentration of particles in a conventional gas. It is, therefore, not surprising that conduction electrons cannot normally be analyzed using classical statistics: in fact, they are subject to Fermi-Dirac statistics (because electrons are fermions).

Recall, from Section 10.5, that the mean number of particles occupying state s (energy ϵ_s) is given by

$$\bar{n}_s = \frac{1}{e^{\beta(\epsilon_s-\mu)} + 1}, \tag{10.134}$$

according to the Fermi-Dirac distribution. Here,

$$\mu \equiv -k\,T\,\alpha \tag{10.135}$$

is termed the *Fermi energy* of the system. This energy is determined by the condition that

$$\sum_r \bar{n}_r = \sum_r \frac{1}{e^{\beta(\epsilon_r-\mu)} + 1} = N, \tag{10.136}$$

where N is the total number of particles contained in the volume V. It is clear, from the previous equation, that the Fermi energy, μ, is generally a function of the temperature, T.

Let us investigate the behavior of the so-called *Fermi function*,

$$F(\epsilon) = \frac{1}{e^{\beta(\epsilon-\mu)} + 1}, \tag{10.137}$$

as ϵ varies. Here, the energy is measured from its lowest possible value $\epsilon = 0$. If the Fermi energy, μ, is such that $\beta\mu \ll -1$ then $\beta(\epsilon-\mu) \gg 1$, and F reduces to the Maxwell-Boltzmann distribution. However, for the case of conduction electrons in a metal, we are interested in the opposite limit, where

$$\beta\mu \equiv \frac{\mu}{kT} \gg 1. \tag{10.138}$$

In this limit, if $0 < \epsilon \ll \mu$ then $\beta(\epsilon-\mu) \ll -1$, so that $F(\epsilon) = 1$. On the other hand, if $\epsilon \gg \mu$ then $\beta(\epsilon-\mu) \gg 1$, so that $F(\epsilon) = \exp[-\beta(\epsilon-\mu)]$ falls off exponentially with increasing ϵ, just like a classical Maxwell-Boltzmann distribution. Note that $F = 1/2$ when $\epsilon = \mu$. The transition region in which F goes from a value close to unity to a value close to zero corresponds to an energy interval of order kT, centered on $\epsilon = \mu$. In fact, $F = 3/4$ when $\epsilon = \mu - (\ln 3)\,kT$, and $F = 1/4$ when $\epsilon = \mu + (\ln 3)\,kT$. The behavior of the Fermi function is illustrated in Figure 10.4.

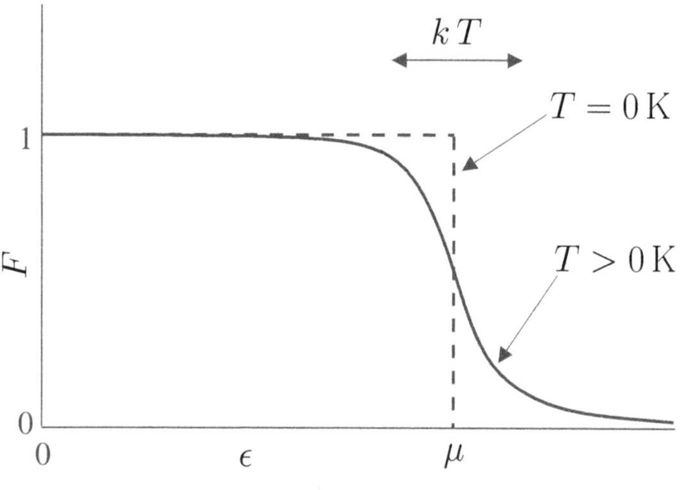

Figure 10.4 The Fermi function.

In the limit as $T \to 0$, the transition region becomes infinitesimally narrow. In this case, $F = 1$ for $\epsilon < \mu$, and $F = 0$ for $\epsilon > \mu$, as illustrated in Figure 10.4. This is an obvious result, because when $T = 0$ the conduction electrons attain their lowest energy, or ground-state, configuration. Because the Pauli exclusion principle requires that there be no more than one electron per single-particle quantum state, the lowest energy configuration is obtained by piling electrons into the lowest available unoccupied states, until all of the electrons are used up. Thus, the last electron added to the pile has a quite considerable energy, $\epsilon = \mu$, because all of the lower energy states are already occupied. Clearly, the exclusion principle implies that a Fermi-Dirac gas possesses a large mean energy, even at absolute zero.

Let us calculate the Fermi energy, $\mu = \mu_F$, of a Fermi-Dirac gas at $T = 0$. The energy of each particle is related to its momentum $\mathbf{p} = \hbar\,\mathbf{k}$ via

$$\epsilon = \frac{p^2}{2\,m} = \frac{\hbar^2\,k^2}{2\,m}, \tag{10.139}$$

where $\mathbf{k}$ is the de Broglie wavevector. Here, m is the electron mass. At $T = 0$, all quantum states whose energy is less than the Fermi energy, μ_F, are filled. The Fermi energy corresponds to a so-called *Fermi momentum*, $p_F = \hbar\,k_F$, which is such that

$$\mu_F = \frac{p_F^2}{2\,m} = \frac{\hbar^2\,k_F^2}{2\,m}. \tag{10.140}$$

Thus, at $T = 0$, all quantum states with $k < k_F$ are filled, and all those with $k > k_F$ are empty.

Now, we know, by analogy with Equation (8.183), that there are $(2\pi)^{-3}\,V$ allowable translational states per unit volume of $\mathbf{k}$-space. The volume of the sphere of radius k_F in $\mathbf{k}$-space is $(4/3)\,\pi\,k_F^3$. It follows that the *Fermi sphere* of radius k_F contains $(4/3)\,\pi\,k_F^3\,(2\pi)^{-3}\,V$ translational states. The number of quantum states inside the sphere is twice this, because electrons possess two possible spin states for every possible translational state. Because the total number of occupied states (i.e., the total number of quantum states inside the Fermi sphere) must equal the total number of particles in the gas, it follows that

$$2\,\frac{V}{(2\pi)^3}\left(\frac{4}{3}\,\pi\,k_F^3\right) = N. \tag{10.141}$$

The previous expression can be rearranged to give

$$k_F = \left(3\pi^2\,\frac{N}{V}\right)^{1/3}. \tag{10.142}$$

Hence,

$$\lambda_F \equiv \frac{2\pi}{k_F} = \frac{2\pi}{(3\pi^2)^{1/3}}\left(\frac{V}{N}\right)^{1/3}, \tag{10.143}$$

which implies that the de Broglie wavelength, λ_F, corresponding to the Fermi energy, is of order the mean separation between particles $(V/N)^{1/3}$. All quantum states with de Broglie wavelengths $\lambda \equiv 2\pi/k > \lambda_F$ are occupied at $T = 0$, whereas all those with $\lambda < \lambda_F$ are empty.

According to Equations (10.140) and (10.142), the Fermi energy at $T = 0$ takes the form

$$\mu_F = \frac{\hbar^2}{2\,m}\left(3\pi^2\,\frac{N}{V}\right)^{2/3}. \tag{10.144}$$

It is easily demonstrated that $\mu_F \gg k\,T$ for conventional metals at room temperature. (See Exercise 10.17.)

The majority of the conduction electrons in a metal occupy a band of completely filled states with energies far below the Fermi energy. In many cases, such electrons have very little effect on the macroscopic properties of the metal. Consider, for example, the contribution of the conduction electrons to the specific heat of the metal. The heat capacity at constant volume, C_V, of these electrons can be calculated from a knowledge of their mean energy, $\overline{E}$, as a function of T: that is,

$$C_V = \left(\frac{\partial\overline{E}}{\partial T}\right)_V. \tag{10.145}$$

If the electrons obeyed classical Maxwell-Boltzmann statistics, so that $F \propto \exp(-\beta\,\epsilon)$ for all electrons, then the equipartition theorem would give

$$\overline{E} = \frac{3}{2}\,N\,k\,T, \tag{10.146}$$

$$C_V = \frac{3}{2}\,N\,k. \tag{10.147}$$

However, the actual situation, in which F has the form shown in Figure 10.4, is very different. A small change in T does not affect the mean energies of the majority of the electrons, with $\epsilon \ll \mu$, because these electrons lie in states that are completely filled, and remain so when the temperature is changed. It follows that these electrons contribute nothing whatsoever to the heat capacity. On the other hand, the relatively small number of electrons, $N_{\rm eff}$, in the energy range of order $k\,T$, centered on the Fermi energy, in which F is significantly different from 0 and 1, do contribute to the specific heat. In the tail end of this region, $F \propto \exp(-\beta\,\epsilon)$, so the distribution reverts to a Maxwell-Boltzmann distribution. Hence, from Equation (10.147), we expect each electron in this region to contribute roughly an amount $(3/2)\,k$ to the heat capacity. Hence, the heat capacity can be written

$$C_V \simeq \frac{3}{2}\,N_{\rm eff}\,k. \tag{10.148}$$

However, because only a fraction $k\,T/\mu$ of the total conduction electrons lie in the tail region of the Fermi-Dirac distribution, we expect

$$N_{\rm eff} \simeq \frac{k\,T}{\mu}\,N. \tag{10.149}$$

It follows that

$$C_V \simeq \frac{3}{2} N k \frac{k T}{\mu}. \tag{10.150}$$

Because $k T \ll \mu$ in conventional metals, the molar specific heat of the conduction electrons is clearly very much less than the classical value $(3/2) R$. This accounts for the fact that the molar specific heat capacities of metals at room temperature are about the same as those of insulators. Before the advent of quantum mechanics, the classical theory predicted incorrectly that the presence of conduction electrons should raise the heat capacities of monovalent metals by 50 percent [i.e., $(3/2) R$] compared to those of insulators.

Note that the specific heat (10.150) is not temperature independent. In fact, using the superscript e to denote the electronic specific heat, the molar specific heat can be written

$$c_V^{(e)} = \gamma T, \tag{10.151}$$

where γ is a (positive) constant of proportionality. At room temperature $c_V^{(e)}$ is completely masked by the much larger specific heat, $c_V^{(L)}$, due to lattice vibrations. However, at very low temperatures $c_V^{(L)} = A T^3$, where A is a (positive) constant of proportionality. (See Section 8.14.) Clearly, at low temperatures, $c_V^{(L)} = A T^3$ approaches zero far more rapidly that the electronic specific heat, as T is reduced. Hence, it should be possible to measure the electronic contribution to the molar specific heat at low temperatures.

The total molar specific heat of a metal at low temperatures takes the form

$$c_V = c_V^{(e)} + c_V^{(L)} = \gamma T + A T^3. \tag{10.152}$$

Hence,

$$\frac{c_V}{T} = \gamma + A T^2. \tag{10.153}$$

It follows that a plot of c_V/T versus T^2 should yield a straight-line whose intercept on the vertical axis gives the coefficient γ. Figure 10.5 shows such a plot. The fact that a good straight-line is obtained verifies that the temperature dependence of the heat capacity predicted by Equation (10.152) is indeed correct.

10.16 Sommerfeld Expansion

Let us examine the conduction electrons in a metal in slightly more detail. In particular, let us try to obtain a more exact expression for the electronic specific

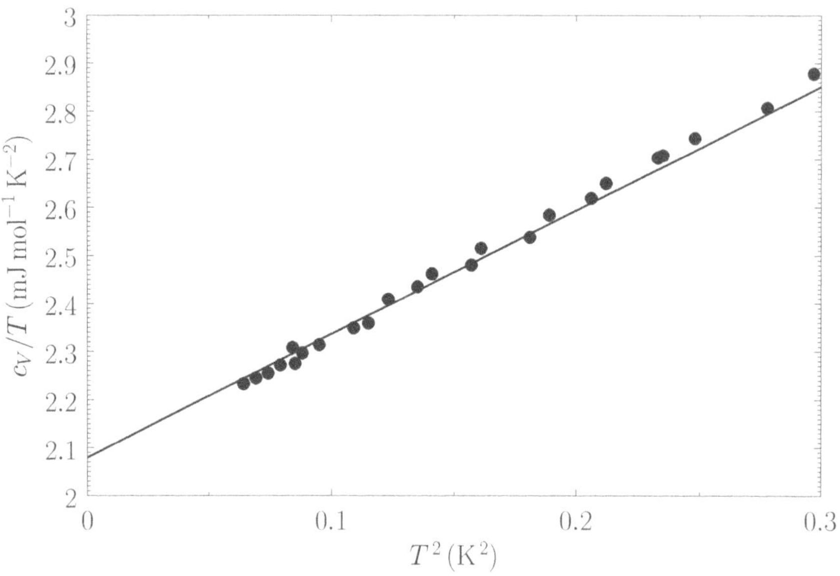

Figure 10.5 The low-temperature heat capacity of potassium, plotted as c_V/T versus T^2. The straight-line shows the fit $c_V/T = 2.08 + 2.57\,T^2$. Data from [Kittel and Kroemer (1980)].

heat. We saw in the previous section that the Fermi energy, μ, is determined by the equation

$$N = \sum_r \frac{1}{e^{\beta(\epsilon_r-\mu)}+1}. \tag{10.154}$$

Likewise, the mean electron energy is

$$\overline{E} = \sum_r \frac{\epsilon_r}{e^{\beta(\epsilon_r-\mu)}+1}. \tag{10.155}$$

Because, in general, the energies of the quantum states are very closely spaced, the sums in the previous two expressions can be replaced by integrals. Now, according to Section 10.12, the number of quantum states per unit volume with wavenumbers in the range k to $k + dk$ is

$$\rho_k(k)\,dk = \frac{k^2}{2\pi^2}\,dk. \tag{10.156}$$

However, the energy of a state with wavenumber k is

$$\epsilon = \frac{\hbar^2\,k^2}{2\,m}, \tag{10.157}$$

where m is the electron mass. Let $\rho(\epsilon)\,d\epsilon$ be the number of electrons whose energies lies in the range ϵ to $\epsilon + d\epsilon$. It follows that

$$\rho(\epsilon)\,d\epsilon = 2\,V\,\rho_k(k)\,\frac{dk}{d\epsilon}\,d\epsilon, \tag{10.158}$$

where the factor of 2 is to take into account the two possible spin states which exist for each translational state. Hence,

$$\rho(\epsilon) = \frac{V}{2\pi^2}\,\frac{(2\,m)^{3/2}}{\hbar^3}\,\epsilon^{1/2}. \tag{10.159}$$

Moreover, Equations (10.154) and (10.155) become

$$N = \int_0^\infty F(\epsilon)\,\rho(\epsilon)\,d\epsilon, \tag{10.160}$$

$$\overline{E} = \int_0^\infty F(\epsilon)\,\epsilon\,\rho(\epsilon)\,d\epsilon, \tag{10.161}$$

where

$$F(\epsilon) = \frac{1}{\mathrm{e}^{\beta\,(\epsilon-\mu)} + 1} \tag{10.162}$$

is the Fermi function.

The integrals on the right-hand sides of Equations (10.160) and (10.161) are both of the general form

$$\int_0^\infty F(\epsilon)\,\varphi(\epsilon)\,d\epsilon, \tag{10.163}$$

where $\varphi(\epsilon)$ is a smoothly varying function of ϵ. Let

$$\psi(\epsilon) = \int_0^\epsilon \varphi(\epsilon')\,d\epsilon'. \tag{10.164}$$

We can integrate Equation (10.163) by parts to give

$$\int_0^\infty F(\epsilon)\,\varphi(\epsilon)\,d\epsilon = [F(\epsilon)\,\psi(\epsilon)]_0^\infty - \int_0^\infty F'(\epsilon)\,\psi(\epsilon)\,d\epsilon, \tag{10.165}$$

which reduces to

$$\int_0^\infty F(\epsilon)\,\varphi(\epsilon)\,d\epsilon = -\int_0^\infty F'(\epsilon)\,\psi(\epsilon)\,d\epsilon, \tag{10.166}$$

because $\varphi(0) = \varphi(\infty)\,F(\infty) = 0$. Here, $F'(\epsilon) \equiv dF/d\epsilon$.

Now, if $k\,T \ll \mu$ then $F(\epsilon)$ is a constant everywhere, apart from a thin region of thickness $k\,T$, centered on $\epsilon = \mu$. (See Figure 10.4.) It follows that $F'(\epsilon)$ is approximately zero everywhere, apart from in this region. Hence, the relatively slowly-varying function $\psi(\epsilon)$ can be Taylor expanded about $\epsilon = \mu$:

$$\psi(\epsilon) = \sum_{m=0,\infty} \frac{1}{m!}\left[\frac{d^m\psi}{d\epsilon^m}\right]_\mu (\epsilon-\mu)^m. \tag{10.167}$$

Thus, Equation (10.166) becomes

$$\int_0^\infty F(\epsilon)\,\varphi(\epsilon)\,d\epsilon = -\sum_{m=0,\infty} \frac{1}{m!}\left[\frac{d^m\psi}{d\epsilon^m}\right]_\mu \int_0^\infty F'(\epsilon)\,(\epsilon-\mu)^m\,d\epsilon. \tag{10.168}$$

From Equation (10.162), we have

$$\int_0^\infty F'(\epsilon)\,(\epsilon-\mu)^m\,d\epsilon = -\int_0^\infty \frac{\beta\,e^{\beta(\epsilon-\mu)}}{[e^{\beta(\epsilon-\mu)}+1]^2}\,(\epsilon-\mu)^m\,d\epsilon, \tag{10.169}$$

which becomes

$$\int_0^\infty F'(\epsilon)\,(\epsilon-\mu)^m\,d\epsilon = -\beta^{-m}\int_{-\beta\mu}^\infty \frac{e^x}{(e^x+1)^2}\,x^m\,dx, \tag{10.170}$$

where $x = \beta(\epsilon-\mu)$. However, because the integrand has a sharp maximum at $x = 0$, and because $\beta\mu \gg 1$, we can replace the lower limit of integration by $-\infty$ with negligible error. Thus, we obtain

$$\int_0^\infty F'(\epsilon)\,(\epsilon-\mu)^m\,d\epsilon = -(k\,T)^m\,I_m, \tag{10.171}$$

where

$$I_m = \int_{-\infty}^\infty \frac{e^x}{(e^x+1)^2}\,x^m\,dx. \tag{10.172}$$

Note that

$$\frac{e^x}{(e^x+1)^2} \equiv \frac{1}{(e^x+1)\,(e^{-x}+1)} \tag{10.173}$$

is an even function of x. It follows, by symmetry, that I_m is zero when m is odd. Moreover,

$$I_0 = \int_{-\infty}^\infty \frac{e^x}{(e^x+1)^2}\,dx = -\left[\frac{1}{e^x+1}\right]_{-\infty}^\infty = 1. \tag{10.174}$$

Finally, it can be demonstrated that

$$I_2 = \frac{\pi^2}{3}. \tag{10.175}$$

(See Exercise 10.3.) Hence, we deduce that

$$\int_0^\infty F(\epsilon)\,\varphi(\epsilon)\,d\epsilon = \int_0^\mu \varphi(\epsilon)\,d\epsilon + \frac{\pi^2}{6}\,(k\,T)^2\left(\frac{d\varphi}{d\epsilon}\right)_\mu + \cdots. \tag{10.176}$$

This expansion is known as the *Sommerfeld expansion*, after its inventor, Arnold Sommerfeld.

Equation (10.160) yields

$$N = \int_0^\mu \rho(\epsilon)\,d\epsilon + \frac{\pi^2}{6}\,(k\,T)^2\left(\frac{d\rho}{d\epsilon}\right)_\mu + \cdots. \tag{10.177}$$

However, it follows from Equation (10.159) that

$$\int_0^{\mu} \rho(\epsilon)\,d\epsilon = \frac{V}{3\pi^2}\,\frac{(2\,m)^{3/2}}{\hbar^3}\,\mu^{3/2}, \tag{10.178}$$

$$\left(\frac{d\rho}{d\epsilon}\right)_{\mu} = \frac{V}{4\pi^2}\,\frac{(2\,m)^{3/2}}{\hbar^3}\,\mu^{-1/2}. \tag{10.179}$$

Hence,

$$N \simeq \frac{V}{3\pi^2}\,\frac{(2\,m)^{3/2}}{\hbar^3}\,\mu^{3/2} + \frac{V}{24}\,\frac{(2\,m)^{3/2}}{\hbar^3}\,(k\,T)^2\,\mu^{-1/2}, \tag{10.180}$$

which can also be written

$$\mu_F^{3/2} \simeq \mu^{3/2} + \frac{\pi^2}{8}\,(k\,T)^2\,\mu^{-1/2}, \tag{10.181}$$

where μ_F is the Fermi energy at $T = 0$. [See Equation (10.144).] The previous equation can be rearranged to give

$$\mu \simeq \mu_F\left[1 + \frac{\pi^2}{8}\left(\frac{k\,T}{\mu}\right)^2\right]^{-2/3}, \tag{10.182}$$

which reduces to

$$\mu \simeq \mu_F\left[1 - \frac{\pi^2}{12}\left(\frac{k\,T}{\mu_F}\right)^2\right], \tag{10.183}$$

assuming that $k\,T/\mu_F \ll 1$. Figure 10.6 shows a comparison between the previous approximate expression for the temperature variation of the Fermi energy of a degenerate electron gas and the numerically-calculated exact value. It can be seen that our approximate expression is surprisingly accurate (at least, for $T \sim \mu_F/k$).

Equation (10.161) yields

$$\overline{E} = \int_0^{\mu} \epsilon\,\rho(\epsilon)\,d\epsilon + \frac{\pi^2}{6}\,(k\,T)^2\left[\frac{d(\epsilon\,\rho)}{d\epsilon}\right]_{\mu} + \cdots. \tag{10.184}$$

However, it follows from Equation (10.159) that

$$\int_0^{\mu} \epsilon\,\rho(\epsilon)\,d\epsilon = \frac{V}{5\pi^2}\,\frac{(2\,m)^{3/2}}{\hbar^3}\,\mu^{5/2}, \tag{10.185}$$

$$\left[\frac{d(\epsilon\,\rho)}{d\epsilon}\right]_{\mu} = \frac{3\,V}{4\pi^2}\,\frac{(2\,m)^{3/2}}{\hbar^3}\,\mu^{1/2}. \tag{10.186}$$

Hence,

$$\overline{E} \simeq \frac{V}{5\pi^2}\,\frac{(2\,m)^{3/2}}{\hbar^3}\,\mu^{5/2} + \frac{V}{8}\,\frac{(2\,m)^{3/2}}{\hbar^3}\,(k\,T)^2\,\mu^{1/2}, \tag{10.187}$$

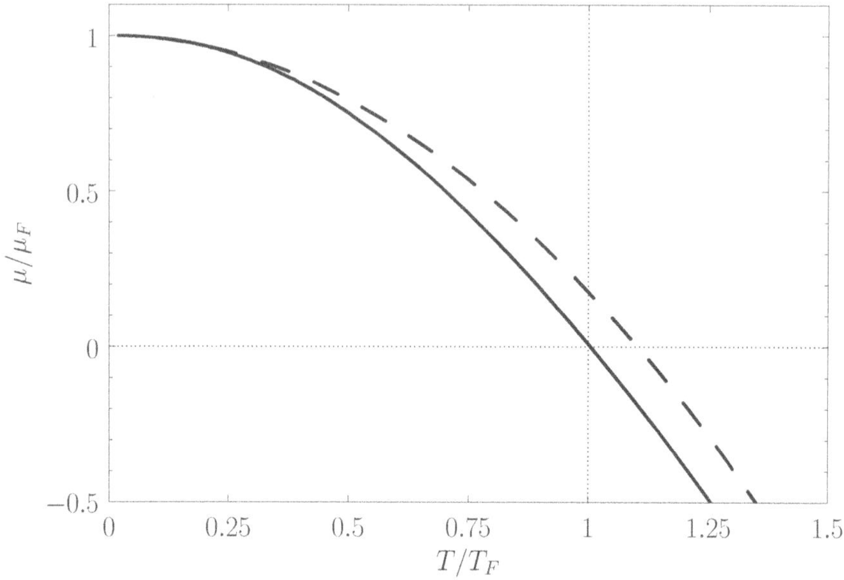

Figure 10.6 The solid curve shows the numerically-calculated exact Fermi energy of a degenerate electron gas as a function of the temperature. The dashed curve shows the analytic approximation $\mu/\mu_F \simeq 1 - (\pi^2/12)\,(T/T_F)^2$. Here, $T_F = \mu_F/k$ is known as the *Fermi temperature.*

which can also be written

$$\overline{E} \simeq \frac{3}{5}\,N\,\mu_F\left[\left(\frac{\mu}{\mu_F}\right)^{5/2} + \frac{5\pi^2}{24}\left(\frac{k\,T}{\mu_F}\right)^2\left(\frac{\mu}{\mu_F}\right)^{1/2}\right]. \tag{10.188}$$

Making use of Equation (10.183), and only retaining terms up to second order in $k\,T/\mu_F$, we obtain

$$\overline{E} \simeq \frac{3}{5}\,N\,\mu_F + \frac{\pi^2}{4}\,N\,\mu_F\left(\frac{k\,T}{\mu_F}\right)^2. \tag{10.189}$$

Hence, the specific heat capacity of the conduction electrons becomes

$$C_V^{(e)} = \frac{d\overline{E}}{dT} = \frac{\pi^2}{2}\,k\,N\,\frac{k\,T}{\mu_F}, \tag{10.190}$$

and the molar specific heat is written

$$c_V^{(e)} = \frac{\pi^2}{2}\left(\frac{k\,T}{\mu_F}\right)R. \tag{10.191}$$

Of course, because $k\,T/\mu_F \ll 1$, this value is much less than the classical estimate, $(3/2)\,R$.

10.17 White-Dwarf Stars

A main-sequence hydrogen-burning star, such as the Sun, is maintained in equilibrium via the balance of the gravitational attraction tending to make it collapse, and the thermal pressure tending to make it expand. Of course, the thermal energy of the star is generated by nuclear reactions occurring deep inside its core. Eventually, however, the star will run out of burnable fuel, and will, therefore, start to collapse, as it radiates away its remaining thermal energy. What is the ultimate fate of such a star?

A burnt-out star is basically a gas of electrons and ions. As the star collapses, its density increases, so the mean separation between its constituent particles decreases. Eventually, the mean separation becomes of order the de Broglie wavelength of the electrons, and the electron gas becomes degenerate. Note, that the de Broglie wavelength of the ions is much smaller than that of the electrons (because of the ions' much larger mass), so the ion gas remains non-degenerate. Now, even at zero temperature, a degenerate electron gas exerts a substantial pressure, because the Pauli exclusion principle prevents the mean electron separation from becoming significantly smaller than the typical de Broglie wavelength. (See Section 10.15 and Exercise 10.18.) Thus, it is possible for a burnt-out star to maintain itself against complete collapse under gravity via the *degeneracy pressure* of its constituent electrons. Such stars are termed *white-dwarfs*. Let us investigate the physics of white-dwarfs in more detail.

The total energy of a white-dwarf star can be written

$$E = K + U, \tag{10.192}$$

where K is the total kinetic energy of the degenerate electrons (the kinetic energy of the ions is negligible), and U is the gravitational potential energy. Let us assume, for the sake of simplicity, that the density of the star is uniform. In this case, the gravitational potential energy takes the form

$$U = -\frac{3}{5}\frac{G M^2}{R}, \tag{10.193}$$

where G is the gravitational constant, M the stellar mass, and R the stellar radius.

Let us assume that the electron gas is highly degenerate, which is equivalent to taking the limit $T \to 0$. In this case, we know, from Section 10.15, that the Fermi momentum can be written

$$p_F = \Lambda\left(\frac{N}{V}\right)^{1/3}, \tag{10.194}$$

where

$$\Lambda = (3\pi^2)^{1/3}\,\hbar. \tag{10.195}$$

Here,

$$V = \frac{4\pi}{3}R^3 \tag{10.196}$$

is the stellar volume, and N the total number of electrons contained in the star. Furthermore, the number of electron states contained in an annular radius of **p**-space lying between radii p and $p + dp$ is

$$dN = \frac{3\,V}{\Lambda^3}\,p^2\,dp. \tag{10.197}$$

(See Exercise 10.20.) Hence, the total kinetic energy of the electron gas can be written

$$K = \frac{3\,V}{\Lambda^3}\int_0^{p_F}\frac{p^2}{2\,m}\,p^2\,dp = \frac{3}{5}\,\frac{V}{\Lambda^3}\,\frac{p_F^5}{2\,m}, \tag{10.198}$$

where m is the electron mass. It follows that

$$K = \frac{3}{5}\,N\,\frac{\Lambda^2}{2\,m}\left(\frac{N}{V}\right)^{2/3}. \tag{10.199}$$

The interior of a white-dwarf star is composed of atoms like C^{12} and O^{16} that contain equal numbers of protons, neutrons, and electrons. Thus,

$$M = 2\,N\,m_p, \tag{10.200}$$

where m_p is the proton mass.

Equations (10.192), (10.193), (10.195), (10.196), (10.199), and (10.200) can be combined to give

$$E = \frac{A}{R^2} - \frac{B}{R}, \tag{10.201}$$

where

$$A = \frac{3}{20}\left(\frac{9\pi}{8}\right)^{2/3}\frac{\hbar^2}{m}\left(\frac{M}{m_p}\right)^{5/3}, \tag{10.202}$$

$$B = \frac{3}{5}\,G\,M^2. \tag{10.203}$$

The equilibrium radius of the star, R_*, is that which minimizes the total energy, E. In fact, it is easily demonstrated that

$$R_* = \frac{2\,A}{B}, \tag{10.204}$$

which yields

$$R_* = \frac{(9\pi)^{2/3}}{8}\,\frac{\hbar^2}{m}\,\frac{1}{G\,m_p^{5/3}\,M^{1/3}}. \tag{10.205}$$

The previous formula can also be written

$$\frac{R_*}{R_\odot} = 0.010\left(\frac{M_\odot}{M}\right)^{1/3}, \tag{10.206}$$

where $R_\odot = 7\times 10^5$ km is the solar radius, and $M_\odot = 2\times 10^{30}$ kg the solar mass. It follows that the radius of a typical solar-mass white-dwarf is about 7000km: that is, about the same as the radius of the Earth. The first white-dwarf to be discovered (in 1862) was the companion of Sirius. Nowadays, thousands of white-dwarfs have been observed, all with properties similar to those described previously.

10.18 Chandrasekhar Limit

One curious feature of white-dwarf stars is that their radius decreases as their mass increases. [See Equation (10.206).] It follows, from Equation (10.199), that the mean energy of the degenerate electrons inside the star increases strongly as the stellar mass increases: in fact, $K \propto M^{7/3}$. Hence, if M becomes sufficiently large then the electrons become relativistic, and the previous analysis needs to be modified. Strictly speaking, the non-relativistic analysis described in the previous section is only valid in the low-mass limit $M \ll M_\odot$. Let us, for the sake of simplicity, consider the ultra-relativistic limit in which $p \gg m\,c$.

The total electron energy (including the rest mass energy) can be written

$$K = \frac{3\,V}{\Lambda^3}\int_0^{p_F}\left(p^2\,c^2 + m^2\,c^4\right)^{1/2} p^2\,dp, \tag{10.207}$$

by analogy with Equation (10.198). Thus,

$$K \simeq \frac{3\,V\,c}{\Lambda^3}\int_0^{p_F}\left(p^3 + \frac{m^2\,c^2}{2}\,p + \cdots\right)dp, \tag{10.208}$$

giving

$$K \simeq \frac{3}{4}\,\frac{V\,c}{\Lambda^3}\left(p_F^4 + m^2\,c^2\,p_F^2 + \cdots\right). \tag{10.209}$$

It follows, from the previous analysis, that the total energy of an ultra-relativistic white-dwarf star can be written in the form

$$E \simeq \frac{A-B}{R} + C\,R, \tag{10.210}$$

where

$$A = \frac{3}{8}\left(\frac{9\pi}{8}\right)^{1/3}\hbar\,c\left(\frac{M}{m_p}\right)^{4/3}, \tag{10.211}$$

$$B = \frac{3}{5}\,G\,M^2, \tag{10.212}$$

$$C = \frac{3}{4}\,\frac{1}{(9\pi)^{1/3}}\,\frac{m^2\,c^3}{\hbar}\left(\frac{M}{m_p}\right)^{2/3}. \tag{10.213}$$

As before, the equilibrium radius R_* is that which minimizes the total energy. E. However, in the ultra-relativistic case, a non-zero value of R_* only exists for $A - B > 0$. When $A - B < 0$, the energy decreases monotonically with decreasing stellar radius. In other words, the degeneracy pressure of the electrons is incapable of halting the collapse of the star under gravity. The criterion that must be satisfied for a relativistic white-dwarf star to be maintained against gravity is that

$$\frac{A}{B} > 1. \tag{10.214}$$

This criterion can be re-written

$$M < M_C, \tag{10.215}$$

where

$$M_C = \frac{15}{64}\,(5\pi)^{1/2}\,\frac{(\hbar\,c/G)^{3/2}}{m_p^2} = 1.72\,M_\odot \tag{10.216}$$

is known as the *Chandrasekhar limit*, after A. Chandrasekhar who first derived it in 1931. A more realistic calculation, which does not assume constant density, yields

$$M_C = 1.4\,M_\odot. \tag{10.217}$$

Thus, if the stellar mass exceeds the Chandrasekhar limit then the star in question cannot become a white-dwarf when its nuclear fuel is exhausted, but, instead, must continue to collapse. What is the ultimate fate of such a star?

10.19 Neutron Stars

At stellar densities that greatly exceed white-dwarf densities, the extreme pressures cause electrons to combine with protons to form neutrons. Thus, any star that collapses to such an extent that its radius becomes significantly less than that characteristic of a white-dwarf is effectively transformed into a gas of neutrons. Eventually, the mean separation between the neutrons becomes comparable with their de Broglie wavelength. At this point, it is possible for the degeneracy pressure of the neutrons to halt the collapse of the star. A star that is maintained against gravity in this manner is called a *neutron star*.

Neutrons stars can be analyzed in a very similar manner to white-dwarf stars. In fact, the previous analysis can be simply modified by letting $m_p \to m_p/2$ and $m \to m_p$. Thus, we conclude that non-relativistic neutrons stars satisfy the mass-radius law:

$$\frac{R_*}{R_\odot} = 0.000011\left(\frac{M_\odot}{M}\right)^{1/3} \tag{10.218}$$

It follows that the radius of a typical solar mass neutron star is a mere 10km. In 1967, Antony Hewish and Jocelyn Bell discovered a class of compact radio sources, called *pulsars*, that emit extremely regular pulses of radio waves. Pulsars have subsequently been identified as rotating neutron stars. To date, many hundreds of these objects have been observed.

When relativistic effects are taken into account, it is found that there is a critical mass above which a neutron star cannot be maintained against gravity. According to our analysis, this critical mass, which is known as the *Oppenheimer-Volkoff limit*, is given by

$$M_{OV} = 4\,M_C = 6.9\,M_\odot. \tag{10.219}$$

A more realistic calculation, which does not assume constant density, does not treat the neutrons as point particles, and takes general relativity into account, gives a somewhat lower value of

$$M_{OV} = 1.5\text{–}2.5\,M_\odot. \tag{10.220}$$

A star whose mass exceeds the Oppenheimer-Volkoff limit cannot be maintained against gravity by degeneracy pressure, and must ultimately collapse to form a black-hole.

10.20 Bose-Einstein Condensation

Consider a gas of weakly-interacting bosons. It is helpful to define the gas's *chemical potential*,

$$\mu = k\,T\,\alpha, \tag{10.221}$$

whose value is determined by the equation

$$N = \sum_r \bar{n}_r = \sum_r \frac{1}{e^{\beta\,(\epsilon_r-\mu)} - 1}. \tag{10.222}$$

Here, N is the total number of particles, and ϵ_r the energy of the single-particle quantum state r. Because, in general, the energies of the quantum states are very closely spaced, the sum in the previous expression can approximated as an integral. Now, according to Section 10.12, the number of quantum states per unit volume with wavenumbers in the range k to $k + dk$ is

$$\rho_k(k)\,dk = \frac{k^2}{2\pi^2}\,dk. \tag{10.223}$$

However, the energy of a state with wavenumber k is

$$\epsilon = \frac{\hbar^2\,k^2}{2\,m}, \tag{10.224}$$

where m is the boson mass. Let $\rho(\epsilon)\,d\epsilon$ be the number of bosons whose energies lies in the range ϵ to $\epsilon + d\epsilon$. It follows that

$$\rho(\epsilon)\,d\epsilon = V\,\rho_k(k)\,\frac{dk}{d\epsilon}\,d\epsilon, \tag{10.225}$$

where V is the volume of the gas. Here, we are assuming that the bosons are spinless, so that there is only one particle state per translational state. Hence,

$$\rho(\epsilon) = \frac{V}{4\pi^2}\,\frac{(2\,m)^{3/2}}{\hbar^3}\,\epsilon^{1/2}, \tag{10.226}$$

and Equation (10.222) becomes

$$N = \int_0^\infty \frac{\rho(\epsilon)}{e^{\beta(\epsilon-\mu)} - 1}\,d\epsilon = \frac{V}{4\pi^2}\,\frac{(2\,m)^{3/2}}{\hbar^3}\int_0^\infty \frac{\epsilon^{1/2}}{e^{\beta(\epsilon-\mu)} - 1}\,d\epsilon. \tag{10.227}$$

However, there is a significant flaw in this formulation. In using the integral approximation, rather than performing the sum, the ground-state, $\epsilon = 0$, has been left out [because $\rho(0) = 0$]. Under ordinary circumstances, this omission does not matter. However, at very low temperatures, bosons tend to condense into the ground-state, and the occupation number of this state becomes very much larger than that of any other state. Under these circumstances, the ground-state must be included in the calculation.

We can overcome the previous difficulty in the following manner. Let there be N_0 bosons in the ground-state, and $N_{\rm ex}$ in the various excited states, so that

$$N = N_0 + N_{\rm ex}. \tag{10.228}$$

Because the ground-state is excluded from expression (10.227), the integral only gives the number of bosons in excited states. In other words,

$$N_{\rm ex} = \frac{V}{4\pi^2}\,\frac{(2\,m)^{3/2}}{\hbar^3}\int_0^\infty \frac{\epsilon^{1/2}}{e^{\beta(\epsilon-\mu)} - 1}\,d\epsilon. \tag{10.229}$$

Now, because the ground-state has zero energy, its mean occupancy number is

$$N_0 = \frac{1}{e^{-\mu/kT} - 1}. \tag{10.230}$$

Moreover, at temperatures very close to absolute zero, we expect $N_0 \simeq N$, which implies that

$$N \simeq \frac{1}{e^{-\mu/kT} - 1}. \tag{10.231}$$

We conclude that

$$-\frac{\mu}{k\,T} \simeq \ln\left(1 + \frac{1}{N}\right) \simeq \frac{1}{N}, \tag{10.232}$$

for large N. Hence, at very low temperatures, we can safely set $\exp(-\mu/k\,T)$ equal to unity in Equation (10.229). Thus, we obtain

$$N_{\rm ex} = \frac{2}{\sqrt{\pi}}\left(\frac{2\pi\,m\,k\,T}{h^2}\right)^{3/2} V \int_0^\infty \frac{x^{1/2}}{{\rm e}^{\,x}-1}\,dx, \tag{10.233}$$

where $x = \epsilon/(k\,T)$. The value of the integral is $\Gamma(3/2)\,\zeta(3/2)$, where $\Gamma(s) = \int_0^\infty x^{s-1}\,{\rm e}^{-x}\,dx$ is a *Gamma function*, and $\zeta(s) = \sum_{n=1,\infty} n^{-s}$ a *Riemann Zeta function*. (See Exercise 10.4.) Furthermore, $\Gamma(3/2) = \sqrt{\pi}/2$, and $\zeta(3/2) = 2.612$. Hence,

$$N_{\rm ex} = \zeta(3/2)\,V\left(\frac{2\pi\,m\,k\,T}{h^2}\right)^{3/2}. \tag{10.234}$$

The so-called *Bose temperature*, T_B, is defined as the temperature above which all the bosons are in excited states. Setting $N_{\rm ex} = N$ and $T = T_B$ in the previous expression, we obtain

$$T_B = \frac{h^2}{2\pi\,m\,k}\left[\frac{N}{\zeta(3/2)\,V}\right]^{2/3}. \tag{10.235}$$

Moreover,

$$\frac{N_{\rm ex}}{N} = \left(\frac{T}{T_B}\right)^{3/2}. \tag{10.236}$$

Thus, the fractional number of bosons in the ground-state is

$$\frac{N_0}{N} = 1 - \left(\frac{T}{T_B}\right)^{3/2}. \tag{10.237}$$

Obviously, the preceding two equations are only valid for $0 < T < T_B$. For $T > T_B$, we have $N_0/N = 0$ and $N_{\rm ex}/N = 1$. Moreover, for $0 < T < T_B$ we have $\mu/k\,T_B = 0$, whereas for $T > T_B$, combining Equations (10.229) (with $N_{\rm ex} = N$) and (10.235) yields

$$\Gamma(3/2)\,\zeta(3/2) = \left(\frac{T}{T_B}\right)^{3/2}\int_0^\infty \frac{x^{1/2}}{{\rm e}^{\,x-\alpha}-1}\,dx, \tag{10.238}$$

where $x = \epsilon/(k\,T)$ and $\alpha = \mu/(k\,T)$. Expanding in powers of ${\rm e}^{-(x-\alpha)}$, we obtain

$$\Gamma(3/2)\,\zeta(3/2) = \left(\frac{T}{T_B}\right)^{3/2}\Gamma(3/2)\sum_{n=1,\infty}\frac{{\rm e}^{\,n\,\alpha}}{n^{3/2}}, \tag{10.239}$$

where $y = n\,x$ (see Exercise 10.4), which reduces to

$$\zeta(3/2) = \left(\frac{T}{T_B}\right)^{3/2}\sum_{n=1,\infty}\frac{{\rm e}^{-n\,\alpha}}{n^{3/2}}. \tag{10.240}$$

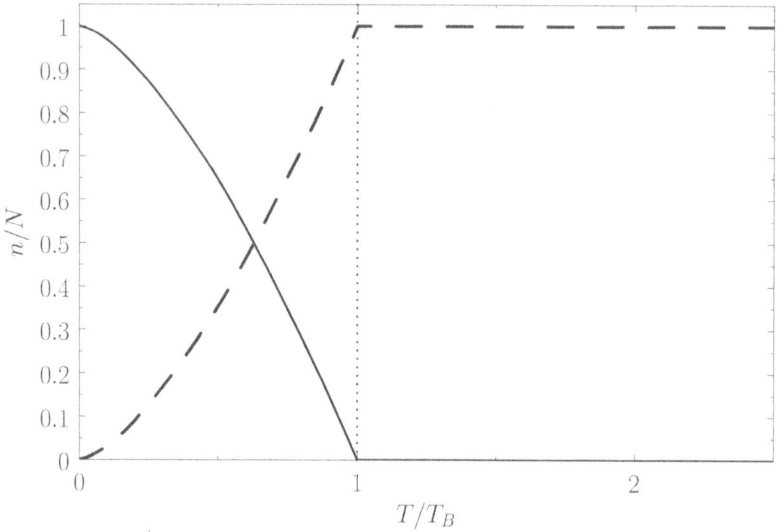

Figure 10.7 Variation with temperature of N_0/N (solid curve) and N_{ex}/N (dashed curve) for a boson gas.

The previous equation can be solved numerically to give $\alpha = \mu/(k\,T_B)$ as a function of T/T_B.

Let us estimate a typical value of the Bose temperature. Consider a boson gas made up of $N_A/10 = 6.02 \times 10^{22}$ He^4 atoms confined to a volume of 1 litre. The mass of a He^4 atom is 6.65×10^{-27} kg. Making use of Equation (10.235), we obtain

$$T_B = \frac{(6.63 \times 10^{-34})^2}{2\pi\,(6.65 \times 10^{-27})\,(1.38 \times 10^{-23})}\left[\frac{6.02 \times 10^{22}}{(2.612)\,(1 \times 10^{-3})}\right]^{2/3}$$
$$= 0.062\,\text{K}. \tag{10.241}$$

At atmospheric pressure, helium liquifies at 4.12 K, long before the Bose temperature is reached. In fact, all real gases liquify before the Bose temperature is reached.

Figure 10.7 shows the variation of N_0/N and N_{ex}/N with T/T_B. A corresponding graph of $\mu/k\,T_B$ [numerically determined from Equation (10.240) when $T > T_B$] is shown in Figure 10.8. The sudden collapse of bosons into the ground-state at temperatures below the Bose temperature is known as *Bose-Einstein condensation*. In 1995, E.A. Cornell and C.E. Wieman led a team of physicists that created a nearly pure condensate by cooling a vapor of rubidium atoms to a temperature of 1.3×10^{-7} K. (Cornell and Wieman were subsequently awarded the Nobel prize in 2001.)

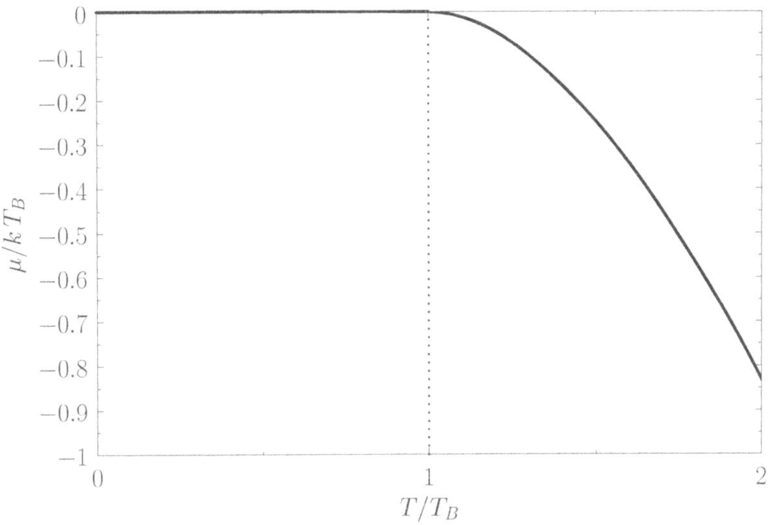

Figure 10.8 Variation with temperature of $\mu/k\,T_B$ for a boson gas.

The mean energy of a boson gas takes the form

$$\overline{E} = \sum_r \frac{\epsilon_r}{e^{\beta(\epsilon_r-\mu)} - 1}. \tag{10.242}$$

As before, we can approximate the sum as an integral, and write

$$\overline{E} = \int_0^\infty \frac{\epsilon\,\rho(\epsilon)}{e^{\beta(\epsilon_r-\mu)} - 1}\,d\epsilon = \frac{V}{4\pi^2}\,\frac{(2\,m)^{3/2}}{\hbar^3}\int_0^\infty \frac{\epsilon^{3/2}}{e^{\beta(\epsilon_r-\mu)} - 1}\,d\epsilon. \tag{10.243}$$

In this case, we do not need to worry about the omission of the ground-state in the integral, because this state makes no contribution to the mean energy (because $\epsilon = 0$ in the ground-state). For temperatures above the Bose temperature, all the bosons are in excited states, and we expect the mean energy to approach the classical value, $(3/2)\,N\,k\,T$. However, below the Bose temperature, a substantial fraction of bosons are in the ground-state, so we expect the mean energy to fall well below the classical value.

As we have seen, the chemical potential of a boson gas is very close to zero for temperatures below the Bose temperature. Hence, setting $\mu = 0$ in the previous expression, and making the substitution $x = \epsilon/(k\,T)$, we obtain

$$\overline{E} = \frac{2}{\sqrt{\pi}}\left(\frac{2\pi\,m\,k\,T}{h^2}\right)^{3/2} V \int_0^\infty \frac{x^{3/2}}{e^x - 1}\,dx. \tag{10.244}$$

The integral is equal to $\Gamma(5/2)\,\zeta(5/2)$, where $\Gamma(5/2) = 3\sqrt{\pi}/4$, and $\zeta(5/2) = 1.34$. (See Exercise 10.4.) Thus, we obtain

$$\overline{E} = \frac{3}{2}\frac{\zeta(5/2)}{\zeta(3/2)}\,N\,k\,T\left(\frac{T}{T_B}\right)^{3/2} = 0.770\,N\,k\,T\left(\frac{T}{T_B}\right)^{3/2}, \tag{10.245}$$

where use has been made of Equation (10.235). Note that $\overline{E} \propto T^{5/2}$ for temperatures below the Bose temperature, but that $\overline{E}$ becomes similar in magnitude to the classical value, $(3/2)\,N\,k\,T$, as $T \rightarrow T_B$. The molar specific heat below the Bose temperature is

$$\frac{c_V}{R} = \frac{1}{N\,k}\frac{d\overline{E}}{dT} = \frac{15}{4}\frac{\zeta(5/2)}{\zeta(3/2)}\left(\frac{T}{T_B}\right)^{3/2} = 1.92\left(\frac{T}{T_B}\right)^{3/2}. \tag{10.246}$$

Likewise, the mean pressure is

$$\bar{p} = \frac{2}{3}\frac{\overline{E}}{V} = \frac{\zeta(5/2)}{\zeta(3/2)}\frac{N\,k\,T}{V}\left(\frac{T}{T_B}\right)^{3/2} = 0.513\,\frac{N\,k\,T}{V}\left(\frac{T}{T_B}\right)^{3/2}. \tag{10.247}$$

(See Exercise 10.7.) In both cases, the quantities become similar in magnitude to their classical values as $T \rightarrow T_B$, but fall far below these values when $T \ll T_B$.

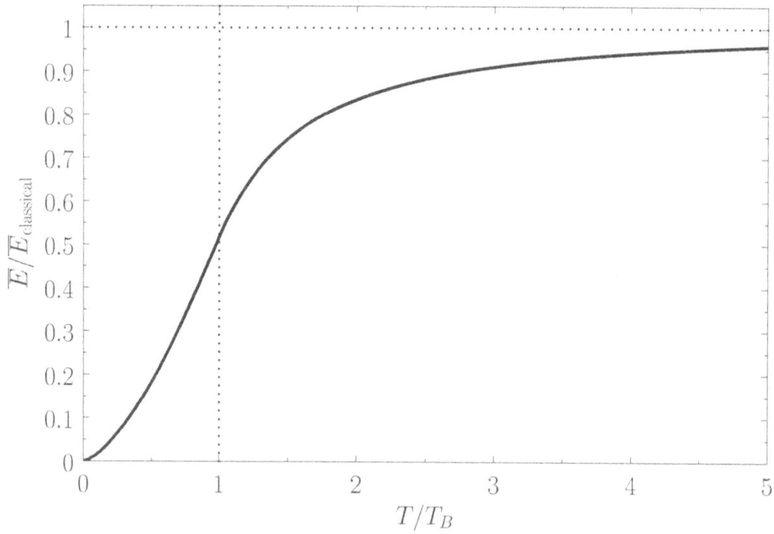

Figure 10.9 Variation with temperature of $\overline{E}/\overline{E}_{\rm classical}$ for a boson gas, where $\overline{E}_{\rm classical} = (3/2)\,N\,k\,T$.

For $T > T_B$, Equation (10.243) becomes

$$\overline{E} = \frac{3}{2}\,N\,k\,T\left(\frac{T}{T_B}\right)^{3/2}\frac{1}{\zeta(3/2)}\sum_{n=1,\infty}\frac{{\rm e}^{n\alpha}}{n^{5/2}} \tag{10.248}$$

(see Exercise 10.4), where α is determined from the numerical solution of Equation (10.240). Figure 10.9 shows how the mean energy of a boson gas varies with temperature, and clearly illustrates that the energy approaches its classical value asymptotically in the limit $T \gg T_B$. Finally, Figure 10.10 shows how the molar heat capacity of a boson gas varies with temperature. It can be seen that the c_V curve has a change in slope at $T = T_B$, reaching a maximum value there of $1.92\,R$. At higher temperatures, c_V approaches the classical value, $(3/2)\,R$, asymptotically.

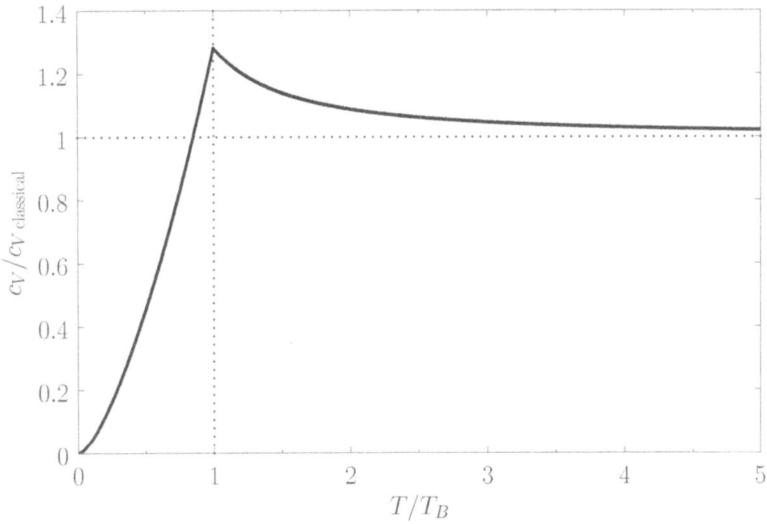

Figure 10.10 Variation with temperature of $c_V/c_{V\,\text{classical}}$ for a boson gas, where $c_{V\,\text{classical}} = (3/2)\,R$.

Exercises

10.1 Let

$$I_n = \int_0^\infty \mathrm{e}^{-x}\,x^n\,dx,$$

where n is a non-negative integer. Integrating by parts, demonstrate that

$$I_n = n\,I_{n-1}.$$

Furthermore, show that

$$I_0 = 1.$$

Hence, deduce that

$$I_n = n!.$$

10.2 Let

$$J_n = \int_0^\infty \frac{x^{n-1}\,dx}{e^x - 1},$$

where n is an integer greater than unity. Demonstrate that

$$J_n = (n-1)!\,\zeta(n),$$

where

$$\zeta(n) = \sum_{k=1,\infty} \frac{1}{k^n}$$

is a Riemann Zeta function. [Hint: $(1-r)^{-1} = \sum_{k=0,\infty} r^k$, provided that $|r| < 1$.] In general, the Zeta function must be evaluated numerically. However, it is possible to prove that $\zeta(2) = \pi^2/6$ and $\zeta(4) = \pi^4/90$.

10.3 Show that

$$\int_{-\infty}^\infty \frac{e^x}{(e^x+1)^2}\,x^2\,dx$$

can be written in the form

$$4\int_0^\infty \frac{x}{e^x+1}\,dx.$$

By expanding the integrand in powers of e^{-x}, demonstrate that

$$\int_0^\infty \frac{x}{e^x+1}\,dx = \int_0^\infty \left(x\,e^{-x} - x\,e^{-2x} + x\,e^{-3x} - x\,e^{-4x} + \cdots\right)dx,$$

and, hence, that

$$\int_0^\infty \frac{x}{e^x+1}\,dx = 1 - \frac{1}{2^2} + \frac{1}{3^2} - \frac{1}{4^2} + \cdots.$$

Rearrange the right-hand side of the previous expression to give

$$\begin{aligned}\int_0^\infty \frac{x}{e^x+1}\,dx &= \left(1 + \frac{1}{2^2} + \frac{1}{3^2} + \frac{1}{4^2} + \cdots\right) - 2\left(\frac{1}{2^2} + \frac{1}{4^2} + \frac{1}{6^2} + \cdots\right)\\ &= \left(1 + \frac{1}{2^2} + \frac{1}{3^2} + \frac{1}{4^2} + \cdots\right) - \frac{2}{2^2}\left(1 + \frac{1}{2^2} + \frac{1}{3^2} + \frac{1}{4^2} + \cdots\right)\\ &= \zeta(2) - \frac{1}{2}\,\zeta(2) = \frac{1}{2}\,\zeta(2) = \frac{\pi^2}{12}.\end{aligned}$$

(See Exercise 10.2.) Thus, deduce that

$$\int_{-\infty}^\infty \frac{e^x}{(e^x+1)^2}\,x^2\,dx = \frac{\pi^2}{3}.$$

10.4 Consider the integral

$$\int_{-\infty}^{\infty} \frac{x^{s-1}}{\mathrm{e}^{x}-1}\,dx,$$

where $s > 1$. By expanding in powers of e^{-x}, show that

$$\int_{-\infty}^{\infty} \frac{x^{s-1}}{e^{x}-1}\,dx = \Gamma(s)\,\zeta(s),$$

where $\Gamma(s) = \int_0^{\infty} x^{s-1}\,\mathrm{e}^{-x}\,dx$ is a Gamma function, and $\zeta(s) = \sum_{n=1,\infty} n^{-s}$ a Riemann Zeta function. Likewise, demonstrate that

$$\int_{-\infty}^{\infty} \frac{x^{s-1}}{e^{x-\alpha}-1}\,dx = \Gamma(s) \sum_{n=1,\infty} \frac{\mathrm{e}^{n\,\alpha}}{n^{s}},$$

where $\alpha < 0$.

10.5 Consider a gas consisting of identical non-interacting particles. The quantum states of a single particle are labeled by the index r. Let the energy of a particle in state r be ϵ_r. Let n_r be the number of particles in quantum state r. The partition function of the gas is thus

$$Z = \sum_R \exp\left(-\beta \sum_r n_r\,\epsilon_r\right),$$

where the first sum is over all allowable values of the n_r, and the second is over all single-particle quantum states. Here, $\beta = 1/(k\,T)$, where T is the absolute temperature.

(a) Demonstrate that

$$\bar{n}_r = -\frac{1}{\beta}\,\frac{\partial \ln Z}{\partial \epsilon_r}.$$

(b) Show that

$$\overline{n_r^2} = \frac{1}{\beta^2\,Z}\,\frac{\partial^2 Z}{\partial \epsilon_r^2} = \frac{1}{\beta^2}\left[\frac{\partial}{\partial \epsilon_r}\left(\frac{\partial \ln Z}{\partial \epsilon_r}\right) + \beta^2\,(\bar{n}_r)^2\right].$$

(c) Hence, deduce that

$$\overline{(\Delta n_r)^2} = \frac{1}{\beta^2}\,\frac{\partial^2 \ln Z}{\partial \epsilon_r^2} = -\frac{1}{\beta}\,\frac{\partial \bar{n}_r}{\partial \epsilon_r}.$$

10.6 Use the results of the previous exercise to show that:

(a) For identical, non-interacting, particles distributed according to the Maxwell-Boltzmann distribution,

$$\frac{\overline{(\Delta n_r)^2}}{\bar{n}_r^2} = \frac{1}{\bar{n}_r}\left(1 - \frac{\bar{n}_r}{\sum_s \bar{n}_s}\right) \simeq \frac{1}{\bar{n}_r}.$$

(b) For photons,

$$\frac{\overline{(\Delta n_r)^2}}{\bar{n}_r^2} = \frac{1}{\bar{n}_r} + 1.$$

(c) For identical, non-interacting, massive particles distributed according to the Bose-Einstein distribution,

$$\frac{\overline{(\Delta n_r)^2}}{\bar{n}_r^2} = \left(\frac{1}{\bar{n}_r} + 1\right)\left(1 + \frac{1}{\beta}\frac{\partial \alpha}{\partial \epsilon_r}\right).$$

Show, also, that

$$\frac{1}{\beta}\frac{\partial \alpha}{\partial \epsilon_r} = -\frac{\bar{n}_r\,(1+\bar{n}_r)}{\sum_s \bar{n}_s\,(1+\bar{n}_s)}.$$

Hence, deduce that

$$\frac{\overline{(\Delta n_r)^2}}{\bar{n}_r^2} = \left(\frac{1}{\bar{n}_r} + 1\right)\left[1 - \frac{\bar{n}_r\,(1+\bar{n}_r)}{\sum_s \bar{n}_s\,(1+\bar{n}_s)}\right] \simeq \frac{1}{\bar{n}_r} + 1.$$

(d) For identical, non-interacting, massive particles distributed according to the Fermi-Dirac distribution,

$$\frac{\overline{(\Delta n_r)^2}}{\bar{n}_r^2} = \left(\frac{1}{\bar{n}_r} - 1\right)\left(1 + \frac{1}{\beta}\frac{\partial \alpha}{\partial \epsilon_r}\right).$$

Show, also, that

$$\frac{1}{\beta}\frac{\partial \alpha}{\partial \epsilon_r} = -\frac{\bar{n}_r\,(1-\bar{n}_r)}{\sum_s \bar{n}_s\,(1-\bar{n}_s)}.$$

Hence, deduce that

$$\frac{\overline{(\Delta n_r)^2}}{\bar{n}_r^2} = \left(\frac{1}{\bar{n}_r} - 1\right)\left[1 - \frac{\bar{n}_r\,(1-\bar{n}_r)}{\sum_s \bar{n}_s\,(1-\bar{n}_s)}\right] \simeq \frac{1}{\bar{n}_r} - 1.$$

Note that, in the case of the Bose-Einstein distribution, the relative dispersion in $\bar{n}_r$ is larger than in the Maxwell-Boltzmann case, whereas in the case of the Fermi-Dirac distribution the relative dispersion is smaller.

10.7 Consider a non-relativistic free particle of mass m in a cubical container of edge-length L, and volume $V = L^3$.

(a) Show that the energy of a general quantum state, r, can be written

$$\epsilon_r = \frac{h^2}{V^{2/3}}\,(n_x^2 + n_y^2 + n_z^2),$$

where n_x, n_y, and n_z are positive integers. Hence, deduce that the contribution to the gas pressure of a particle in this state is

$$p_r = -\frac{\partial \epsilon_r}{\partial V} = \frac{2}{3}\frac{\epsilon_r}{V}.$$

(b) Demonstrate that the mean pressure of a gas of weakly-interacting non-relativistic particles is

$$\bar{p} = \frac{2}{3}\frac{\overline{E}}{V},$$

where $\overline{E}$ is its mean total kinetic energy, irrespective of whether the particles obey classical, Bose-Einstein, or Fermi-Dirac statistics.

10.8 As an electron moves in a molecule, there exists, at any instance in time, a separation of positive and negative charges within the molecule. The molecule therefore possesses a time-varying electric dipole moment, $\mathbf{p}$. Assuming that the molecule is located at the origin, its instantaneous dipole moment generates an electric field

$$\mathbf{E}(\mathbf{r}) = \frac{3\,(\mathbf{p}\cdot\mathbf{r})\,\mathbf{r} - r^2\,\mathbf{p}}{4\pi\,\epsilon_0\,r^5},$$

where $\mathbf{r}$ is the vector displacement from the origin. A second molecule, located at position $\mathbf{r}$, relative to the first, develops an induced dipole moment, $\mathbf{p}' = \alpha\,\mathbf{E}(\mathbf{r})$, in response to the electric field of the first. Here, α is the molecular polarizability. Moreover, the electric field generates a force $\mathbf{f}' = \nabla[\mathbf{p}'\cdot\mathbf{E}(\mathbf{r})]$ acting on the second molecule. (Of course, an equal and opposite force acts on the first molecule.) Use these facts to deduce that the mean (i.e., averaged over time, and all possible orientations of $\mathbf{p}$) intermolecular potential between a pair of neutral molecules is attractive, and takes the form

$$u(r) = -\frac{\alpha\,\langle p^2\rangle}{16\pi^2\,\epsilon_0^2\,r^6},$$

where $\langle\cdots\rangle$ denotes a time average.

10.9 Consider a van der Waals gas whose equation of state is

$$\bar{p} = \frac{R\,T}{v-b} - \frac{a}{v^2}.$$

Here, $a > 0$, > 0 are constants, and v is the molar volume. Show that the various thermodynamic potentials of the gas are:

$$F = -\nu R T \left\{ \ln \left[\frac{n_Q}{N_A} (v - b) \right] + 1 \right\} - \frac{\nu a}{v},$$

$$S = \nu R \left\{ \ln \left[\frac{n_Q}{N_A} (v - b) \right] + \frac{5}{2} \right\},$$

$$\overline{E} = \frac{3}{2} \nu R T - \frac{\nu a}{v},$$

$$H = \nu R T \left(\frac{v}{v - b} + \frac{3}{2} \right) + \frac{2 \nu a}{v},$$

$$G = \nu R T \left\{ - \ln \left[\frac{n_Q}{N_A} (v - b) \right] + \frac{v}{v - b} - 1 \right\} - \frac{2 \nu a}{v}.$$

Here, $n_Q = (2\pi\, m\, k\, T/h^2)^{3/2}$, where m is the molecular mass, is known as the *quantum concentration*, and is the particle concentration at which the de Broglie wavelength is equal to the mean inter-particle spacing. Obviously, the previous expression are only valid when $N_A/v \ll n_Q$.

10.10 Show that the energy density of radiation inside a radiation-filled cavity whose walls are held at absolute temperature T is

$$\bar{u} = \frac{\pi^2}{15} \frac{(k\, T)^4}{(c\, \hbar)^3}.$$

Demonstrate that the mean pressure that the radiation exerts on the walls of the cavity is

$$\bar{p} = \frac{1}{3}\, \bar{u}.$$

[Hint: Modify and reuse the analysis of Exercise 10.7.]

10.11 Apply the thermodynamic relation $T\, dS = d\overline{E} + \bar{p}\, dV$ to a photon gas. Here, we can write $\overline{E} = V\, \bar{u}$, where $\bar{u}(T)$ is the mean energy density of the radiation (which is independent of the volume). The radiation pressure is $\bar{p} = \bar{u}/3$.

(a) Considering S as a function of T and V, show that

$$\left(\frac{\partial S}{\partial V} \right)_T = \frac{4}{3} \frac{\bar{u}}{T},$$

$$\left(\frac{\partial S}{\partial T} \right)_V = \frac{V}{T} \frac{d\bar{u}}{dT}.$$

(b) Demonstrate that the mathematical identity $(\partial^2 S/\partial V\, \partial T) = (\partial^2 S/\partial T\, \partial V)$ leads to a differential equation for $\bar{u}(T)$ that can be integrated to give

$$\bar{u} = a\, T^4,$$

where a is a constant.

(c) Hence, deduce that

$$S = \frac{4}{3}\,\frac{\overline{E}}{T}.$$

10.12 Show that the total number of photons in a radiation-filled cavity of volume V, whose walls are held at temperature T, is

$$N = 16\pi\,\zeta(3)\,V\left(\frac{k\,T}{h\,c}\right)^{3},$$

where

$$\zeta(n) = \frac{1}{(n-1)!}\int_0^\infty \frac{x^{n-1}\,dx}{{\rm e}^{\,x}-1}$$

is a Riemann Zeta function. (See Exercise 10.2.) Note that $\zeta(3) \simeq 1.202$. Demonstrate that the mean energy per photon and the entropy per photon are

$$\frac{\overline{E}}{N} = \frac{\pi^4}{30\,\zeta(3)}\,k\,T = 2.70\,k\,T,$$

$$\frac{S}{N} = \frac{2\,\pi^4}{45\,\zeta(3)}\,k = 3.60\,k,$$

respectively. Note that the entropy per photon is a constant, independent of the temperature.

10.13 Electromagnetic radiation at temperature T fills a cavity of volume V. If the cavity is thermally insulated, and expands quasi-statically, show that

$$T \propto \frac{1}{V^{1/3}}.$$

(Neglect the heat capacity of the cavity walls.)

10.14 (a) The partition function for a photon gas is

$$\ln Z = -\sum_r \ln\left(1-{\rm e}^{-\beta\,\epsilon_r}\right),$$

where $\epsilon_r = \hbar\,\omega_r$, and ω_r is the frequency of an individual photon state. The spacing between the allowed frequencies can be made arbitrarily small by increasing the volume, V, of the container. Hence, the sum can be approximated as an integral. Now, the number of photon states per unit volume whose frequencies lie between ω and $\omega+d\omega$ is $\sigma(\omega)\,d\omega$, where

$$\sigma(\omega) = 2\,\frac{\omega^2}{2\pi^2\,c^3}.$$

(See Section 10.12.) Thus, we can write

$$\ln Z = -V \int_0^\infty \sigma(\omega)\, \ln\left(1 - e^{-\beta\hbar\omega}\right) d\omega.$$

Show that

$$\ln Z = -8\pi\, V \left(\frac{k\,T}{h\,c}\right)^3 \int_0^\infty y^2 \ln(1 - e^{-y})\, dy.$$

Integrating by parts, demonstrate that

$$\ln Z = \frac{8\pi}{3}\, V \left(\frac{k\,T}{h\,c}\right)^3 \int_0^\infty \frac{y^3}{e^y - 1}\, dy.$$

Hence, deduce that

$$\ln Z = \frac{8\pi^5}{45} \left(\frac{k\,T}{h\,c}\right)^3 V.$$

[Hint: See Exercise 10.2.]

(b) Using the standard results

$$\overline{E} = -\frac{\partial \ln Z}{\partial \beta},$$

$$\bar{p} = \frac{1}{\beta}\frac{\partial \ln Z}{\partial V},$$

$$S = k\,(\ln Z + \beta\,\overline{E}),$$

show that

$$\overline{E} = \frac{8\pi^5}{15}\frac{(k\,T)^4}{(h\,c)^3}\, V,$$

$$\bar{p} = \frac{\overline{E}}{3\,V},$$

$$S = \frac{4}{3}\frac{\overline{E}}{T}.$$

10.15 Show that the power per unit area radiated by a black-body of temperature T peaks at angular frequency ω_c, where $\hbar\,\omega_c = y\,k\,T$, and y is the solution of the transcendental equation

$$y = 3\,(1 - e^{-y}).$$

Solve this equation by iteration [i.e., $y_{n+1} = 3\,(1 - e^{-y_n})$, where y_n is the nth guess], and, thereby, show that

$$\hbar\,\omega_c = 2.82\,k\,T.$$

10.16 A black (non-reflective) plane at temperature T_u is parallel to a black plane at temperature $T_l < T_u$. The net energy flux density in vacuum between the two planes is $J_U = \sigma (T_u^4 - T_l^4)$, where σ is the Stefan-Boltzmann constant. A third black plane is inserted between the other two, and is allowed to come to a steady-state temperature T_m. Find T_m in terms of T_u and T_l, and show that the net energy flux is cut in half because of the presence of this plane. This is the principle of the *heat shield*, and is widely used to reduce radiant heat transfer.

10.17 Consider the conduction electrons in silver, which is monovalent (one free electron per atom), has a mass density $10.5 \times 10^3\,\mathrm{kg\,m^{-3}}$, and an atomic weight of 107. Show that the Fermi energy (at $T = 0$) is

$$\mu_F = 5.6\,\mathrm{eV},$$

and the equivalent temperature—the so-called *Fermi temperature*—is

$$T_F = \frac{\mu_F}{k} = 65{,}000\,\mathrm{K}.$$

Thus, if $T \ll T_F$ (as is definitely the case at room temperature) then the conduction electrons are highly degenerate. Furthermore, show that at room temperature ($T = 300\,\mathrm{K}$), the contribution of the conduction electrons to the molar specific heat is only

$$\frac{c_V^{(e)}}{R} = 2.2 \times 10^{-2}.$$

10.18 Demonstrate that the mean pressure of the conduction electrons in a metal can be written

$$\bar{p} = \frac{2}{5}\,\frac{N\,k\,T_F}{V}\left[1 + \frac{5\pi^2}{12}\left(\frac{T}{T_F}\right)^2 + \cdots\right],$$

where T_F is the Fermi temperature. (Hint: See Exercise 10.17.) Show that for silver (see previous exercise),

$$\bar{p} = 2.1 \times 10^{10}\,\mathrm{Pa}.$$

10.19 Show that the contribution of the conduction electrons in a metal to the isothermal compressibility is

$$\kappa_T = \frac{3}{5\,\bar{p}},$$

where $\bar{p}$ is the electron's mean pressure. Estimate κ_T for silver. (See the previous exercise.) Compare your estimate to the experimental value $0.99 \times 10^{-11}\,\mathrm{Pa}^{-1}$.

10.20 Justify Equation (10.197).

10.21 A system of N bosons of mass m and zero spin is in a container of volume V, at an absolute temperature T. The number of particles is

$$N = \frac{V}{4\pi^2}\left(\frac{2m}{\hbar^2}\right)\int_0^\infty \frac{\epsilon^{1/2}}{e^{(x-\mu)/kT}-1}\,d\epsilon.$$

[See Equation (10.227).] In the limit in which the gas is dilute, $\exp(-\mu/kT) \gg 1$, and the Bose-Einstein distribution becomes the Maxwell-Boltzmann distribution. Evaluate the integral in this approximation, and show that

$$\exp\left(-\frac{\mu}{kT}\right) = \left(\frac{d}{\lambda}\right)^3,$$

where $\lambda = h/(2\pi m k T)^{1/2}$ is the particles' de Broglie wavelength, and $d = (V/N)^{1/3}$ the mean inter-particle distance. Hence, deduce that, in the classical limit, the average distance between particles is much larger than the de Broglie wavelength.

Appendix A

Physical Constants

Planck constant :	$h = 6.62606957 \times 10^{-34}$ J s
Reduced Planck constant :	$\hbar = h/(2\pi) = 1.05457173 \times 10^{-34}$ J s
Electron mass :	$m_e = 9.10938291 \times 10^{-31}$ kg
Electron charge (magnitude) :	$e = 1.60217657 \times 10^{-19}$ C
Proton mass :	$m_p = 1.67262178 \times 10^{-27}$ kg
Electric permittivity of free space :	$\epsilon_0 = 8.85418782 \times 10^{-12}$ F m^{-1}
Magnetic permeability of free space :	$\mu_0 = 4\pi \times 10^{-7}$ H m^{-1}
Speed of light in vacuum :	$c = 2.99792458 \times 10^{8}$ m s^{-1}
Bohr magneton :	$\mu_B = 9.27400968 \times 10^{-24}$ J T^{-1}
Molar ideal gas constant :	$R = 8.3144621$ J K^{-1}
Boltzmann constant :	$k = 1.3806488 \times 10^{-23}$ J K^{-1}
Avogadro number :	$N_A = R/k = 6.0221409 \times 10^{23}$
Absolute zero :	$T_0 = 273.15°$ C
Stefan-Boltzmann constant :	$\sigma = 5.670367 \times 10^{-8}$ W m^{-2} K^{-4}
Gravitational constant :	$G = 6.67408 \times 10^{-11}$ m^3 kg^{-1} s^{-2}

Appendix B

Classical Mechanics

B.1 Generalized Coordinates

Let the q_i, for $i = 1, \mathcal{F}$, be a set of coordinates that uniquely specifies the instantaneous configuration of some classical dynamical system. Here, it is assumed that each of the q_i can vary independently. The q_i might be Cartesian coordinates, or polar coordinates, or angles, or some mixture of all three types of coordinate, and are, therefore, termed *generalized coordinates*. A dynamical system whose instantaneous configuration is fully specified by $\mathcal{F}$ independent generalized coordinates is said to have $\mathcal{F}$ *degrees of freedom*. For instance, the instantaneous position of a particle moving freely in three dimensions is completely specified by its three Cartesian coordinates, x, y, and z. Moreover, these coordinates are clearly independent of one another. Hence, a dynamical system consisting of a single particle moving freely in three dimensions has three degrees of freedom. If there are two freely moving particles then the system has six degrees of freedom, and so on.

Suppose that we have a dynamical system consisting of N particles moving freely in three dimensions. This is an $\mathcal{F} = 3N$ degree of freedom system whose instantaneous configuration can be specified by $\mathcal{F}$ Cartesian coordinates. Let us denote these coordinates the x_j, for $j = 1, \mathcal{F}$. Thus, x_1, x_2, x_3 are the Cartesian coordinates of the first particle, x_4, x_5, x_6 the Cartesian coordinates of the second particle, et cetera. Suppose that the instantaneous configuration of the system can also be specified by $\mathcal{F}$ generalized coordinates, which we shall denote the q_i, for $i = 1, \mathcal{F}$. Thus, the q_i might be the spherical coordinates of the particles. In general, we expect the x_j to be functions of the q_i. In other words,

$$x_j = x_j(q_1, q_2, \cdots, q_{\mathcal{F}}, t), \tag{B.1}$$

for $j = 1, \mathcal{F}$. Here, for the sake of generality, we have included the possibility that the functional relationship between the x_j and the q_i might depend on the time, t, explicitly. This would be the case if the dynamical system were subject to time-varying constraints. For instance, a system consisting of a particle constrained to move on a surface that is itself moving. Finally, by the chain rule, the variation of the x_j caused by a variation of the q_i (at constant t) is given by

$$\delta x_j = \sum_{i=1,\mathcal{F}} \frac{\partial x_j}{\partial q_i}\,\delta q_i, \tag{B.2}$$

for $j = 1, \mathcal{F}$.

B.2 Generalized Forces

The work done on the dynamical system when its Cartesian coordinates change by δx_j is simply

$$\delta W = \sum_{j=1,\mathcal{F}} f_j\,\delta x_j. \tag{B.3}$$

Here, the f_j are the Cartesian components of the forces acting on the various particles making up the system. Thus, f_1, f_2, f_3 are the components of the force acting on the first particle, f_4, f_5, f_6 the components of the force acting on the second particle, et cetera. Using Equation (B.2), we can also write

$$\delta W = \sum_{j=1,\mathcal{F}} f_j \sum_{i=1,\mathcal{F}} \frac{\partial x_j}{\partial q_i}\,\delta q_i. \tag{B.4}$$

The previous expression can be rearranged to give

$$\delta W = \sum_{i=1,\mathcal{F}} Q_i\,\delta q_i, \tag{B.5}$$

where

$$Q_i = \sum_{j=1,\mathcal{F}} f_j\,\frac{\partial x_j}{\partial q_i}. \tag{B.6}$$

Here, the Q_i are termed *generalized forces*. More explicitly, Q_i is termed the force *conjugate* to the coordinate q_i. Note that a generalized force does not necessarily have the dimensions of force. However, the product $Q_i\,q_i$ must have the dimensions of work. Thus, if a particular q_i is a Cartesian coordinate then the associated Q_i is a force. Conversely, if a particular q_i is an angle then the associated Q_i is a torque.

Suppose that the dynamical system in question is energy conserving. It follows that

$$f_j = -\frac{\partial U}{\partial x_j}, \tag{B.7}$$

for $j = 1, \mathcal{F}$, where $U(x_1, x_2, \cdots, x_{\mathcal{F}}, t)$ is the system's potential energy. Hence, according to Equation (B.6),

$$Q_i = -\sum_{j=1,\mathcal{F}} \frac{\partial U}{\partial x_j}\frac{\partial x_j}{\partial q_i} = -\frac{\partial U}{\partial q_i}, \tag{B.8}$$

for $i = 1, \mathcal{F}$.

B.3 Lagrange's Equation

The Cartesian equations of motion of our dynamical system take the form

$$m_j\,\ddot{x}_j = f_j, \tag{B.9}$$

for $j = 1, \mathcal{F}$, where m_1, m_2, m_3 are each equal to the mass of the first particle, m_4, m_5, m_6 are each equal to the mass of the second particle, et cetera. Furthermore, the kinetic energy of the system can be written

$$K = \frac{1}{2}\sum_{j=1,\mathcal{F}} m_j\,\dot{x}_j^2. \tag{B.10}$$

Now, because $x_j = x_j(q_1, q_2, \cdots, q_{\mathcal{F}}, t)$, we can write

$$\dot{x}_j = \sum_{i=1,\mathcal{F}} \frac{\partial x_j}{\partial q_i}\,\dot{q}_i + \frac{\partial x_j}{\partial t}, \tag{B.11}$$

for $j = 1, \mathcal{F}$. Hence, it follows that $\dot{x}_j = \dot{x}_j(\dot{q}_1, \dot{q}_2, \cdots, \dot{q}_{\mathcal{F}}, q_1, q_2, \cdots, q_{\mathcal{F}}, t)$. According to the previous equation,

$$\frac{\partial \dot{x}_j}{\partial \dot{q}_i} = \frac{\partial x_j}{\partial q_i}, \tag{B.12}$$

where we are treating the $\dot{q}_i$ and the q_i as independent variables.

Multiplying Equation (B.12) by $\dot{x}_j$, and then differentiating with respect to time, we obtain

$$\frac{d}{dt}\left(\dot{x}_j\,\frac{\partial \dot{x}_j}{\partial \dot{q}_i}\right) = \frac{d}{dt}\left(\dot{x}_j\,\frac{\partial x_j}{\partial q_i}\right) = \ddot{x}_j\,\frac{\partial x_j}{\partial q_i} + \dot{x}_j\,\frac{d}{dt}\left(\frac{\partial x_j}{\partial q_i}\right). \tag{B.13}$$

Now,

$$\frac{d}{dt}\left(\frac{\partial x_j}{\partial q_i}\right) = \sum_{k=1,\mathcal{F}} \frac{\partial^2 x_j}{\partial q_i\,\partial q_k}\,\dot{q}_k + \frac{\partial^2 x_j}{\partial q_i\,\partial t}. \tag{B.14}$$

Furthermore,

$$\frac{1}{2}\frac{\partial \dot{x}_j^2}{\partial \dot{q}_i} = \dot{x}_j\,\frac{\partial \dot{x}_j}{\partial \dot{q}_i}, \tag{B.15}$$

and

$$\frac{1}{2}\frac{\partial \dot{x}_j^2}{\partial q_i} = \dot{x}_j\,\frac{\partial \dot{x}_j}{\partial q_i} = \dot{x}_j\,\frac{\partial}{\partial q_i}\left(\sum_{k=1,\mathcal{F}}\frac{\partial x_j}{\partial q_k}\,\dot{q}_k + \frac{\partial x_j}{\partial t}\right)$$
$$= \dot{x}_j\left(\sum_{k=1,\mathcal{F}}\frac{\partial^2 x_j}{\partial q_i\,\partial q_k}\,\dot{q}_k + \frac{\partial^2 x_j}{\partial q_i\,\partial t}\right)$$
$$= \dot{x}_j\,\frac{d}{dt}\left(\frac{\partial x_j}{\partial q_i}\right), \tag{B.16}$$

where use has been made of Equation (B.14). Thus, it follows from Equations (B.13), (B.15), and (B.16) that

$$\frac{d}{dt}\left(\frac{1}{2}\frac{\partial \dot{x}_j^2}{\partial \dot{q}_i}\right) = \ddot{x}_j\,\frac{\partial x_j}{\partial q_i} + \frac{1}{2}\frac{\partial \dot{x}_j^2}{\partial q_i}. \tag{B.17}$$

Let us take the previous equation, multiply by m_j, and then sum over all j. We obtain

$$\frac{d}{dt}\left(\frac{\partial K}{\partial \dot{q}_i}\right) = \sum_{j=1,\mathcal{F}} f_j\,\frac{\partial x_j}{\partial q_i} + \frac{\partial K}{\partial q_i}, \tag{B.18}$$

where use has been made of Equations (B.9) and (B.10). Thus, it follows from Equation (B.6) that

$$\frac{d}{dt}\left(\frac{\partial K}{\partial \dot{q}_i}\right) = Q_i + \frac{\partial K}{\partial q_i}. \tag{B.19}$$

Finally, making use of Equation (B.8), we get

$$\frac{d}{dt}\left(\frac{\partial K}{\partial \dot{q}_i}\right) = -\frac{\partial U}{\partial q_i} + \frac{\partial K}{\partial q_i}. \tag{B.20}$$

It is helpful to introduce a function L, called the *Lagrangian*,] which is defined as the difference between the kinetic and potential energies of the dynamical system under investigation:

$$L = K - U. \tag{B.21}$$

Because the potential energy, U, is clearly independent of the $\dot{q}_i$, it follows from Equation (B.20) that

$$\frac{d}{dt}\left(\frac{\partial L}{\partial \dot{q}_i}\right) - \frac{\partial L}{\partial q_i} = 0, \tag{B.22}$$

for $i = 1, \mathcal{F}$. This equation is known as *Lagrange's equation*.

According to the previous analysis, if we can express the kinetic and potential energies of our dynamical system solely in terms of our generalized coordinates, and their time derivatives, then we can immediately write down the equations of motion of the system, expressed in terms of the generalized coordinates, using Lagrange's equation, (B.22). Unfortunately, this scheme only works for energy-conserving systems.

B.4 Generalized Momenta

Consider the motion of a single particle moving in one dimension. The kinetic energy is

$$K = \frac{1}{2}\, m\, \dot{x}^2, \tag{B.23}$$

where m is the mass of the particle, and x its displacement. The particle's linear momentum is $p = m\,\dot{x}$. However, this can also be written

$$p = \frac{\partial K}{\partial \dot{x}} = \frac{\partial L}{\partial \dot{x}}, \tag{B.24}$$

because $L = K - U$, and the potential energy, U, is independent of $\dot{x}$.

Consider a dynamical system described by $\mathcal{F}$ generalized coordinates, q_i, for $i = 1, \mathcal{F}$. By analogy with the previous expression, we can define *generalized momenta* of the form

$$p_i = \frac{\partial L}{\partial \dot{q}_i}, \tag{B.25}$$

for $i = 1, \mathcal{F}$. Here, p_i is sometimes called the momentum *conjugate* to the coordinate q_i. Hence, Lagrange's equation, (B.22), can be written

$$\frac{dp_i}{dt} = \frac{\partial L}{\partial q_i}, \tag{B.26}$$

for $i = 1, \mathcal{F}$. Note that a generalized momentum does not necessarily have the dimensions of linear momentum.

Suppose that the Lagrangian, L, does not depend explicitly on some coordinate q_k. It follows from Equation (B.26) that

$$\frac{dp_k}{dt} = \frac{\partial L}{\partial q_k} = 0. \tag{B.27}$$

Hence,

$$p_k = \text{constant}. \tag{B.28}$$

The coordinate q_k is said to be *ignorable* in this case. Thus, we conclude that the generalized momentum associated with an ignorable coordinate is a constant of the motion.

B.5 Calculus of Variations

It is a well-known fact, first enunciated by Archimedes, that the shortest distance between two points in a plane is a straight-line. However, suppose that we wish to demonstrate this result from first principles. Let us consider the length, l, of various curves, $y(x)$, that run between two fixed points, A and B, in a plane. Now, l takes the form

$$l = \int_A^B [dx^2 + dy^2]^{1/2} = \int_a^b [1 + y'^2(x)]^{1/2}\, dx, \tag{B.29}$$

where $y' \equiv dy/dx$. Note that l is a function of the function $y(x)$. In mathematics, a function of a function is termed a *functional.*

In order to find the shortest path between points A and B, we need to minimize the functional l with respect to small variations in the function $y(x)$, subject to the constraint that the end points, A and B, remain fixed. In other words, we need to solve

$$\delta l = 0. \tag{B.30}$$

The meaning of the previous equation is that if $y(x) \to y(x) + \delta y(x)$, where $\delta y(x)$ is small, then the first-order variation in l, denoted δl, vanishes. In other words, $l \to l + O(\delta y^2)$. The particular function $y(x)$ for which $\delta l = 0$ obviously yields an extremum of l (i.e., either a maximum or a minimum). Hopefully, in the case under consideration, it yields a minimum of l.

Consider a general functional of the form

$$I = \int_a^b F(y, y', x)\, dx, \tag{B.31}$$

where the end points of the integration are fixed. Suppose that $y(x) \to y(x)+\delta y(x)$. The first-order variation in I is written

$$\delta I = \int_a^b \left(\frac{\partial F}{\partial y}\, \delta y + \frac{\partial F}{\partial y'}\, \delta y' \right) dx, \tag{B.32}$$

where $\delta y' = d(\delta y)/dx$. Setting δI to zero, we obtain

$$\int_a^b \left(\frac{\partial F}{\partial y}\, \delta y + \frac{\partial F}{\partial y'}\, \delta y' \right) dx = 0. \tag{B.33}$$

This equation must be satisfied for all possible small perturbations, $\delta y(x)$.

Integrating the second term in the integrand of the previous equation by parts, we get

$$\int_a^b \left[\frac{\partial F}{\partial y} - \frac{d}{dx}\left(\frac{\partial F}{\partial y'} \right) \right] \delta y\, dx + \left[\frac{\partial F}{\partial y'}\, \delta y \right]_a^b = 0. \tag{B.34}$$

If the end points are fixed then $\delta y = 0$ at $x = a$ and $x = b$. Hence, the last term on the left-hand side of the previous equation is zero. Thus, we obtain

$$\int_a^b \left[\frac{\partial F}{\partial y} - \frac{d}{dx}\left(\frac{\partial F}{\partial y'}\right)\right] \delta y\, dx = 0. \tag{B.35}$$

The previous equation must be satisfied for all small perturbations $\delta y(x)$. The only way in which this is possible is for the expression enclosed in square brackets in the integral to be zero. Hence, the functional I attains an extremum value whenever

$$\frac{d}{dx}\left(\frac{\partial F}{\partial y'}\right) - \frac{\partial F}{\partial y} = 0. \tag{B.36}$$

This condition is known as the *Euler-Lagrange equation.*

Let us consider some special cases. Suppose that F does not explicitly depend on y. It follows that $\partial F/\partial y = 0$. Hence, the Euler-Lagrange equation, (B.36), simplifies to

$$\frac{\partial F}{\partial y'} = \text{constant}. \tag{B.37}$$

Next, suppose that F does not depend explicitly on x. Multiplying Equation (B.36) by y', we obtain

$$y'\,\frac{d}{dx}\left(\frac{\partial F}{\partial y'}\right) - y'\,\frac{\partial F}{\partial y} = 0. \tag{B.38}$$

However,

$$\frac{d}{dx}\left(y'\,\frac{\partial F}{\partial y'}\right) = y'\,\frac{d}{dx}\left(\frac{\partial F}{\partial y'}\right) + y''\,\frac{\partial F}{\partial y'}. \tag{B.39}$$

Thus, we get

$$\frac{d}{dx}\left(y'\,\frac{\partial F}{\partial y'}\right) = y'\,\frac{\partial F}{\partial y} + y''\,\frac{\partial F}{\partial y'}. \tag{B.40}$$

If F is not an explicit function of x then the right-hand side of the previous equation is the total derivative of F, namely dF/dx. Hence, we obtain

$$\frac{d}{dx}\left(y'\,\frac{\partial F}{\partial y'}\right) = \frac{dF}{dx}, \tag{B.41}$$

which yields

$$y'\,\frac{\partial F}{\partial y'} - F = \text{constant}. \tag{B.42}$$

Returning to the case under consideration, we have $F = (1+y'^2)^{1/2}$, according to Equations (B.29) and (B.31). Hence, F is not an explicit function of y, so Equation (B.37) yields

$$\frac{\partial F}{\partial y'} = \frac{y'}{\sqrt{1+y'^2}} = c, \tag{B.43}$$

where c is a constant. So,

$$y' = \frac{c}{\sqrt{1-c^2}} = \text{constant.} \tag{B.44}$$

Of course, $y' = \text{constant}$ is the equation of a straight-line. Thus, the shortest distance between two fixed points in a plane is indeed a straight-line.

B.6 Conditional Variation

Suppose that we wish to find the function, $y(x)$, that maximizes or minimizes the functional

$$I = \int_a^b F(y, y', x)\,dx, \tag{B.45}$$

subject to the constraint that the value of

$$J = \int_a^b G(y, y', x)\,dx \tag{B.46}$$

remains constant. We can achieve our goal by finding an extremum of the new functional $K = I + \lambda J$, where $\lambda(x)$ is an undetermined function. We know that $\delta J = 0$, because the value of J is fixed, so if $\delta K = 0$ then $\delta I = 0$ as well. In other words, finding an extremum of K is equivalent to finding an extremum of I. Application of the Euler-Lagrange equation yields

$$\frac{d}{dx}\left(\frac{\partial F}{\partial y'}\right) - \frac{\partial F}{\partial y} + \left[\frac{d}{dx}\left(\frac{\partial[\lambda G]}{\partial y'}\right) - \frac{\partial[\lambda G]}{\partial y}\right] = 0. \tag{B.47}$$

In principle, the previous equation, together with the constraint (B.46), yields the functions $\lambda(x)$ and $y(x)$. Incidentally, λ is generally termed a *Lagrange multiplier*. If F and G have no explicit x-dependence then λ is usually a constant.

As an example, consider the following famous problem. Suppose that a uniform chain of fixed length l is suspended by its ends from two equal-height fixed points that are a distance a apart, where $a < l$. What is the equilibrium configuration of the chain?

Suppose that the chain has the uniform density per unit length ρ. Let the x- and y-axes be horizontal and vertical, respectively, and let the two ends of the chain lie at $(\pm a/2, 0)$. The equilibrium configuration of the chain is specified by the function $y(x)$, for $-a/2 \leq x \leq +a/2$, where $y(x)$ is the vertical distance of the chain below its end points at horizontal position x. Of course, $y(-a/2) = y(+a/2) = 0$.

The stable equilibrium state of a conservative dynamical system is one that minimizes the system's potential energy. The potential energy of the chain is written

$$U = -\rho\,g \int y\,ds = -\rho\,g \int_{-a/2}^{a/2} y\,[1 + y'^{\,2}]^{1/2}\,dx, \tag{B.48}$$

where $ds = [dx^2 + dy^2]^{1/2}$ is an element of length along the chain, and g the acceleration due to gravity. Hence, we need to minimize U with respect to small variations in $y(x)$. However, the variations in $y(x)$ must be such as to conserve the fixed length of the chain. Hence, our minimization procedure is subject to the constraint that

$$l = \int ds = \int_{-a/2}^{a/2} [1 + y'^{\,2}]^{1/2}\,dx \tag{B.49}$$

remains constant.

It follows, from the previous discussion, that we need to minimize the functional

$$K = U + \lambda\,l = \int_{-a/2}^{a/2} (-\rho\,g\,y + \lambda)\,[1 + y'^{\,2}]^{1/2}\,dx, \tag{B.50}$$

where λ is an, as yet, undetermined constant. Because the integrand in the functional does not depend explicitly on x, we have from Equation (B.42) that

$$y'^{\,2}\,(-\rho\,g\,y + \lambda)\,[1 + y'^{\,2}]^{-1/2} - (-\rho\,g\,y + \lambda)\,[1 + y'^{\,2}]^{1/2} = k, \tag{B.51}$$

where k is a constant. This expression reduces to

$$y'^{\,2} = \left(\lambda' + \frac{y}{h}\right)^2 - 1, \tag{B.52}$$

where $\lambda' = \lambda/k$, and $h = -k/\rho\,g$.

Let

$$\lambda' + \frac{y}{h} = -\cosh z. \tag{B.53}$$

Making this substitution, Equation (B.52) yields

$$\frac{dz}{dx} = -h^{-1}. \tag{B.54}$$

Hence,

$$z = -\frac{x}{h} + c, \tag{B.55}$$

where c is a constant. It follows from Equation (B.53) that

$$y(x) = -h\left[\lambda' + \cosh\left(-\frac{x}{h} + c\right)\right]. \tag{B.56}$$

The previous solution contains three undetermined constants, h, λ', and c. We can eliminate two of these constants by application of the boundary conditions $y(\pm a/2) = 0$. This yields

$$\lambda' + \cosh\left(\mp\frac{a}{2\,h} + c\right) = 0. \tag{B.57}$$

Hence, $c = 0$, and $\lambda' = -\cosh(a/2\,h)$. It follows that

$$y(x) = h\left[\cosh\left(\frac{a}{2\,h}\right) - \cosh\left(\frac{x}{h}\right)\right]. \tag{B.58}$$

The final unknown constant, h, is determined via the application of the constraint (B.49). Thus,

$$l = \int_{-a/2}^{a/2} [1 + y'^{\,2}]^{1/2}\,dx = \int_{-a/2}^{a/2} \cosh\left(\frac{x}{h}\right)dx = 2\,h\,\sinh\left(\frac{a}{2\,h}\right). \tag{B.59}$$

Hence, the equilibrium configuration of the chain is given by the curve (B.58), which is known as a *catenary*, where the parameter h satisfies

$$\frac{l}{2\,h} = \sinh\left(\frac{a}{2\,h}\right). \tag{B.60}$$

B.7 Multi-Function Variation

Suppose that we wish to maximize or minimize the functional

$$I = \int_a^b F(y_1, y_2, \cdots, y_{\mathcal{F}}, y_1', y_2', \cdots, y_{\mathcal{F}}', x)\,dx. \tag{B.61}$$

Here, the integrand F is now a functional of the $\mathcal{F}$ independent functions $y_i(x)$, for $i = 1, \mathcal{F}$. A fairly straightforward extension of the analysis in Section B.5 yields $\mathcal{F}$ separate Euler-Lagrange equations,

$$\frac{d}{dx}\left(\frac{\partial F}{\partial y_i'}\right) - \frac{\partial F}{\partial y_i} = 0, \tag{B.62}$$

for $i = 1, \mathcal{F}$, which determine the $\mathcal{F}$ functions $y_i(x)$. If F does not explicitly depend on the function y_k then the kth Euler-Lagrange equation simplifies to

$$\frac{\partial F}{\partial y_k'} = \text{constant}. \tag{B.63}$$

Likewise, if F does not explicitly depend on x then all $\mathcal{F}$ Euler-Lagrange equations simplify to

$$y_i'\,\frac{\partial F}{\partial y_i'} - F = \text{constant}, \tag{B.64}$$

for $i = 1, \mathcal{F}$.

B.8 Hamilton's Principle

We can specify the instantaneous configuration of a conservative dynamical system with $\mathcal{F}$ degrees of freedom in terms of $\mathcal{F}$ independent generalized coordinates, q_i, for $i = 1, \mathcal{F}$. Let $K(q_1, q_2, \cdots, q_{\mathcal{F}}, \dot{q}_1, \dot{q}_2, \cdots, \dot{q}_{\mathcal{F}}, t)$ and $U(q_1, q_2, \cdots, q_{\mathcal{F}}, t)$ represent the kinetic and potential energies of the system, respectively, expressed in terms of these generalized coordinates. The Lagrangian of the system is defined

$$L(q_1, q_2, \cdots, q_{\mathcal{F}}, \dot{q}_1, \dot{q}_2, \cdots, \dot{q}_{\mathcal{F}}, t) = K - U. \tag{B.65}$$

Finally, the $\mathcal{F}$ Lagrangian equations of motion of the system take the form

$$\frac{d}{dt}\left(\frac{\partial L}{\partial \dot{q}_i}\right) - \frac{\partial L}{\partial q_i} = 0, \tag{B.66}$$

for $i = 1, \mathcal{F}$.

Note that the previous equations of motion have exactly the same mathematical form as the Euler-Lagrange equations, (B.62). Indeed, it is clear that the $\mathcal{F}$ Lagrangian equations of motion, (B.66), can all be derived from a single equation: namely,

$$\delta \int_{t_1}^{t_2} L(q_1, q_2, \cdots, q_{\mathcal{F}}, \dot{q}_1, \dot{q}_2, \cdots, \dot{q}_{\mathcal{F}}, t)\, dt = 0. \tag{B.67}$$

In other words, the motion of the system in a given time interval is such as to maximize or minimize the time integral of the Lagrangian, which is known as the *action integral*. Thus, the laws of Newtonian dynamics can be summarized in a single statement:

> The motion of a dynamical system in a given time interval is such as to maximize or minimize the action integral.

(In practice, the action integral is almost always minimized.) This statement is known as *Hamilton's principle*, and was first formulated in 1834 by the Irish mathematician William Hamilton.

B.9 Hamilton's Equations

Consider a dynamical system with $\mathcal{F}$ degrees of freedom that is described by the generalized coordinates q_i, for $i = 1, \mathcal{F}$. Suppose that neither the kinetic energy, K, nor the potential energy, U, depend explicitly on the time, t. In conventional dynamical systems, the potential energy is generally independent of the $\dot{q}_i$, whereas the kinetic energy takes the form of a homogeneous quadratic function of the $\dot{q}_i$. In other words,

$$K = \sum_{i,j=1,\mathcal{F}} m_{ij}\, \dot{q}_i\, \dot{q}_j, \tag{B.68}$$

where the m_{ij} depend on the q_i, but not on the $\dot{q}_i$. It is easily demonstrated from the previous equation that

$$\sum_{i=1,\mathcal{F}} \dot{q}_i \frac{\partial K}{\partial \dot{q}_i} = 2\,K. \tag{B.69}$$

Recall, from Section B.4, that generalized momentum conjugate to the ith generalized coordinate is defined

$$p_i = \frac{\partial L}{\partial \dot{q}_i} = \frac{\partial K}{\partial \dot{q}_i}, \tag{B.70}$$

where $L = K - U$ is the Lagrangian of the system, and we have made use of the fact that U is independent of the $\dot{q}_i$. Consider the function

$$H = \sum_{i=1,\mathcal{F}} \dot{q}_i\,p_i - L = \sum_{i=1,\mathcal{F}} \dot{q}_i\,p_i - K + U. \tag{B.71}$$

If all of the conditions discussed previously are satisfied then Equations (B.69) and (B.70) yield

$$H = K + U. \tag{B.72}$$

In other words, the function H is equal to the total energy of the system.

Consider the variation of the function H. We have

$$\delta H = \sum_{i=1,\mathcal{F}} \left(\delta\dot{q}_i\,p_i + \dot{q}_i\,\delta p_i - \frac{\partial L}{\partial \dot{q}_i}\,\delta\dot{q}_i - \frac{\partial L}{\partial q_i}\,\delta q_i\right). \tag{B.73}$$

The first and third terms in the bracket cancel, because $p_i = \partial L/\partial\dot{q}_i$. Furthermore, because Lagrange's equation can be written $\dot{p}_i = \partial L/\partial q_i$ (see Section B.4), we obtain

$$\delta H = \sum_{i=1,\mathcal{F}} (\dot{q}_i\,\delta p_i - \dot{p}_i\,\delta q_i)\,. \tag{B.74}$$

Suppose, now, that we can express the total energy of the system, H, solely as a function of the q_i and the p_i, with no explicit dependence on the $\dot{q}_i$. In other words, suppose that we can write $H = H(q_i, p_i)$. When the energy is written in this fashion it is generally termed the *Hamiltonian* of the system. The variation of the Hamiltonian function takes the form

$$\delta H = \sum_{i=1,\mathcal{F}} \left(\frac{\partial H}{\partial p_i}\,\delta p_i + \frac{\partial H}{\partial q_i}\,\delta q_i\right). \tag{B.75}$$

A comparison of the previous two equations yields

$$\dot{q}_i = \frac{\partial H}{\partial p_i}, \tag{B.76}$$

$$\dot{p}_i = -\frac{\partial H}{\partial q_i}, \tag{B.77}$$

for $i = 1, \mathcal{F}$. These $2\mathcal{F}$ first-order differential equations are known as *Hamilton's equations*. Hamilton's equations are often a useful alternative to Lagrange's equations, which take the form of $\mathcal{F}$ second-order differential equations.

Appendix C

Wave Mechanics

C.1 Introduction

According to classical physics (i.e., physics prior to the 20th century), particles and waves are distinct classes of physical entity that possess markedly different properties. For instance, particles are discrete, which means that they cannot be arbitrarily divided. In other words, it makes sense to talk about one electron, or two electrons, but not about a third of an electron. Waves, on the other hand, are continuous, which means that they can be arbitrarily divided. In other words, given a wave whose amplitude has a certain value, it makes sense to talk about a similar wave whose amplitude is one third, or any other fraction whatsoever, of this value. Particles are also highly localized in space. For example, atomic nuclei have very small radii of order 10^{-15} m, whereas electrons act like point particles (i.e., they have no discernible spatial extent). Waves, on the other hand, are non-localized in space. In fact, a wave is defined as a disturbance that is periodic in space, with some finite periodicity length (i.e., wavelength). Hence, it is fairly meaningless to talk about a disturbance being a wave unless it extends over a region of space that is at least a few wavelengths in size.

The classical scenario, just described, in which particles and waves are distinct phenomena, had to be significantly modified in the early decades of the 20th century. During this time period, physicists discovered, much to their surprise, that, under certain circumstances, waves act as particles, and particles act as waves. This bizarre behavior is known as *wave-particle duality*. For instance, the photoelectric effect (see Section C.2) shows that electromagnetic waves sometimes act like swarms of massless particles called photons. Moreover, the phenomenon of electron diffraction by atomic lattices (see Section C.3) implies that electrons sometimes possess wave-like properties. Wave-particle duality usually only manifests itself on atomic and sub-atomic lengthscales [i.e., on lengthscales less than, or of order, 10^{-10} m. (See Section C.3).] The classical picture remains valid on significantly longer lengthscales. Thus, on macroscopic lengthscales, waves only act like waves, particles only act like particles, and there is no wave-particle duality. However, on atomic lengthscales, classical mechanics, which governs the macroscopic behavior of massive particles, and classical electrodynamics, which governs the macroscopic behavior of electromagnetic fields—neither of which take wave-particle duality into account—must be replaced by new theories. The theories in question are called *quantum mechanics* and *quantum electrodynamics*, respectively. In this appendix, we shall discuss a simple version of quantum mechanics in which the microscopic dynamics of massive particles (i.e., particles with finite mass) is described entirely in terms of wavefunctions. This particular version of quantum mechanics is known as *wave mechanics*.

C.2 Photoelectric Effect

The so-called *photoelectric effect*, by which a polished metal surface emits electrons when illuminated by visible or ultra-violet light, was discovered by Heinrich Hertz in 1887. The following facts regarding this effect can be established via careful observation. First, a given surface only emits electrons when the frequency of the light with which it is illuminated exceeds a certain threshold value that is a property of the metal. Second, the current of photoelectrons, when it exists, is proportional to the intensity of the light falling on the surface. Third, the energy of the photoelectrons is independent of the light intensity, but varies linearly with the light frequency. These facts are inexplicable within the framework of classical physics.

In 1905, Albert Einstein proposed a radical new theory of light in order to account for the photoelectric effect. According to this theory, light of fixed angular frequency ω consists of a collection of indivisible discrete packages, called *quanta*,[1] whose energy is

$$E = \hbar\,\omega. \tag{C.1}$$

Here, $\hbar = 1.055 \times 10^{-34}$ J s is a new constant of nature, known as *Planck's constant*. (Strictly speaking, it is Planck's constant divided by 2π). Incidentally, $\hbar$ is called Planck's constant, rather than Einstein's constant, because Max Planck first introduced the concept of the quantization of light, in 1900, when trying to account for the electromagnetic spectrum of a black body (i.e., a perfect emitter and absorber of electromagnetic radiation).

[1] Plural of *quantum*: Latin neuter of *quantus*: how much.

Suppose that the electrons at the surface of a piece of metal lie in a potential well of depth W. In other words, the electrons have to acquire an energy W in order to be emitted from the surface. Here, W is generally called the *workfunction* of the surface, and is a property of the metal. Suppose that an electron absorbs a single quantum of light, otherwise known as a *photon*. Its energy therefore increases by $\hbar\,\omega$. If $\hbar\,\omega$ is greater than W then the electron is emitted from the surface with the residual kinetic energy

$$K = \hbar\,\omega - W. \tag{C.2}$$

Otherwise, the electron remains trapped in the potential well, and is not emitted. Here, we are assuming that the probability of an electron absorbing two or more photons is negligibly small compared to the probability of it absorbing a single photon (as is, indeed, the case for relatively low-intensity illumination). Incidentally, we can determine Planck's constant, as well as the workfunction of the metal, by plotting the kinetic energy of the emitted photoelectrons as a function of the wave frequency. This plot is a straight-line whose slope is $\hbar$, and whose intercept with the ω axis is $W/\hbar$. Finally, the number of emitted electrons increases with the intensity of the light because, the more intense the light, the larger the flux of photons onto the surface. Thus, Einstein's quantum theory of light is capable of accounting for all three of the previously mentioned observational facts regarding the photoelectric effect. In the following, we shall assume that the central component of Einstein's theory—namely, Equation (C.1)—is a general result that applies to all particles, not just photons.

C.3 Electron Diffraction

In 1927, George Paget Thomson discovered that if a beam of electrons is made to pass through a thin metal film then the regular atomic array in the metal acts as a sort of diffraction grating, so that when a photographic film, placed behind the metal, is developed, an interference pattern is discernible. This implies that electrons have wave-like properties. Moreover, the electron wavelength, λ, or, alternatively, the wavenumber, $k = 2\pi/\lambda$, can be deduced from the spacing of the maxima in the interference pattern. Thomson found that the momentum, p, of an electron is related to its wavenumber, k, according to the following simple relation:

$$p = \hbar\,k. \tag{C.3}$$

The associated wavelength, $\lambda = 2\pi/k$, is known as the *de Broglie wavelength*, because this relation was first hypothesized by Louis de Broglie in 1926. In the following, we shall assume that Equation (C.3) is a general result that applies to all particles, not just electrons.

It turns out that wave-particle duality only manifests itself on lengthscales less than, or of order, the de Broglie wavelength. Under normal circumstances, this wavelength is fairly small. For instance, the de Broglie wavelength of an electron is

$$\lambda_e = 1.2 \times 10^{-9}\,[E(\mathrm{eV})]^{-1/2}\,\mathrm{m}, \tag{C.4}$$

where the electron energy is conveniently measured in units of electron-volts (eV). (An electron accelerated from rest through a potential difference of 1000 V acquires an energy of 1000 eV, and so on. Electrons in atoms typically have energies in the range 10 to 100 eV.) Moreover, the de Broglie wavelength of a proton is

$$\lambda_p = 2.9 \times 10^{-11}\,[E(\mathrm{eV})]^{-1/2}\,\mathrm{m}. \tag{C.5}$$

C.4 Representation of Waves via Complex Numbers

In mathematics, the symbol i is conventionally used to represent the square-root of minus one: that is, the solution of $\mathrm{i}^2 = -1$. A real number, x (say), can take any value in a continuum of values lying between $-\infty$ and $+\infty$. On the other hand, an *imaginary number* takes the general form $\mathrm{i}\,y$, where y is a real number. It follows that the square of a real number is a positive real number, whereas the square of an imaginary number is a negative real number. In addition, a general *complex number* is written

$$z = x + \mathrm{i}\,y, \tag{C.6}$$

where x and y are real numbers. In fact, x is termed the real part of z, and y the imaginary part of z. This is written mathematically as $x = \mathrm{Re}(z)$ and $y = \mathrm{Im}(z)$. Finally, the *complex conjugate* of z is defined $z^* = x - \mathrm{i}\,y$.

Just as we can visualize a real number as a point on an infinite straight-line, we can visualize a complex number as a point in an infinite plane. The coordinates of the point in question are the real and imaginary parts of the number: that is, $z \equiv (x, y)$. The distance, $r = (x^2 + y^2)^{1/2}$, of the representative point from the origin is termed the *modulus* of the corresponding complex number, z. This is written mathematically as $|z| = (x^2 + y^2)^{1/2}$. Incidentally, it follows that $z\,z^* = x^2 + y^2 = |z|^2$. The angle, $\theta = \tan^{-1}(y/x)$, that the straight-line joining the representative point to the origin subtends with the real axis is termed the *argument* of the corresponding complex number, z. This is written mathematically as $\arg(z) = \tan^{-1}(y/x)$. It follows from standard trigonometry that $x = r\,\cos\theta$, and $y = r\,\sin\theta$. Hence, $z = r\,\cos\theta + \mathrm{i}\,r\sin\theta$.

Complex numbers are often used to represent waves and wavefunctions. All such representations ultimately depend on a fundamental mathematical identity, known as *Euler's theorem*, which takes the form

$$e^{i\phi} \equiv \cos\phi + i\,\sin\phi, \tag{C.7}$$

where ϕ is a real number. Incidentally, given that $z = r\,\cos\theta + i\,r\,\sin\theta = r\,(\cos\theta + i\,\sin\theta)$, where z is a general complex number, $r = |z|$ its modulus, and $\theta = \arg(z)$ its argument, it follows from Euler's theorem that any complex number, z, can be written

$$z = r\,e^{i\theta}, \tag{C.8}$$

where $r = |z|$ and $\theta = \arg(z)$ are real numbers.

A one-dimensional wavefunction takes the general form

$$\psi(x, t) = A\,\cos(\omega\,t - k\,x - \phi), \tag{C.9}$$

where $A > 0$ is the wave amplitude, ϕ the phase angle, k the wavenumber, and ω the angular frequency. Consider the complex wavefunction

$$\psi(x, t) = \psi_0\,e^{-i\,(\omega\,t - k\,x)}, \tag{C.10}$$

where ψ_0 is a complex constant. We can write

$$\psi_0 = A\,e^{i\phi}, \tag{C.11}$$

where A is the modulus, and ϕ the argument, of ψ_0. Hence, we deduce that

$$\begin{aligned} \mathrm{Re}\left[\psi_0\,e^{-i\,(\omega\,t-k\,x)}\right] = \mathrm{Re}\left[A\,e^{i\phi}\,e^{-i\,(\omega\,t-k\,x)}\right] &= \mathrm{Re}\left[A\,e^{-i\,(\omega\,t-k\,x-\phi)}\right] \\ &= A\,\mathrm{Re}\left[e^{-i\,(\omega\,t-k\,x-\phi)}\right]. \end{aligned} \tag{C.12}$$

Thus, it follows from Euler's theorem, and Equation (C.9), that

$$\mathrm{Re}\left[\psi_0\,e^{-i\,(\omega\,t-k\,x)}\right] = A\,\cos(\omega\,t - k\,x - \phi) = \psi(x, t). \tag{C.13}$$

In other words, a general one-dimensional real wavefunction, (C.9), can be represented as the real part of a complex wavefunction of the form (C.10). For ease of notation, the "take the real part" aspect of the previous expression is usually omitted, and our general one-dimension wavefunction is simply written

$$\psi(x, t) = \psi_0\,e^{-i\,(\omega\,t - k\,x)}. \tag{C.14}$$

The main advantage of the complex representation, (C.14), over the more straightforward real representation, (C.9), is that the former enables us to combine the amplitude, A, and the phase angle, ϕ, of the wavefunction into a single complex amplitude, ψ_0.

C.5 Schrödinger's Equation

The basic premise of wave mechanics is that a massive particle of energy E and linear momentum p, moving in the x-direction (say), can be represented by a one-dimensional *complex wavefunction* of the form

$$\psi(x,t) = \psi_0\,e^{-i\,(\omega\,t-k\,x)}, \tag{C.15}$$

where the complex amplitude, ψ_0, is arbitrary, while the wavenumber, k, and the angular frequency, ω, are related to the particle momentum, p, and energy, E, via the fundamental relations (C.3) and (C.1), respectively. The previous one-dimensional wavefunction is the solution of a one-dimensional wave equation that determines how the wavefunction evolves in time. As described in the following, we can guess the form of this wave equation by drawing an analogy with classical physics.

A classical particle of mass m, moving in a one-dimensional potential $U(x)$, satisfies the energy conservation equation

$$E = K + U, \tag{C.16}$$

where

$$K = \frac{p^2}{2\,m} \tag{C.17}$$

is the particle's kinetic energy. Hence,

$$E\,\psi = (K+U)\,\psi \tag{C.18}$$

is a valid, but not obviously useful, wave equation.

However, it follows from Equations (C.1) and (C.15) that

$$\frac{\partial\psi}{\partial t} = -i\,\omega\,\psi_0\,e^{-i\,(\omega\,t-k\,x)} = -i\,\frac{E}{\hbar}\,\psi, \tag{C.19}$$

which can be rearranged to give

$$E\,\psi = i\,\hbar\,\frac{\partial\psi}{\partial t}. \tag{C.20}$$

Likewise, from Equations (C.3) and (C.15),

$$\frac{\partial\psi}{\partial x} = i\,k\,\psi_0\,e^{-i\,(k\,x-\omega\,t)} = i\,\frac{p}{\hbar}\,\psi, \tag{C.21}$$

$$\frac{\partial^2\psi}{\partial x^2} = -k^2\,\psi_0\,e^{-i\,(k\,x-\omega\,t)} = -\frac{p^2}{\hbar^2}\,\psi, \tag{C.22}$$

which can be rearranged to give

$$K\,\psi = \frac{p^2}{2\,m}\,\psi = -\frac{\hbar^2}{2\,m}\,\frac{\partial^2\psi}{\partial x^2}. \tag{C.23}$$

Thus, combining Equations (C.18), (C.20), and (C.23), we obtain

$$\mathrm{i}\,\hbar\,\frac{\partial\psi}{\partial t} = -\frac{\hbar^2}{2\,m}\,\frac{\partial^2\psi}{\partial x^2} + U(x)\,\psi. \tag{C.24}$$

This equation, which is known as *Schrödinger's equation*—because it was first formulated by Erwin Schrödinder in 1926—is the fundamental equation of wave mechanics.

For a massive particle moving in free space (i.e., $U = 0$), the complex wave-function (C.15) is a solution of Schrödinger's equation, (C.24), provided

$$\omega = \frac{\hbar}{2\,m}\,k^2. \tag{C.25}$$

The previous expression can be thought of as the dispersion relation for matter waves in free space. The associated *phase velocity* is

$$v_p = \frac{\omega}{k} = \frac{\hbar\,k}{2\,m} = \frac{p}{2\,m}, \tag{C.26}$$

where use has been made of Equation (C.3). However, this phase velocity is only half the classical velocity, $v = p/m$, of a massive (non-relativistic) particle.

Incidentally, Equation (C.21) suggests that

$$p \equiv -\mathrm{i}\,\hbar\,\frac{\partial}{\partial x} \tag{C.27}$$

in quantum mechanics, whereas Equation (C.24) suggests that the most general form of Schrödinger's equation is

$$\mathrm{i}\,\hbar\,\frac{\partial\psi}{\partial t} = H\,\psi, \tag{C.28}$$

where

$$H(x, p) \equiv \frac{p^2}{2\,m} + U(x) = -\frac{\hbar^2}{2\,m}\,\frac{\partial^2}{\partial x^2} + U(x) \tag{C.29}$$

is the Hamiltonian of the system. (See Section B.9.)

C.6 Probability Interpretation of Wavefunction

After many false starts, physicists in the early 20th century eventually came to the conclusion that the only physical interpretation of a particle wavefunction that is consistent with experimental observations is probabilistic in nature. To be more exact, if $\psi(x, t)$ is the complex wavefunction of a given particle, moving in one dimension along the x-axis, then the probability of finding the particle between x and $x + dx$ at time t is

$$P(x, t) = |\psi(x, t)|^2\,dx. \tag{C.30}$$

We can interpret

$$P(t) = \int_{-\infty}^{\infty} |\psi(x,t)|^2\,dx \tag{C.31}$$

as the probability of the particle being found anywhere between $x = -\infty$ and $x = +\infty$ at time t. This follows, via induction, from the fundamental result in probability theory that the probability of the occurrence of one or other of two mutually exclusive events (such as the particle being found in two non-overlapping regions) is the sum (or integral) of the probabilities of the individual events. (See Section 2.3.) Assuming that the particle exists, it is certain that it will be found somewhere between $x = -\infty$ and $x = +\infty$ at time t. Because a certain event has probability 1 (see Section 2.2), our probability interpretation of the wavefunction is only tenable provided

$$\int_{-\infty}^{\infty} |\psi(x,t)|^2\,dx = 1 \tag{C.32}$$

at all times. A wavefunction that satisfies the previous condition—which is known as the *normalization condition*—is said to be properly normalized.

Suppose that we have a wavefunction, $\psi(x,t)$, which is such that it satisfies the normalization condition (C.32) at time $t = 0$. Furthermore, let the wavefunction evolve in time according to Schrödinger's equation, (C.24). Our probability interpretation of the wavefunction only makes sense if the normalization condition remains satisfied at all subsequent times. This follows because if the particle is certain to be found somewhere on the x-axis (which is the interpretation put on the normalization condition) at time $t = 0$ then it is equally certain to be found somewhere on the x-axis at a later time (because we are not considering any physical process by which particles can be created or destroyed). Thus, it is necessary for us to demonstrate that Schrödinger's equation preserves the normalization of the wavefunction.

Taking Schrödinger's equation, and multiplying it by ψ^* (the complex conjugate of the wavefunction), we obtain

$$\mathrm{i}\,\hbar\,\frac{\partial\psi}{\partial t}\,\psi^* = -\frac{\hbar^2}{2\,m}\,\frac{\partial^2\psi}{\partial x^2}\,\psi^* + U(x)\,|\psi|^2. \tag{C.33}$$

The complex conjugate of the previous expression yields

$$-\mathrm{i}\,\hbar\,\frac{\partial\psi^*}{\partial t}\,\psi = -\frac{\hbar^2}{2\,m}\,\frac{\partial^2\psi^*}{\partial x^2}\,\psi + U(x)\,|\psi|^2. \tag{C.34}$$

Here, use has been made of the readily demonstrated results $(\psi^*)^* = \psi$ and $\mathrm{i}^* = -\mathrm{i}$, as well as the fact that $U(x)$ is real. Taking the difference between the previous two expressions, we obtain

$$\mathrm{i}\,\hbar\left(\frac{\partial\psi}{\partial t}\,\psi^* + \frac{\partial\psi^*}{\partial t}\,\psi\right) = -\frac{\hbar^2}{2\,m}\left(\frac{\partial^2\psi}{\partial x^2}\,\psi^* - \frac{\partial^2\psi^*}{\partial x^2}\,\psi\right), \tag{C.35}$$

which can be written

$$\mathrm{i}\,\hbar\,\frac{\partial|\psi|^2}{\partial t} = -\frac{\hbar^2}{2\,m}\,\frac{\partial}{\partial x}\left(\frac{\partial\psi}{\partial x}\,\psi^* - \frac{\partial\psi^*}{\partial x}\,\psi\right). \tag{C.36}$$

Integrating in x, we get

$$\mathrm{i}\,\hbar\,\frac{d}{dt}\int_{-\infty}^{\infty}|\psi|^2\,dx = -\frac{\hbar^2}{2\,m}\left[\frac{\partial\psi}{\partial x}\,\psi^* - \frac{\partial\psi^*}{\partial x}\,\psi\right]_{-\infty}^{\infty}. \tag{C.37}$$

Finally, assuming that the wavefunction is localized in space: that is,

$$|\psi(x,t)| \to 0 \quad \text{as} \quad |x| \to \infty, \tag{C.38}$$

we obtain

$$\frac{d}{dt}\int_{-\infty}^{\infty}|\psi|^2\,dx = 0. \tag{C.39}$$

It follows, from the preceding analysis, that if a localized wavefunction is properly normalized at $t = 0$ (i.e., if $\int_{-\infty}^{\infty}|\psi(x,0)|^2\,dx = 1$) then it will remain properly normalized as it evolves in time according to Schrödinger's equation. Incidentally, a wavefunction that is not localized cannot be properly normalized, because its normalization integral $\int_{-\infty}^{\infty}|\psi|^2\,dx$ is necessarily infinite. For such a wavefunction, $|\psi(x,t)|^2\,dx$ gives the relative, rather than the absolute, probability of finding the particle between x and $x + dx$ at time t. In other words, [cf., Equation (C.30)]

$$P(x,t) \propto |\psi(x,t)|^2\,dx. \tag{C.40}$$

C.7 Wave Packets

As we have seen, the wavefunction of a massive particle of momentum p and energy E, moving in free space along the x-axis, can be written

$$\psi(x,t) = \bar{\psi}\,\mathrm{e}^{-\mathrm{i}\,(\omega\,t - k\,x)}, \tag{C.41}$$

where $k = p/\hbar$, $\omega = E/\hbar$, and $\bar{\psi}$ is a complex constant. Here, ω and k are linked via the matter wave dispersion relation, (C.25). Expression (C.41) represents a plane wave that propagates in the x-direction with the phase velocity $v_p = \omega/k$. However, it follows from Equation (C.26) that this phase velocity is only half of the classical velocity of a massive particle.

According to the discussion in the previous section, the most reasonable physical interpretation of the wavefunction is that $|\psi(x,t)|^2\,dx$ is proportional to (assuming that the wavefunction is not properly normalized) the probability of finding the particle between x and $x + dx$ at time t. However, the modulus squared of the wavefunction (C.41) is $|\bar{\psi}|^2$, which is a constant that depends on neither

x nor t. In other words, the previous wavefunction represents a particle that is equally likely to be found anywhere on the x-axis at all times. Hence, the fact that this wavefunction propagates at a phase velocity that does not correspond to the classical particle velocity has no observable consequences.

How can we write the wavefunction of a particle that is localized in x? In other words, a particle that is more likely to be found at some positions on the x-axis than at others. It turns out that we can achieve this goal by forming a linear combination of plane waves of different wavenumbers: that is,

$$\psi(x,t) = \int_{-\infty}^{\infty} \bar{\psi}(k)\,\mathrm{e}^{-\mathrm{i}\,(\omega\,t-k\,x)}\,dk. \tag{C.42}$$

Here, $\bar{\psi}(k)$ represents the complex amplitude of plane waves of wavenumber k within this combination. In writing the previous expression, we are relying on the assumption that matter waves are superposable. In other words, it is possible to add two valid wave solutions to form a third valid wave solution. The ultimate justification for this assumption is that matter waves satisfy the linear wave equation (C.24).

There is a fundamental mathematical theorem, known as *Fourier's theorem*, which states that if

$$f(x) = \int_{-\infty}^{\infty} \bar{f}(k)\,\mathrm{e}^{\,\mathrm{i}\,k\,x}\,dk, \tag{C.43}$$

then

$$\bar{f}(k) = \frac{1}{2\pi}\int_{-\infty}^{\infty} f(x)\,\mathrm{e}^{-\mathrm{i}\,k\,x}\,dx. \tag{C.44}$$

Here, $\bar{f}(k)$ is known as the Fourier transform of the function $f(x)$. We can use Fourier's theorem to find the k-space function, $\bar{\psi}(k)$, that generates any given x-space wavefunction, $\psi(x)$, at a given time.

For instance, suppose that at $t = 0$ the wavefunction of our particle takes the form

$$\psi(x,0) \propto \exp\left[\mathrm{i}\,k_0\,x - \frac{(x-x_0)^2}{4\,(\Delta x)^2}\right]. \tag{C.45}$$

Thus, the initial probability distribution for the particle's x-coordinate is Gaussian (see Section 2.9):

$$|\psi(x,0)|^2 \propto \exp\left[-\frac{(x-x_0)^2}{2\,(\Delta x)^2}\right]. \tag{C.46}$$

It follows that a measurement of the particle's position is most likely to yield the value x_0, and very unlikely to yield a value which differs from x_0 by more than $3\,\Delta x$. Thus, Equation (C.45) is the wavefunction of a particle that is initially localized in some region of x-space, centered on $x = x_0$, whose width is of order Δx. This type of wavefunction is known as a *wave packet*.

According to Equation (C.42),

$$\psi(x,0) = \int_{-\infty}^{\infty} \bar{\psi}(k)\,\mathrm{e}^{\,\mathrm{i}\,k\,x}\,dk. \tag{C.47}$$

Hence, we can employ Fourier's theorem to invert this expression to give

$$\bar{\psi}(k) \propto \int_{-\infty}^{\infty} \psi(x,0)\,\mathrm{e}^{-\mathrm{i}\,k\,x}\,dx. \tag{C.48}$$

Making use of Equation (C.45), we obtain

$$\bar{\psi}(k) \propto \mathrm{e}^{-\mathrm{i}\,(k-k_0)\,x_0}\int_{-\infty}^{\infty} \exp\left[-\mathrm{i}\,(k-k_0)\,(x-x_0) - \frac{(x-x_0)^2}{4\,(\Delta x)^2}\right] dx. \tag{C.49}$$

Changing the variable of integration to $y = (x-x_0)/(2\,\Delta x)$, the previous expression reduces to

$$\bar{\psi}(k) \propto \mathrm{e}^{-\mathrm{i}\,k\,x_0-\beta^2/4}\int_{-\infty}^{\infty} \mathrm{e}^{-(y-y_0)^2}\,dy, \tag{C.50}$$

where $\beta = 2\,(k-k_0)\,\Delta x$ and $y_0 = -\mathrm{i}\,\beta/2$. The integral in the previous equation is now just a number, as can easily be seen by making the second change of variable $z = y - y_0$. Hence, we deduce that

$$\bar{\psi}(k) \propto \exp\left[-\mathrm{i}\,k\,x_0 - \frac{(k-k_0)^2}{4\,(\Delta k)^2}\right], \tag{C.51}$$

where

$$\Delta k = \frac{1}{2\,\Delta x}. \tag{C.52}$$

If $|\psi(x,0)|^2\,dx$ is proportional to the probability of a measurement of the particle's position yielding a value in the range x to $x + dx$ at time $t = 0$ then it stands to reason that $|\bar{\psi}(k)|^2\,dk$ is proportional to the probability of a measurement of the particle's wavenumber yielding a value in the range k to $k + dk$. (Recall that $p = \hbar\,k$, so a measurement of the particle's wavenumber, k, is equivalent to a measurement of the particle's momentum, p). According to Equation (C.51),

$$|\bar{\psi}(k)|^2 \propto \exp\left[-\frac{(k-k_0)^2}{2\,(\Delta k)^2}\right]. \tag{C.53}$$

This probability distribution is a Gaussian in k-space. [See Equation (C.46).] Hence, a measurement of k is most likely to yield the value k_0, and very unlikely to yield a value that differs from k_0 by more than $3\,\Delta k$.

We have just seen that a wave packet with a Gaussian probability distribution of characteristic width Δx in x-space [see Equation (C.46)] is equivalent to a wave packet with a Gaussian probability distribution of characteristic width Δk in k-space [see Equation (C.53)], where

$$\Delta x \Delta k = \frac{1}{2}. \tag{C.54}$$

This illustrates an important property of wave packets. Namely, in order to construct a packet that is highly localized in x-space (i.e., with small Δx) we need to combine plane waves with a very wide range of different k-values (i.e., with large Δk). Conversely, if we only combine plane waves whose wavenumbers differ by a small amount (i.e., if Δk is small) then the resulting wave packet is highly extended in x-space (i.e., Δx is large).

According to standard wave theory, a wave packet made up of a superposition of plane waves that is strongly peaked around some central wavenumber k_0 propagates at the *group velocity*,

$$v_g = \frac{d\omega(k_0)}{dk}, \tag{C.55}$$

rather than the phase velocity, $v_p = (\omega/k)_{k_0}$, assuming that all of the constituent plane waves satisfy a dispersion relation of the form $\omega = \omega(k)$. For the case of matter waves, the dispersion relation is (C.25). Thus, the associated group velocity is

$$v_g = \frac{\hbar k_0}{m} = \frac{p}{m}, \tag{C.56}$$

where $p = \hbar k_0$. This velocity is identical to the classical velocity of a (non-relativistic) massive particle. We conclude that the matter wave dispersion relation (C.25) is perfectly consistent with classical physics, as long as we recognize that particles must be identified with wave packets (which propagate at the group velocity) rather than plane waves (which propagate at the phase velocity).

Standard wave theory also implies that the spatial extent of a wave packet of initial extent $(\Delta x)_0$ grows, as the packet evolves in time, like

$$\Delta x \simeq (\Delta x)_0 + \frac{d^2\omega(k_0)}{dk^2} \frac{t}{(\Delta x)_0}, \tag{C.57}$$

where k_0 is the packet's central wavenumber. Thus, it follows from the matter wave dispersion relation, (C.25), that the width of a particle wave packet grows in time as

$$\Delta x \simeq (\Delta x)_0 + \frac{\hbar}{m} \frac{t}{(\Delta x)_0}. \tag{C.58}$$

For example, if an electron wave packet is initially localized in a region of atomic dimensions (i.e., $\Delta x \sim 10^{-10}$ m) then the width of the packet doubles in about 10^{-16} s.

C.8 Heisenberg's Uncertainty Principle

According to the analysis contained in the previous section, a particle wave packet that is initially localized in x-space, with characteristic width Δx, is also localized in k-space, with characteristic width $\Delta k = 1/(2\,\Delta x)$. However, as time progresses, the width of the wave packet in x-space increases [see Equation (C.58)], while that of the packet in k-space stays the same [because $\bar{\psi}(k)$ is given by Equation (C.48) at all times]. Hence, in general, we can say that

$$\Delta x\,\Delta k \gtrsim \frac{1}{2}. \tag{C.59}$$

Furthermore, we can interpret Δx and Δk as characterizing our uncertainty regarding the values of the particle's position and wavenumber, respectively.

A measurement of a particle's wavenumber, k, is equivalent to a measurement of its momentum, p, because $p = \hbar\,k$. Hence, an uncertainty in k of order Δk translates to an uncertainty in p of order $\Delta p = \hbar\,\Delta k$. It follows, from the previous inequality, that

$$\Delta x\,\Delta p \gtrsim \frac{\hbar}{2}. \tag{C.60}$$

This is the famous *Heisenberg uncertainty principle*, first proposed by Werner Heisenberg in 1927. According to this principle, it is impossible to simultaneously measure the position and momentum of a particle (exactly). Indeed, a good knowledge of the particle's position implies a poor knowledge of its momentum, and vice versa. The uncertainty principle is a direct consequence of representing particles as waves.

It is apparent, from Equation (C.58), that a particle wave packet of initial spatial extent $(\Delta x)_0$ spreads out in such a manner that its spatial extent becomes

$$\Delta x \sim \frac{\hbar\,t}{m\,(\Delta x)_0} \tag{C.61}$$

at large t. It is readily demonstrated that this spreading of the wave packet is a consequence of the uncertainty principle. Indeed, because the initial uncertainty in the particle's position is $(\Delta x)_0$, it follows that the uncertainty in its momentum is of order $\hbar/(\Delta x)_0$. This translates to an uncertainty in velocity of $\Delta v = \hbar/[m\,(\Delta x)_0]$. Thus, if we imagine that part of the wave packet propagates at $v_0 + \Delta v/2$, and another part at $v_0 - \Delta v/2$, where v_0 is the mean propagation velocity, then it follows that the wave packet will spread out as time progresses. Indeed, at large t, we expect the width of the wave packet to be

$$\Delta x \sim \Delta v\,t \sim \frac{\hbar\,t}{m\,(\Delta x)_0}, \tag{C.62}$$

which is identical to Equation (C.61). Evidently, the spreading of a particle wave packet, as time progresses, should be interpreted as representing an increase in our uncertainty regarding the particle's position, rather than an increase in the spatial extent of the particle itself.

C.9 Stationary States

Consider separable solutions to Schrödinger's equation of the form

$$\psi(x,t) = \psi(x)\,\mathrm{e}^{-\mathrm{i}\,\omega\,t}. \tag{C.63}$$

According to Equation (C.20), such solutions have definite energies, $E = \hbar\,\omega$. For this reason, they are usually written

$$\psi(x,t) = \psi(x)\,\mathrm{e}^{-\mathrm{i}\,(E/\hbar)\,t}. \tag{C.64}$$

The probability of finding the particle between x and $x + dx$ at time t is

$$P(x,t) = |\psi(x,t)|^2\,dx = |\psi(x)|^2\,dx. \tag{C.65}$$

This probability is time independent. For this reason, states whose wavefunctions are of the form (C.64) are known as *stationary states*. Moreover, $\psi(x)$ is called a stationary wavefunction. Substituting (C.64) into Schrödinger's equation, (C.24), we obtain the following differential equation for $\psi(x)$:

$$-\frac{\hbar^2}{2\,m}\frac{d^2\psi}{dx^2} + U(x)\,\psi = E\,\psi. \tag{C.66}$$

This equation is called the *time-independent Schrödinger equation*. Of course, the most general form of this equation is

$$H\,\psi = E\,\psi, \tag{C.67}$$

where H is the Hamiltonian. (See Section C.5.)

Consider a particle trapped in a one-dimensional square potential well, of infinite depth, which is such that

$$U(x) = \begin{cases} 0 & 0 \leq x \leq a \\ \infty & \text{otherwise} \end{cases}. \tag{C.68}$$

The particle is excluded from the region $x < 0$ or $x > a$, so $\psi = 0$ in this region (i.e., there is zero probability of finding the particle outside the well). Within the well, a particle of definite energy E has a stationary wavefunction, $\psi(x)$, that satisfies

$$-\frac{\hbar^2}{2\,m}\frac{d^2\psi}{dx^2} = E\,\psi. \tag{C.69}$$

The boundary conditions are

$$\psi(0) = \psi(a) = 0. \tag{C.70}$$

This follows because $\psi = 0$ in the region $x < 0$ or $x > a$, and $\psi(x)$ must be continuous [because a discontinuous wavefunction would generate a singular term (i.e., the term involving $d^2\psi/dx^2$) in the time-independent Schrödinger equation, (C.66), that could not be balanced, even by an infinite potential].

Let us search for solutions to Equation (C.69) of the form

$$\psi(x) = \psi_0 \sin(k\,x), \tag{C.71}$$

where ψ_0 is a constant. It follows that

$$\frac{\hbar^2 k^2}{2\,m} = E. \tag{C.72}$$

The solution (C.71) automatically satisfies the boundary condition $\psi(0) = 0$. The second boundary condition, $\psi(a) = 0$, leads to a quantization of the wavenumber: that is,

$$k = n\,\frac{\pi}{a}, \tag{C.73}$$

where $n = 1, 2, 3$, et cetera. (A "quantized" quantity is one that can only take certain discrete values.) Here, the integer n is known as a *quantum number*. According to Equation (C.72), the energy is also quantized. In fact, $E = E_n$, where

$$E_n = n^2\,\frac{\hbar^2\,\pi^2}{2\,m\,a^2}. \tag{C.74}$$

Thus, the allowed wavefunctions for a particle trapped in a one-dimensional square potential well of infinite depth are

$$\psi_n(x,t) = A_n\,\sin\left(n\,\pi\,\frac{x}{a}\right)\,\exp\left(-{\rm i}\,n^2\,\frac{E_1}{\hbar}\,t\right), \tag{C.75}$$

where n is a positive integer, and A_n a constant. We cannot have $n = 0$, because, in this case, we obtain a null wavefunction: that is, $\psi = 0$, everywhere. Furthermore, if n takes a negative integer value then it generates exactly the same wavefunction as the corresponding positive integer value (assuming $A_{-n} = -A_n$).

The constant A_n, appearing in the previous wavefunction, can be determined from the constraint that the wavefunction be properly normalized. For the case under consideration, the normalization condition (C.32) reduces to

$$\int_0^a |\psi(x)|^2\,dx = 1. \tag{C.76}$$

It follows from Equation (C.75) that $|A_n|^2 = 2/a$. Hence, the properly normalized version of the wavefunction (C.75) is

$$\psi_n(x,t) = \left(\frac{2}{a}\right)^{1/2}\,\sin\left(n\,\pi\,\frac{x}{a}\right)\,\exp\left(-{\rm i}\,n^2\,\frac{E_1}{\hbar}\,t\right). \tag{C.77}$$

At first sight, it seems rather strange that the lowest possible energy for a particle trapped in a one-dimensional potential well is not zero, as would be the case in classical mechanics, but rather $E_1 = \hbar^2\,\pi^2/(2\,m\,a^2)$. In fact, as explained in the following, this residual energy is a direct consequence of Heisenberg's uncertainty principle. A particle trapped in a one-dimensional well of width a is likely to be found anywhere inside the well. Thus, the uncertainty in the particle's position is $\Delta x \sim a$. It follows from the uncertainty principle, (C.60), that

$$\Delta p \gtrsim \frac{\hbar}{2\,\Delta x} \sim \frac{\hbar}{a}. \tag{C.78}$$

In other words, the particle cannot have zero momentum. In fact, the particle's momentum must be at least $p \sim \hbar/a$. However, for a free particle, $E = p^2/2\,m$. Hence, the residual energy associated with the particle's residual momentum is

$$E \sim \frac{p^2}{m} \sim \frac{\hbar^2}{m\,a^2} \sim E_1. \tag{C.79}$$

This type of residual energy, which often occurs in quantum mechanical systems, and has no equivalent in classical mechanics, is called *zero point energy*.

C.10 Three-Dimensional Wave Mechanics

Up to now, we have only discussed wave mechanics for a particle moving in one dimension. However, the generalization to a particle moving in three dimensions is fairly straightforward. A massive particle moving in three dimensions has a complex wavefunction of the form [cf., Equation (C.15)]

$$\psi(x, y, z, t) = \psi_0\,{\rm e}^{-{\rm i}\,(\omega\,t - {\bf k}\cdot{\bf r})}, \tag{C.80}$$

where ψ_0 is a complex constant, and ${\bf r} = (x,\,y,\,z)$. Here, the wavevector, ${\bf k}$, and the angular frequency, ω, are related to the particle momentum, ${\bf p}$, and energy, E, according to [cf., Equations (C.3) and (C.27)]

$${\bf p} = \hbar\,{\bf k} = -{\rm i}\,\hbar\,\nabla, \tag{C.81}$$

and [cf., Equation (C.1)]

$$E = \hbar\,\omega, \tag{C.82}$$

respectively. Generalizing the analysis of Section C.5, the three-dimensional version of Schrödinger's equation is [cf., Equations (C.24) and (C.29)]

$${\rm i}\,\hbar\,\frac{\partial\psi}{\partial t} = H\,\psi \equiv -\frac{\hbar^2}{2\,m}\,\nabla^2\psi + U({\bf r})\,\psi, \tag{C.83}$$

where

$$H = \frac{p^2}{2\,m} + U({\bf r}) = -\frac{\hbar^2}{2\,m}\,\nabla^2 + U({\bf r}) \tag{C.84}$$

is the Hamiltonian, and the differential operator

$$\nabla^2 \equiv \frac{\partial^2}{\partial x^2} + \frac{\partial^2}{\partial y^2} + \frac{\partial^2}{\partial z^2} \tag{C.85}$$

is known as the *Laplacian*. The interpretation of a three-dimensional wavefunction is that the probability of simultaneously finding the particle between x and $x + dx$, between y and $y + dy$, and between z and $z + dz$, at time t is [cf., Equation (C.30)]

$$P(x, y, z, t) = |\psi(x, y, z, t)|^2 \, dx \, dy \, dz. \tag{C.86}$$

Moreover, the normalization condition for the wavefunction becomes [cf., Equation (C.32)]

$$\int_{-\infty}^{\infty}\int_{-\infty}^{\infty}\int_{-\infty}^{\infty} |\psi(x, y, z, t)|^2 \, dx \, dy \, dz = 1. \tag{C.87}$$

It can be demonstrated that Schrödinger's equation, (C.83), preserves the normalization condition, (C.87), of a localized wavefunction. Heisenberg's uncertainty principle generalizes to [cf., Equation (C.60)]

$$\Delta x \, \Delta p_x \gtrsim \frac{\hbar}{2}, \tag{C.88}$$

$$\Delta y \, \Delta p_y \gtrsim \frac{\hbar}{2}, \tag{C.89}$$

$$\Delta z \, \Delta p_z \gtrsim \frac{\hbar}{2}. \tag{C.90}$$

Finally, a stationary state of energy E is written [cf., Equation (C.64)]

$$\psi(x, y, z, t) = \psi(x, y, z) \, e^{-i (E/\hbar) t}, \tag{C.91}$$

where the stationary wavefunction, $\psi(x, y, z)$, satisfies [cf., Equation (C.66)]

$$H \psi \equiv -\frac{\hbar^2}{2 m} \nabla^2 \psi + U(\mathbf{r}) \, \psi = E \, \psi. \tag{C.92}$$

As an example of a three-dimensional problem in wave mechanics, consider a particle trapped in a square potential well of infinite depth, such that

$$U(x, y, z) = \begin{cases} 0 & 0 \leq x \leq a, \ 0 \leq y \leq a, \ 0 \leq z \leq a \\ \infty & \text{otherwise} \end{cases}. \tag{C.93}$$

Within the well, the stationary wavefunction, $\psi(x, y, z)$, satisfies

$$-\frac{\hbar^2}{2 m} \nabla^2 \psi = E \, \psi, \tag{C.94}$$

subject to the boundary conditions

$$\psi(0, y, z) = \psi(x, 0, z) = \psi(x, y, 0) = 0, \tag{C.95}$$

and

$$\psi(a, y, z) = \psi(x, a, z) = \psi(x, y, a) = 0, \tag{C.96}$$

because $\psi = 0$ outside the well. Let us try a separable wavefunction of the form

$$\psi(x, y, z) = \psi_0 \, \sin(k_x \, x) \, \sin(k_y \, y) \, \sin(k_z \, z). \tag{C.97}$$

This expression automatically satisfies the boundary conditions (C.95). The remaining boundary conditions, (C.96), are satisfied provided

$$k_x = n_x \, \frac{\pi}{a}, \tag{C.98}$$

$$k_y = n_y \, \frac{\pi}{a}, \tag{C.99}$$

$$k_z = n_z \, \frac{\pi}{a}, \tag{C.100}$$

where the quantum numbers n_x, n_y, and n_z are (independent) positive integers. Note that a stationary state generally possesses as many quantum numbers as there are degrees of freedom of the system (in this case, three). Substitution of the wavefunction (C.97) into Equation (C.94) yields

$$E = \frac{\hbar^2}{2\,m} \, (k_x^2 + k_y^2 + k_z^2). \tag{C.101}$$

Thus, it follows from Equations (C.98)–(C.100) that the particle energy is quantized, and that the allowed *energy levels* are

$$E_{n_x,n_y,n_z} = \frac{\hbar^2}{2\,m\,a^2} \, (n_x^2 + n_y^2 + n_z^2). \tag{C.102}$$

The properly normalized [see Equation (C.87)] stationary wavefunctions corresponding to these energy levels are

$$\psi_{n_x,n_y,n_z}(x, y, z) = \left(\frac{2}{a}\right)^{3/2} \sin\left(n_x \, \pi \, \frac{x}{a}\right) \sin\left(n_y \, \pi \, \frac{y}{a}\right) \sin\left(n_z \, \pi \, \frac{z}{a}\right). \tag{C.103}$$

As is the case for a particle trapped in a one-dimensional potential well, the lowest energy level for a particle trapped in a three-dimensional well is not zero, but rather

$$E_{1,1,1} = 3\,E_1. \tag{C.104}$$

Here,

$$E_1 = \frac{\hbar^2}{2\,m\,a^2} \tag{C.105}$$

is the *ground state* (i.e., the lowest energy state) energy in the one-dimensional case. It follows from Equation (C.102) that distinct permutations of n_x, n_y, and

n_z that do not alter the value of $n_x^2 + n_y^2 + n_z^2$ also do not alter the energy. In other words, in three dimensions, it is possible for distinct wavefunctions to be associated with the same energy level. In this situation, the energy level is said to be *degenerate*. The ground state energy level, $3\,E_1$, is non-degenerate, because the only combination of $(n_x,\, n_y,\, n_z)$ that gives this energy is (1, 1, 1). However, the next highest energy level, $6\,E_1$, is degenerate, because it is obtained when $(n_x,\, n_y,\, n_y)$ take the values (2, 1, 1), or (1, 2, 1), or (1, 1, 2). In fact, a non-degenerate energy level corresponds to a case where the three quantum numbers (i.e., n_x, n_y, and n_z) all have the same value, whereas a three-fold degenerate energy level corresponds to a case where only two of the quantum numbers have the same value, and, finally, a six-fold degenerate energy level corresponds to a case where the quantum numbers are all different.

C.11 Simple Harmonic Oscillator

The classical Hamiltonian of a simple harmonic oscillator is

$$H = \frac{p^2}{2\,m} + \frac{1}{2}\,K\,x^2, \tag{C.106}$$

where $K > 0$ is the so-called force constant of the oscillator. Assuming that the quantum-mechanical Hamiltonian has the same form as the classical Hamiltonian, the time-independent Schrödinger equation for a particle of mass m and energy E moving in a simple harmonic potential becomes

$$\frac{d^2\psi}{dx^2} = \frac{2\,m}{\hbar^2}\left(\frac{1}{2}\,K\,x^2 - E\right)\psi. \tag{C.107}$$

Let $\omega = \sqrt{K/m}$, where ω is the oscillator's classical angular frequency of oscillation. Furthermore, let

$$y = \sqrt{\frac{m\,\omega}{\hbar}}\,x, \tag{C.108}$$

and

$$\epsilon = \frac{2\,E}{\hbar\,\omega}. \tag{C.109}$$

Equation (C.107) reduces to

$$\frac{d^2\psi}{dy^2} - \left(y^2 - \epsilon\right)\psi = 0. \tag{C.110}$$

We need to find solutions to the previous equation that are bounded at infinity. In other words, solutions that satisfy the boundary condition $\psi \to 0$ as $|y| \to \infty$.

Consider the behavior of the solution to Equation (C.110) in the limit $|y| \gg 1$. As is easily seen, in this limit, the equation simplifies somewhat to give

$$\frac{d^2\psi}{dy^2} - y^2\,\psi \simeq 0. \tag{C.111}$$

The approximate solutions to the previous equation are

$$\psi(y) \simeq A(y)\,\mathrm{e}^{\pm y^2/2}, \tag{C.112}$$

where $A(y)$ is a relatively slowly-varying function of y. Clearly, if $\psi(y)$ is to remain bounded as $|y| \to \infty$ then we must chose the exponentially decaying solution. This suggests that we should write

$$\psi(y) = h(y)\,\mathrm{e}^{-y^2/2}, \tag{C.113}$$

where we would expect $h(y)$ to be an algebraic, rather than an exponential, function of y.

Substituting Equation (C.113) into Equation (C.110), we obtain

$$\frac{d^2h}{dy^2} - 2\,y\,\frac{dh}{dy} + (\epsilon - 1)\,h = 0. \tag{C.114}$$

Let us attempt a power-law solution of the form

$$h(y) = \sum_{i=0,\infty} c_i\,y^i. \tag{C.115}$$

Inserting this test solution into Equation (C.114), and equating the coefficients of y^i, we obtain the recursion relation

$$c_{i+2} = \frac{(2\,i - \epsilon + 1)}{(i+1)\,(i+2)}\,c_i. \tag{C.116}$$

Consider the behavior of $h(y)$ in the limit $|y| \to \infty$. The previous recursion relation simplifies to

$$c_{i+2} \simeq \frac{2}{i}\,c_i. \tag{C.117}$$

Hence, at large $|y|$, when the higher powers of y dominate, we have

$$h(y) \sim C \sum_j \frac{y^{2j}}{j!} \sim C\,\exp\left(+y^2\right). \tag{C.118}$$

It follows that $\psi(y) = h(y)\,\exp\left(-y^2/2\right)$ varies as $\exp\left(+y^2/2\right)$ as $|y| \to \infty$. This behavior is unacceptable, because it does not satisfy the boundary condition $\psi \to 0$ as $|y| \to \infty$. The only way in which we can prevent ψ from blowing up as $|y| \to \infty$ is to demand that the power series (C.115) terminate at some finite value of i. This implies, from the recursion relation (C.116), that

$$\epsilon = 2\,n + 1, \tag{C.119}$$

where n is a non-negative integer. Note that the number of terms in the power series (C.115) is $n + 1$. Finally, using Equation (C.109), we obtain

$$E = (n + 1/2)\,\hbar\,\omega, \tag{C.120}$$

for $n = 0, 1, 2, \cdots$.

Hence, we conclude that a particle moving in a harmonic potential has quantized energy levels that are equally spaced. The spacing between successive energy levels is $\hbar\,\omega$, where ω is the classical oscillation frequency. Furthermore, the lowest energy state ($n = 0$) possesses the finite energy $(1/2)\,\hbar\,\omega$. This is sometimes called *zero-point energy*. It is easily demonstrated that the (normalized) wavefunction of the lowest-energy state takes the form

$$\psi_0(x) = \frac{e^{-x^2/2\,d^2}}{\pi^{1/4}\,\sqrt{d}}, \tag{C.121}$$

where $d = \sqrt{\hbar/m\,\omega}$.

C.12 Angular Momentum

In classical mechanics, the vector angular momentum, $\mathbf{L}$, of a particle of position vector $\mathbf{r}$ and linear momentum $\mathbf{p}$ is defined as

$$\mathbf{L} = \mathbf{r} \times \mathbf{p}. \tag{C.122}$$

In other words,

$$L_x = y\,p_z - z\,p_y, \tag{C.123}$$

$$L_y = z\,p_x - x\,p_z, \tag{C.124}$$

$$L_z = x\,p_y - y\,p_x. \tag{C.125}$$

In quantum mechanics, the operators, p_i, that represent the Cartesian components of linear momentum, are represented as the spatial differential operators $-i\,\hbar\,\partial/\partial x_i$. [See Equation (C.81)]. It follows that:

$$L_x = -i\,\hbar\left(y\,\frac{\partial}{\partial z} - z\,\frac{\partial}{\partial y}\right), \tag{C.126}$$

$$L_y = -i\,\hbar\left(z\,\frac{\partial}{\partial x} - x\,\frac{\partial}{\partial z}\right), \tag{C.127}$$

$$L_z = -i\,\hbar\left(x\,\frac{\partial}{\partial y} - y\,\frac{\partial}{\partial x}\right). \tag{C.128}$$

In addition, let

$$L^2 = L_x^2 + L_y^2 + L_z^2 \tag{C.129}$$

be the magnitude-squared of the angular momentum vector.

It is most convenient to work in terms of the standard spherical coordinates, r, θ, and ϕ. These are defined with respect to our usual Cartesian coordinates as follows:

$$x = r\,\sin\theta\,\cos\phi, \tag{C.130}$$

$$y = r\,\sin\theta\,\sin\phi, \tag{C.131}$$

$$z = r\,\cos\theta. \tag{C.132}$$

We deduce, after some tedious analysis, that

$$\frac{\partial}{\partial x} = \sin\theta\,\cos\phi\,\frac{\partial}{\partial r} + \frac{\cos\theta\,\cos\phi}{r}\,\frac{\partial}{\partial\theta} - \frac{\sin\phi}{r\,\sin\theta}\,\frac{\partial}{\partial\phi}, \tag{C.133}$$

$$\frac{\partial}{\partial y} = \sin\theta\,\sin\phi\,\frac{\partial}{\partial r} + \frac{\cos\theta\,\sin\phi}{r}\,\frac{\partial}{\partial\theta} + \frac{\cos\phi}{r\,\sin\theta}\,\frac{\partial}{\partial\phi}, \tag{C.134}$$

$$\frac{\partial}{\partial z} = \cos\theta\,\frac{\partial}{\partial r} - \frac{\sin\theta}{r}\,\frac{\partial}{\partial\theta}. \tag{C.135}$$

Making use of the definitions (C.126)–(C.129), after more tedious algebra, we obtain

$$L_x = -\mathrm{i}\,\hbar\left(-\sin\phi\,\frac{\partial}{\partial\theta} - \cos\phi\,\cot\theta\,\frac{\partial}{\partial\phi}\right), \tag{C.136}$$

$$L_y = -\mathrm{i}\,\hbar\left(\cos\phi\,\frac{\partial}{\partial\theta} - \sin\phi\,\cot\theta\,\frac{\partial}{\partial\phi}\right), \tag{C.137}$$

$$L_z = -\mathrm{i}\,\hbar\,\frac{\partial}{\partial\phi}, \tag{C.138}$$

as well as

$$L^2 = -\hbar^2\left[\frac{1}{\sin\theta}\frac{\partial}{\partial\theta}\left(\sin\theta\,\frac{\partial}{\partial\theta}\right) + \frac{1}{\sin^2\theta}\frac{\partial^2}{\partial\phi^2}\right]. \tag{C.139}$$

We, thus, conclude that all of our angular momentum operators can be represented as differential operators involving the angular spherical coordinates, θ and ϕ, but not involving the radial coordinate, r.

Let us search for an angular wavefunction, $Y_{l,m}(\theta,\phi)$, that is a simultaneous eigenstate of L^2 and L_z. In other words,

$$L^2\,Y_{l,m} = l\,(l+1)\,\hbar^2\,Y_{l,m}, \tag{C.140}$$

$$L_z\,Y_{l,m} = m\,\hbar\,Y_{l,m}, \tag{C.141}$$

where l and m are dimensionless constants. We also want the wavefunction to satisfy the normalization constraint

$$\oint |Y_{l,m}|^2\,d\Omega = 1, \tag{C.142}$$

where $d\Omega = \sin\theta\, d\theta\, d\phi$ is an element of solid angle, and the integral is over all solid angle. Thus, we are searching for well-behaved angular functions that simultaneously satisfy,

$$\left[\frac{1}{\sin\theta}\frac{\partial}{\partial\theta}\left(\sin\theta\,\frac{\partial}{\partial\theta}\right)+\frac{1}{\sin^2\theta}\frac{\partial^2}{\partial\phi^2}\right]Y_{l,m} = -l\,(l+1)\,Y_{l,m}, \tag{C.143}$$

$$\frac{\partial Y_{l,m}}{\partial\phi} = \mathrm{i}\,m\,Y_{l,m}, \tag{C.144}$$

$$\int_0^{\pi}\int_0^{2\pi}|Y_{l,m}|^2\,\sin\theta\,d\theta\,d\phi = 1. \tag{C.145}$$

As is well known, the requisite functions are the so-called *spherical harmonics*,

$$Y_{l,m}(\theta,\phi) = (-1)^m\left[\frac{2\,l+1}{4\pi}\,\frac{(l-m)!}{(l+m)!}\right]^{1/2}P_{l,m}(\cos\theta)\,\mathrm{e}^{\,\mathrm{i}\,m\phi}, \tag{C.146}$$

for $m \geq 0$. Here, the $P_{l,m}$ are known as *associated Legendre polynomials*, and are written

$$P_{l,m}(u) = (-1)^{l+m}\,\frac{(l+m)!}{(l-m)!}\,\frac{\left(1-u^2\right)^{-m/2}}{2^l\,l!}\left(\frac{d}{du}\right)^{l-m}\left(1-u^2\right)^l, \tag{C.147}$$

for $m \geq 0$. Note that

$$Y_{l,-m} = (-1)^m\,Y^{*}_{l,m}. \tag{C.148}$$

Finally, the constant l is constrained to take non-negative integer values, whereas the constant m is constrained to take integer values in the range $-l \leq m \leq +l$. Thus, l is a quantum number that determines the value of L^2. In fact, $L^2 = l\,(l+1)\,\hbar^2$. Likewise, m is a quantum number that determines the value of L_z. In fact, $L_z = m\,\hbar$.

The classical Hamiltonian of an extended object spinning with constant angular momentum $\mathbf{L}$ about one of its principal axes of rotation is

$$H = \frac{L^2}{2\,I}, \tag{C.149}$$

where I is the associated principal moment of inertia. Let us assume that the quantum-mechanical Hamiltonian of an object spinning about a principal axis of rotation has the same form. We can solve the energy eigenvalue problem,

$$H\,\psi = E\,\psi, \tag{C.150}$$

by writing $\psi(r,\theta,\phi) = R(r)\,Y_{l,m}(\theta,\phi)$, where $R(r)$ is arbitrary. It immediately follows from Equation (C.140) that

$$E = \frac{l\,(l+1)\,\hbar^2}{2\,I}. \tag{C.151}$$

In other words, the energy levels are quantized in terms of the quantum number, l, that specifies the value of L^2.

The classical Hamiltonian of an extended object spinning about a general axis is

$$H = \frac{L_x^2}{2\,I_x} + \frac{L_y^2}{2\,L_y} + \frac{L_z^2}{2\,L_z}, \tag{C.152}$$

where the Cartesian axes are aligned along the body's principal axes of rotation, and I_x, I_y, I_z are the corresponding principal moments of inertia. Suppose that the body is axially symmetric about the z-axis. It follows that $I_x = I_y$. The previous Hamiltonian can be written

$$H = \frac{L^2}{2\,I_x} + \left(\frac{1}{2\,I_z} - \frac{1}{2\,I_x}\right) L_z^2. \tag{C.153}$$

Let us assume that the quantum-mechanical Hamiltonian of an axisymmetric object spinning about an arbitrary axis has the same form as the previous Hamiltonian. We can solve the energy eigenvalue problem, $H\,\psi = E\,\psi$, by writing $\psi(r,\theta,\phi) = R(r)\,Y_{l,m}(\theta,\phi)$, where $R(r)$ is arbitrary. It immediately follows from Equations (C.140) and (C.141) that

$$E = \frac{[l\,(l+1) - m^2]\,\hbar^2}{2\,I_x} + \frac{m^2\,\hbar^2}{2\,I_z}. \tag{C.154}$$

In other words, the energy levels are quantized in terms of the quantum number, l, that specifies the value of L^2, as well as the quantum number, m, that determines the value of L_z. If $I_z \ll I_x$ (in other words, if the object is highly elongated along its symmetry axis) then the spacing between energy levels corresponding to different values of m becomes much greater than the spacing between energy levels corresponding to different values of l. In this case, it is plausible that the system remains in the $m = 0$ state (because it cannot acquire sufficient energy to reach the $m = 1$ state), so that

$$E = \frac{l\,(l+1)\,\hbar^2}{2\,I_x}. \tag{C.155}$$

Bibliography

Arajs, S. and Colvin, R. V. (1961). Paramagnetism of polycrystalline gadolinium, terbium, and dysprosium metals, *Journal of Applied Physics* **32**, p. 336S.

Carter, A. H. (2001). *Classical and Statistical Thermodynamics* (Prentice-Hall, Upper Saddle River, NY).

Henry, W. E. (1952). Spin paramagnetism of Cr^{+++}, Fe^{+++}, and Gd^{+++} at liquid helium temperatures and in strong magnetic fields, *Physical Review* **88**, p. 561.

Kaul, R. and Thompson, E. D. (1969). Temperature dependence of the magnetic moment and hyperfine constant of nickel, *Journal of Applied Physics* **40**, p. 1383.

Khinchin, A. I. (1949). *Mathematical Foundations of Statistical Mechanics* (Dover, New York NY).

Kittel, C. (1956). *Introduction to Solid-State Physics*, 2nd edn. (John Wiley & Sons, New York NY).

Kittel, C. and Kroemer, H. (1980). *Thermal Physics* (W. H. Freeman & Co., New York NY).

Mather, J. C. *et al.* (1990). A preliminary measurement of the cosmic microwave background spectrum by the cosmic background explorer (cobe) satellite, *Astrophysical Journal Letters* **354**, p. L37.

Partington, J. R. and Shilling, W. G. (1924). *The Specific Heats of Gases* (Benn, London UK).

Reif, F. (1965). *Fundamentals of Statistical and Thermal Physics* (McGraw-Hill International, Auckland, New Zealand).

Rumble, J. R., Lide, D. R., and Bruno, T. J. (2017). *CRC Handbook of Chemistry and Physics*, 98th edn. (CRC Press, Boca Raton FL).

Schroeder, D. V. (2000). *An Introduction to Thermal Physics* (Addison Wesley Longman, San Francisco CA).

Seitz, F. (1943). *The Physics of Metals* (McGraw-Hill, New York NY).

Tolman, R. C. (1938). *The Principles of Statistical Mechanics* (Oxford University Press, Oxford UK).

Index

www.ingramcontent.com/pod-product-compliance
Lightning Source LLC
LaVergne TN
LVHW011703200926
848484LV00004B/1060
* 9 7 8 9 8 1 1 2 2 3 3 5 8 *